INTERNET and COMMUNICATIONS

This new book series presents the latest research and technological developments in the field of Internet and multimedia systems and applications. We remain committed to publishing high-quality reference and technical books written by experts in the field.

If you are interested in writing, editing, or contributing to a volume in this series, or if you have suggestions for needed books, please contact Dr. Borko Furht at the following address:

Borko Furht, Ph.D.
Department Chairman and Professor
Computer Science and Engineering
Florida Atlantic University
777 Glades Road
Boca Raton, FL 33431 U.S.A.

E-mail: borko@cse.fau.edu

Edgar H. Callaway, Jr.

Wireless Sensor Networks

Architectures and Protocols

AUERBACH PUBLICATIONS

A CRC Press Company
Boca Raton London New York Washington, D.C.

Library of Congress Cataloging-in-Publication Data

Callaway, Edgar H.
Wireless sensor networks : architectures and protocols / Edgar H. Callaway.
p. cm.
Includes bibliographical references and index.
ISBN 0-8493-1823-8 (alk. paper)
1. Sensor networks. 2. Wireless LANs. I. Title.

Tk7872.D48C35 2003
004.6'8—dc21   2003051886

Visit the Auerbach PublicationsWeb site at www.auerbach-publications.com

© 2004 by CRC Press LLC
Auerbach is an imprint of CRC Press LLC

No claim to original U.S. Government works
International Standard Book Number 0-8493-1823-8
Library of Congress Card Number 2003051886
Printed in the United States of America 3 4 5 6 7 8 9 0
Printed on acid-free paper

Dedication

To Jan

Contents

Contents

Acknowledgments

This book is an extension of my dissertation,[1] completed at Florida Atlantic University (Boca Raton) in 2002 under the guidance of Dr. Ravi Shankar. If, as Newton said, a body remains at rest until acted upon by an external force, his was the force that moved me to return to school and complete my education; and for that, I am grateful. I also thank the other members of my committee, Dr. Valentine Aalo, Dr. Raymond Barrett, Dr. Borko Furht, Dr. Sam Hsu, and Dr. Fred Martin, for consenting to serve on the committee and for their constructive criticism of my work.

Much of the research for this book was done while I was employed at the Florida Communication Research Laboratory, Plantation, a division of Motorola Labs. I am indebted to the laboratory directors, Dr. Larry Dworsky and Dr. Chip Shanley, and my immediate superior, Dr. Bob O'Dea, for their support, which went far beyond corporate standard operating procedure. I also benefited from many useful technical discussions with my coworkers, including Anthony Allen, Monique Bourgeois, Priscilla Chen, Dr. Neiyer Correal, Dr. Lance Hester, Dr. Jian Huang, Dr. Yan Huang, Masahiro Maeda, Qicai Shi, Bob Stengel, and especially Paul Gorday, Sumit Talwalkar, and David Taubenheim for their consultations on the Signal Processing Worksystem (SPW).

I thank Gary Pace of Motorola's Semiconductor Products Sector, Boynton Beach, Florida, for the differential amplifier circuit discussed in Chapter 7, and for inculcating in me the value of low-voltage, low-power design. In addition to many of the aforementioned individuals, Dan Brueske, Barbara Doutre, Dr. Antonio Faraone, Latonia Gordon, and Dr. Kai Siwiak reviewed various early drafts of this book; any remaining errors of omission or commission, however, are mine.

Special thanks are due to Joan Lange, Kim Searer, and Martha Mitchell, corporate librarians who went to extraordinary efforts to track down obscure references for me on short notice.

Acknowledgments

My parents, Pat and Ed, deserve special thanks for emphasizing the value of education to a young man more often interested in other things. I also recognize my greatest debt of gratitude: to my wife, Jan. Without her support and understanding, this book would not have been possible.

Edgar H. Callaway, Jr.

Note

1. Edgar H. Callaway, Jr., A Communication Protocol for Wireless Sensor Networks, Ph.D. dissertation, Florida Atlantic University, Boca Raton, FL, August 2002.

Chapter 1
Introduction to Wireless Sensor Networks

1.1 APPLICATIONS AND MOTIVATION[1*]

In recent years, the desire for connectivity has caused an exponential growth in wireless communication. Wireless data networks, in particular, have led this trend due to the increasing exchange of data in Internet services such as the World Wide Web, e-mail, and data file transfers. The capabilities needed to deliver such services are characterized by an increasing need for data throughput in the network; applications now under development, such as wireless multimedia distribution in the home, indicate that this trend will continue. Wireless Local Area Networks (WLANs) provide an example of this phenomenon. The original (1997) Institute of Electrical and Electronic Engineers (IEEE) WLAN standard, 802.11, had a gross data rate of 2 megabits per second (Mb/s);[2,3] the most popular variant now is 802.11b, with a rate of 11 Mb/s;[4] and 802.11a, with a rate of 54 Mb/s, is now entering the market.[5] Wireless Personal Area Networks (WPANs), defined as networks employing no fixed infrastructure and having communication links less than 10 meters in length centered on an individual, form another example: the HomeRF 1.0 specification, released in January 1999 by the Home RF [sic] Working Group, has a raw data rate of 800 kb/s with an optional 1.6 Mb/s mode;[6] the Bluetooth™ 1.0 specification, released in July 1999 by the Bluetooth Special Interest Group (SIG) and later standardized as IEEE 802.15.1,[7] has a raw data rate of 1 Mb/s;[8,9] and IEEE 802.15.3, released in June 2003, has a maximum raw data rate of 55 Mb/s.[10] Both the 802.11 and 802.15 organizations have begun the definition of protocols with data throughputs greater than 100 Mb/s.

Other potential wireless network applications exist, however. These applications, which have relaxed throughput requirements and are often measured in a few bits per day, include industrial control and monitoring; home automation and consumer electronics; security and military sensing;

---

* Portions reprinted from Reference 1. © 2001, IEEE.

asset tracking and supply chain management; intelligent agriculture; and health monitoring.[11] Because most of these low-data-rate applications involve sensing of one form or another, networks supporting them have been called wireless sensor networks, or Low-Rate WPANs (LR-WPANs), because they require short-range links without a preexisting infrastructure. An overview of applications for wireless sensor networks follows.

1.1.1 Industrial Control and Monitoring

A large, industrial facility typically has a relatively small control room, surrounded by a relatively large physical plant. The control room has indicators and displays that describe the state of the plant (the state of valves, the condition of equipment, the temperature and pressure of stored materials, etc.), as well as input devices that control actuators in the physical plant (valves, heaters, etc.) that affect the observed state of the plant. The sensors describing the state of the physical plant, their displays in the control room, the control input devices, and the actuators in the plant are often all relatively inexpensive when compared with the cost of the armored cable that must be used to communicate between them in a wired installation. Significant cost savings may be achieved if an inexpensive wireless means were available to provide this communication. Because the information being communicated is state information, it often changes slowly. Thus, in normal operation, the required data throughput of the network is relatively low, but the required reliability of the network is very high. A wireless sensor network of many nodes, providing multiple message routing paths of multihop communication, can meet these requirements.

An example of wireless industrial control is the control of commercial lighting. Much of the expense in the installation of lights in a large building concerns the control of the lights — where the wired switches will be, which lights will be turned on and off together, dimming of the lights, etc. A flexible wireless system can employ a handheld controller that can be programmed to control a large number of lights in a nearly infinite variety of ways, while still providing the security needed by a commercial installation.

A further example is the use of wireless sensor networks for industrial safety applications. Wireless sensor networks may employ sensors to detect the presence of noxious, poisonous, or otherwise dangerous materials, providing early detection and identification of leaks or spills of chemicals or biological agents before serious damage can result (and before the material can reach the public). Because the wireless networks may employ distributed routing algorithms, have multiple routing paths, and can be self-healing and self-maintaining, they can be resilient in the face of an explosion or other damage to the industrial plant, providing officials with critical plant status information under very difficult conditions.

The monitoring and control of rotating or otherwise moving machinery is another area suitable for wireless sensor networks. In such applications, wired sensor and actuators are often impractical, yet it may be important to monitor the temperature, vibration, lubrication flow, etc. of the rotating components of the machine to optimize the time between maintenance periods, when the machine must be taken off-line. To do this, it is important that the wireless sensor system be capable of operating for the full interval between maintenance periods; to do otherwise defeats the purpose of the sensors. This, in turn, requires the use of a wireless sensor network with very low energy requirements. The sensor node often must be physically small and inexpensive as well. Wireless sensor networks may be of particular use in the prediction of component failure for aircraft, where these attributes may be used to particular advantage.[12]

Still another application in this area for wireless sensor networks is the heating, ventilating, and air conditioning (HVAC) of buildings. HVAC systems are typically controlled by a small number of strategically located thermostats and humidistats. The number of these thermostats and humidistats is limited, however, by the costs associated with their wired connection to the rest of the HVAC system. In addition, the air handlers and dampers that directly control the room environment are also wired; for the same reasons, their numbers are also limited.

The heat load generated by people in a building is quite dynamic, however. Diurnal, hebdomadal, seasonal, and annual variations occur. These variations are associated with the distribution of people in the building throughout the day, week, season, and year; important changes also affect the heat load of the building at more irregular intervals. For example, when organizations reorganize and remodel, space previously used for offices may be used by heat-generating laboratory or manufacturing equipment. Changes to the building itself must also be considered: interior walls may be inserted, moved, or removed; windows, curtains, and awnings may be added or removed, etc. Due to all these possible variations and, as nearly anyone who works in an office building can attest, improvement is needed.

The root cause of such unsatisfactory HVAC function is that the control system lacks sufficient information about the environment in the building to maintain a comfortable environment for all. Because they do not require the expense of wired sensors and actuators, wireless sensor networks may be employed to greatly increase the information about the building environment available to the HVAC control system, and to greatly decrease the granularity of its response. Wireless thermostats and humidistats may be placed in several places around each room to provide detailed information to the control system. Similarly, wireless bypass dampers and volume dampers can be used in great number to fine-tune the response of the HVAC system to any situation. Should everyone in an office area move to the conference room for a meeting, for example, the system can respond by closing the volume dampers in the office area, while opening the volume dampers in the conference room. Should the group leave the building, the HVAC system may instruct the wireless bypass dampers to respond to the change in total building heat load. Should the group return during a driving rainstorm, the humidistat in the umbrella and coat closet could detect the increased humidity in that closet. The HVAC system could then place especially dry air there, without affecting the occupants elsewhere in the building.

The wireless HVAC system can also solve one of the great problems facing the HVAC engineer: balancing heating and air conditioning. It is often the case that heat sources are not uniformly distributed throughout a building. In the home, for example, kitchens tend to be warm, due to the heat of cooking, while bedrooms tend to be cool. In winter, more heated air needs to be sent to the bedroom, where it is cooler, and less heated air needs to be sent to the kitchen, where it is warmer. In summer, however, just the opposite is true — more cooled air needs to be sent to the kitchen, where it is warmer, and less cooled air needs to be sent to the bedroom, where it is cooler. This difference between the air distribution of heating and air conditioning is a difficult and expensive problem to solve with wired control systems, because a volume damper to each room in the house must be independently controlled. Often, the dampers are placed in a single, fixed position, leaving some areas perpetually cold and others perpetually warm. With wireless sensors and actuators in the HVAC system, however, the problem becomes trivial; the damper(s) to each room can be controlled by the sensor(s) in each room, leading to perfect system balance at any time of the year.

Such a wireless HVAC system has other advantages. Close monitoring of system performance enables problems to be identified and corrected before occupant complaints arise. In addition to the living-area sensors, wireless sensors may be placed inside air ducts (to monitor the performance of heat exchange apparatus, for example) without requiring maintenance personnel to make manual measurements atop ladders. In addition, sensors may be placed in attics and crawlspaces that contain ductwork; anomalous temperatures in such areas may indicate costly leaks of heated or cooled air. For these reasons, total building HVAC costs should drop, while occupant comfort would increase when wireless sensors and actuators are employed.

1.1.2 Home Automation and Consumer Electronics

The home is a very large application space for wireless sensor networks.[13] Many of the industrial applications just described have parallels in the home. For example, a home HVAC system equipped with wireless thermostats and dampers can keep the rooms on the sunny side of the house comfortable — without chilling the occupants on the shady side of the house — more effectively than a home equipped with only a single, wired thermostat. However, many other opportunities are available.

One application is the "universal" remote control, a personal digital assistant (PDA)-type device that can control not only the television, DVD player, stereo, and other home electronic equipment, but the lights, curtains, and locks that are also equipped with a wireless sensor network connection. With the universal remote control, one may control the house from the comfort of one's armchair. Its most intriguing potential, however, comes from the combination of multiple services, such as having the curtains close automatically when the television is turned on, or perhaps automatically muting the home entertainment system when a call is received on the telephone or the doorbell rings. With the scale and personal computer both connected via a wireless sensor network, one's weight may be automatically recorded without the need for manual intervention (and the possibility of stretching the truth "just this once").

A major use of wireless sensor networks in the home is expected to be for personal computer peripherals, such as wireless keyboards and mice. Such applications take advantage of the low cost and low power consumption that are the sine qua non of wireless sensor networks. Another application in the home is sensor-based information appliances that transparently interact and work symbiotically together as well as with the home occupant.[14] These networks are an extension of the information appliances proposed by Norman.[15]

Toys represent another large market for wireless sensor networks. The list of toys that can be enhanced or enabled by wireless sensor networks is limited only by one's imagination, and range from conventional radio-controlled cars and boats to computer games employing wireless joysticks and controllers. A particularly intriguing field is personal computer (PC)-enhanced toys, which employ the computing power of a nearby computer to enrich the behavior of the toy itself. For example, speech recognition and synthesis may be performed by placing the microphone and speaker in the toy, along with the appropriate analog-to-digital and digital-to-analog converters, but employing a wireless connection to the computer, which performs the recognition and synthesis functions. By not placing the relatively expensive yet limited speech recognition and synthesis circuits in the toy, and using the (much more powerful) computing power already present in the computer, the cost of the toy may be significantly reduced, while greatly improving the capabilities and performance of the toy. It is also possible to give the toy complex behavior that is not practical to implement in other technologies.[16]

Another major home application is an extension of the Remote Keyless Entry (RKE) feature found on many automobiles. With wireless sensor networks, wireless locks, door and window sensors, and wireless light controls, the homeowner may have a device similar to a key fob with a button. When this button is pressed, the device locks all the doors and windows in the home, turns off most indoor lights (save a few night lights), turns on outdoor security lights, and sets the home's HVAC system to nighttime (sleeping) mode. The user receives a reassuring "beep" once this is all done successfully, and sleeps soundly, knowing that the home is secure. Should a door be left open, or some other problem exists, a small display on the device indicates the source of the trouble. The network may even employ a full home security system to detect a broken window or other trouble.

Outside of the home, the location-aware capabilities of wireless sensor networks are suitable for a diverse collection of consumer-related activities, including tourism[17] and shopping.[18,19] In these applications, location can be used to provide context-specific information to the consumer. In the case of the tourism guide, the user is provided only information relevant to his present view; in the case of the shopping guide, the user is provided information relevant to the products before him, including sale items and special discounts and offers.

1.1.3 Security and Military Sensing

The wireless security system described above for the home can be augmented for use in industrial security applications. Such systems, employing proprietary communication protocols, have existed for several years.[20] They can support multiple sensors relevant to industrial security, including passive infrared, magnetic door opening, smoke, and broken glass sensors, and sensors for direct human intervention (the "panic button" sensor requesting immediate assistance).

As with many technologies, some of the earliest proposed uses of wireless sensor networks were for military applications.[21] One of the great benefits of using wireless sensor networks is that they can be used to replace guards and sentries around defensive perimeters, keeping soldiers out of harm's way. In this way, they can serve the same function as antipersonnel mines, without the attendant hazard mines represent to allied personnel during the battle (or the civilian population afterward). In addition to such defensive applications, deployed wireless sensor networks can be used to locate and identify targets for potential attack, and to support the attack by locating friendly troops and unmanned vehicles. They may be equipped with acoustic microphones, seismic vibration sensors, magnetic sensors, ultrawideband radar, and other sensors.[22]

Wireless sensor networks can be small, unobtrusive, and camouflaged to resemble native rock, trees, or even roadside litter. By their nature, multihop networks are redundant. These networks have distributed control and routing algorithms (i.e., without a single point of failure), a feature that makes them difficult to destroy in battle.[23] The use of spread spectrum techniques, combined with the bursty transmission format common to many wireless sensor networks (to optimize battery life), can give them a low probability of detection by electronic means. The relative location determination capability of many ad hoc wireless sensor networks can enable the network nodes to be used as elements of a retrodirective array[24] of randomly distributed radiating elements; such an array can be used to provide exfiltration of the sensor network data.[25,26] The relative location information is used to align the relative carrier phase of the signals transmitted by each node; with this information, the exfiltrated data may be transmitted not just in the direction of the incoming signal, but in any desired direction. Beamforming techniques can also be applied to the sensors themselves, to enhance their sensitivity and improve detection probabilities.[27]

Wireless sensor networks can also be effective in the monitoring and control of civilian populations with the use of optical, audio, chemical, biological, and radiological sensors to track individuals and groups. The control of wireless sensor networks and the data they produce in a free society, while an important public policy discussion,[28] is outside the scope of this text.

1.1.4 Asset Tracking and Supply Chain Management

A very large unit volume application of wireless sensor networks is expected to be asset tracking and supply chain management.

Asset tracking can take many forms. One example is the tracking of shipping containers in a large port. Such port facilities may have tens of thousands of containers, some of which are empty and in storage, while others are bound for many different destinations. The containers are stacked, both on land and on ship. An important factor in the shipper's productivity (and profitability) is how efficiently the containers can be organized so that they can be handled the fewest number of times and with the fewest errors. For example, it is important that the containers next needed be on top of a nearby stack instead of at the bottom of a stack 1 km away. An error in the location record of any container can be disastrous; a "lost" container can be found only by an exhaustive search of a very large facility. Wireless sensor networks can be used to advantage in such a situation; by placing sensors on each container, its location can always be determined.

Similar situations involving large numbers of items that must be tracked occur in rail yards, where thousands of railroad cars of all types must be organized, and in the manufacture of durable goods, such as cars and trucks, that may sit in large lots or warehouses after manufacture, but before delivery to a retailer.

A related application is that of supply chain management. An item in a large warehouse, but with its precise location unknown, is practically lost because it is unavailable to be used or sold. This represents inventory shrinkage, even though the item is physically on the premises, and is therefore a business expense. In a manner similar to that of the asset tracking application described previously, wireless sensor networks can be used to reduce this cost; however, additional benefits may be obtained. In a large distribution chain, one of the most vexing problems facing the distributor is to quickly and accurately identify the location of material to be sold. Knowing where a product is can mean the difference between making or not making a sale, but knowing the status of the entire supply chain — from raw materials through components to final product — can help a business operate more efficiently. For example, transferring excess product from Division X (where it is selling slowly) to Division Y (where it is selling briskly) can help a company avoid the purchase of component parts to manufacture more product for Division Y. Wireless sensor networks placed along the supply chain enable everyone in the business to make better decisions because more information about product in the supply chain is available.

This information can also be used as a competitive advantage; by being able to tell a customer exactly where his product is (or even where the component parts of his product are) in the supply chain, the customer's confidence of on-time delivery (and opinion of the seller's competence) rises. This has already been used extensively in the package shipping industry, so much so that customers expect this service as a matter of course — a shipper that cannot tell a customer where his package is at any given time is rarely reused.

The use of wireless sensor networks for the tracking of nuclear materials has already been demonstrated in the Authenticated Tracking and Monitoring System (ATMS).[29,30] The ATMS employs wireless sensors (including the state of the door seal, as well as infrared, smoke, radiation, and temperature sensors) within a shipping container (e.g., a railroad car) to monitor the state of its contents. Notification of sensor events are wirelessly transmitted within the shipping container to a mobile processing unit, connected to both a Global Positioning System (GPS) receiver and an International Maritime Satellite (INMARSAT) transceiver. Through the INMARSAT system, the location and status of each shipment may be monitored anywhere in the world.

1.1.5 Intelligent Agriculture and Environmental Sensing

A textbook example of the use of wireless sensor networks in agriculture is the rain gauge. Large farms and ranches may cover several square miles, and they may receive rain only sporadically and only on some portions of the farm. Irrigation is expensive, so it is important to know which fields have received rain, so that irrigation may be omitted, and which fields have not and must be irrigated. Such an application is ideal for wireless sensor networks. The amount of data sent over the network can be very low (as low as one bit — "yes or no" — in response to the "Did it rain today?" query), and the message latency can be on the order of minutes. Yet, costs must be low, and power consumption must be low enough for the entire network to last an entire growing season.

The wireless sensor network is capable of much more than just soil moisture measurements, however, because the network can be fitted with a near-infinite variety of chemical and biological sensors. The data that is provided by such a network is capable of providing the farmer with a graphical view of soil moisture; temperature; the need for pesticides, herbicides, and fertilizers; received sunshine; and many other quantities. This type of application is especially important in vineyards, where subtle environmental changes may have large effects on the value of the crop and how it is processed.

The location determination features of many wireless sensor networks also may be used in advanced control systems to enable more automation of farming equipment.

Many applications of wireless sensor networks are also used on ranches. Ranchers may use wireless sensor networks in the location determination of animals within the ranch and, with sensors placed on each animal, determine the need for treatments to prevent parasites. Dairy farmers may use wireless sensors to determine the onset of estrus in cattle, a labor-intensive manual process at present. Hog and chicken farmers typically have many animals in cooled, ventilated barns. Should the temperature rise excessively, many thousands of animals may be lost. Wireless sensor networks can be used to monitor the temperature throughout the barn, keeping the animals safe.

Wireless sensor networks may also be used for low-power sensing of environmental contaminants such as mercury.[31] Integrated microcantilever sensors sensitive to particular contaminants can achieve parts-per-trillion sensitivities. These microelectromechanical (MEMS) sensors may be integrated with a wireless transceiver in a standard complementary metal oxide semiconductor (CMOS) process, providing a very low-cost solution to the monitoring of chemical and biological agents.

1.1.6 Health Monitoring

A market for wireless sensor networks that is expected to grow quickly is the field of health monitoring. "Health monitoring" is usually defined as "monitoring of non-life-critical health information," to differentiate it from medical telemetry, although the definition is broad and nonspecific, and some medical telemetry applications can be considered for wireless sensor networks.

Two general classes of health monitoring applications are available for wireless sensor networks. One class is athletic performance monitoring, for example, tracking one's pulse and respiration rate via wearable sensors and sending the information to a personal computer for later analysis.[32] The other class is at-home health monitoring, for example, personal weight management.[33] The patient's weight may be wirelessly sent to a personal computer for storage. Other examples are daily blood sugar monitoring and recording by a diabetic, and remote monitoring of patients with chronic disorders.[34]

The use of wireless sensor networks in health monitoring is expected to accelerate due to the development of biological sensors compatible with conventional CMOS integrated circuit processes.[35] These sensors, which can detect enzymes, nucleic acids, and other biologically important materials, can be very small and inexpensive, leading to many applications in pharmaceuticals and medical care.

A developing field in the health monitoring market is that of implanted medical devices. In the United States, the Federal Communications Commission (FCC) established regulations governing the Medical Implant Communications Service, in January 2000, "for transmitting data in support of diagnostic or therapeutic functions associated with implanted medical devices."[36] These types of systems can be used for a number of purposes, from monitoring cardiac pacemakers to specialized drug delivery systems.

A developing field related to both health monitoring and security is that of disaster relief. For example, the wireless sensors of the HVAC system in a collapsed multistory building (perhaps the result of an earthquake) can provide victim location information to rescue workers if acoustic sensors, activated automatically by accelerometers or manually by emergency personnel, are included. Water and gas sensors also could be used to give rescuers an understanding of the conditions beneath them in the rubble. Even if no additional sensors were included, the identities and pre- and post-collapse locations of the surviving network nodes can be used to help workers understand how the building collapsed, where air pockets or other survivable areas may be, and can be used by forensic investigators to make future buildings safer.

Wireless disaster relief systems, in the form of avalanche rescue beacons, are already on the market. Avalanche rescue beacons, which continuously transmit signals that rescuers can use to locate the wearer in time of emergency, are used by skiers and other mountaineers in avalanche-prone areas. The present systems have their limitations, however; principal among these is that they provide only location information, and give no information about the health of the victim. In a large avalanche, when emergency personnel can detect several beacons, they have no way to decide who should be assisted first. It was recently proposed that these systems be enhanced by the addition of health sensors, including oximeters and thermometers, so that would-be rescuers would be able to perform triage in a large avalanche, identifying those still alive under the snow.[37]

1.2 NETWORK PERFORMANCE OBJECTIVES

To meet the requirements of the applications just described, a successful wireless sensor network design must have several unique features. The need for these features leads to a combination of interesting technical issues not found in other wireless networks.

1.2.1 Low Power Consumption

Wireless sensor network applications typically require network components with average power consumption that is substantially lower than currently provided in implementations of existing wireless networks such as Bluetooth. For instance, devices for certain types of industrial and medical sensors, smart tags, and badges, powered from small coin cell batteries, should last from several months to many years. Applications involving the monitoring and control of industrial equipment require exceptionally long battery life so that the existing maintenance schedules of the monitored equipment are not compromised. Other applications, such as environmental monitoring over large areas, may require a very large number of devices that make frequent battery replacement impractical. Also, certain applications cannot employ a battery at all; network nodes in these applications must get their energy by mining or scavenging energy from the environment.[38,39] An example of this type is the wireless automobile tire pressure sensor, for which it is desirable to obtain energy from the mechanical or thermal energy present in the tire instead of a battery that may need to be replaced before the tire does.

In addition to average power consumption, primary power sources with limited average power sourcing capability often have limited peak power sourcing capabilities as well; this factor should also be considered in the system design.

1.2.2 Low Cost

Cost plays a fundamental role in applications adding wireless connectivity to inexpensive or disposable products, and for applications with a large number of nodes in the network, such as wireless supermarket price tags. These potential applications require wireless links of low complexity that are low in cost relative to the total product cost.

To meet this objective, the communication protocol and network design must avoid the need for high-cost components, such as discrete filters, by employing relaxed analog tolerances wherever possible, and minimize silicon area by minimizing protocol complexity and memory requirements. In addition, however, it should be recognized that one of the largest costs of many networks is administration and maintenance. To be a true low-cost system, the network should be ad hoc and capable of self-configuration and self-maintenance. An "ad hoc" network in this context is defined to be a network without a predetermined physical distribution, or logical topology, of the nodes. "Self-configuration" is defined to be the ability of network nodes to detect the presence of other nodes and to organize into a structured, functioning network without human intervention. "Self-maintenance" is defined to be the ability of the network to detect, and recover from, faults appearing in either network nodes or communication links, again without human intervention.

To facilitate the volume production expected of such systems and devices, thereby minimizing the cost of the wireless network components, the development of a standardized communication protocol is necessary. Recently, the IEEE 802 Local Area Network/Metropolitan Area Network Standards Committee (LMSC) created Working Group 15 to develop a set of standards for WPANs.[40] To address the need for low-power, low-cost wireless networking, in December 2000, the IEEE New Standards Committee (NesCom) officially sanctioned a new task group in Working Group 15 to begin the development of a standard for Low-Rate WPANs (LR-WPANs), to be called 802.15.4.[41] The goal of Task Group 4, as defined in its Project Authorization Request, was to provide a standard having ultra-low complexity, cost, and power for low-data-rate wireless connectivity among inexpensive, fixed, portable, and moving devices. Location awareness was considered a unique capability of the standard. The scope of Task Group 4, as for all IEEE 802 wireless standards,[42] was limited to the creation of specifications of the Physical (PHY) layer and Media Access Control (MAC) sublayer of the Data Link Layer in the International Standards Organization (ISO) Open Systems Interconnection (OSI) reference model;[43,44] further it was required that the standard be compatible with the 802.2 Logical Link Control layer standard. The 802.15.4 standard was approved in May 2003.

1.2.3 Worldwide Availability

Many of the proposed applications of wireless sensor networks, such as wireless luggage tags and shipping container location systems, implicitly require that the network be capable of operation worldwide. Further, to maximize production, marketing, sales, and distribution efficiency of products that may have wireless sensor network devices embedded in them, and avoid the establishment of regional variants that must be individually monitored through (perhaps separate) distribution chains, it is desirable to produce devices capable of worldwide operation. Although, in theory, this capability may be obtained by employing Global Positioning System (GPS) or Global Navigation Satellite System (GLONASS) receivers in each network node and adjusting node behavior according to its location, the cost of adding a second receiver, plus the additional performance flexibility required to meet the varying worldwide requirements, makes this approach economically unviable. It is, therefore, desirable to employ a single band worldwide — one that has minimal variation in government regulatory requirements from country to country — to maximize the total available market for wireless sensor networks.

1.2.4 Network Type

A conventional star network employing a single master and one or more slave devices may satisfy many applications. Because the transmit power of the network devices is limited by government regulation and battery life concerns, however, this network design limits the physical area a network may serve to the range of a single device (the master). When additional range is needed, network types that support multi-hop routing (e.g., mesh or cluster types) must be employed; the additional memory and computational cost for routing tables or algorithms, in addition to network maintenance overhead, must be supported without excessive cost or power consumption. It should be recognized that for many applications, wireless sensor networks are of relatively large order (e.g., > 256 nodes); device density may also be high (e.g., in active supermarket price tag applications). This book will focus on wireless sensor networks capable of multi-hop routing through dense networks.

1.2.5 Security

The security of wireless sensor networks has two facets of equal importance — how secure the network actually is and how secure the network is *perceived* to be by users and (especially) potential users. The perception of security is important because users have a natural concern when their data (whatever it may be) is transmitted over the air for anyone to receive. Often, an application employing wireless sensor networks replaces an earlier wired version in which users could physically see the wires or cables carrying their information, and know, with reasonable certainty, that no one else was receiving their information or injecting false information for them to receive. The wireless application must work to regain that confidence to attain the wide market needed to lower costs.

Security is more than just message encryption, however. In fact, in many applications, encryption (keeping a message secret, or private) is not an important security goal of wireless sensor networks. Often, the most important security goals are to ensure that any message received has not been modified in any way and is from the sender it purports to be. That is, if one has a wireless light and light switch in a home, there is often little to be gained by encrypting the commands "turn light on" and "turn light off." Not only does any potential eavesdropper know that only two possible commands are likely (in this simple example), but he or she may also be able to see the light shining out the home window from his or her position in the street. Having secret commands in this application is, therefore, of relatively little importance.

What is of greater importance, however, is that the malicious eavesdropper in the street not be able to inject false or modified messages into the wireless sensor network, perhaps causing the light to turn on and off at random, for example. This requires a second type of security, message authentication and integrity checking, which is performed by appending a message- and sender-dependent Message Integrity Code (MIC) to the transmitted message. (In the security field, the MIC is often termed the Message Authentication Code (MAC), but MIC is used in this text to avoid possible confusion with the Media Access Control layer of the OSI protocol stack.) The desired recipient and sender share a key, which is used by the sender to generate the MIC as well as by the recipient to confirm the integrity of the message and the identity of the sender. To avoid "replay attacks," in which an eavesdropper records a message and retransmits it later, a message counter or timer is included in the calculation of the MIC. In this way, no two authentic messages — even containing the same data — are identical.

Regarding security, the wireless sensor network designer faces three difficulties:

- The length of the MIC, as well as the security plan in general, must be balanced with the typical length of data to be transmitted, and the desire for short transmitted messages. Although a 16-byte (128-bit) MIC is often cited as necessary for the most secure systems, it becomes unwieldy when single-bit data is being passed (e.g., turn on, turn off). The designer must be able to balance the security needs of some users with the low-power requirements of the network. Note that this may involve choices of MIC length, as well as combinations of message authentication, integrity checking, and encryption — and must be performed automatically, as part of a self-organizing network.

• To minimize the cost of the network devices, the security features must be capable of implementation with inexpensive hardware, with a minimum addition of logic gates, random access memory (RAM), and read-only memory (ROM). In addition, the computational power (i.e., microcomputer clock speed, number of available coprocessors, etc.) available in most network devices is very limited. This combination of low gate count, small memory requirements, and low executed instruction count limits the types of security algorithms that can be used.

• Finally, perhaps the most difficult problem to solve in general is key distribution. Many methods are available, including several types of public key cryptography employing dedicated key loading devices and various types of direct user intervention. All have their advantages and disadvantages when used in a given application; the wireless sensor network designer must select the appropriate one for the application at hand.

Wireless sensor networks have additional requirements, including the need for scalability to very large networks, fault tolerance, and the need to operate in a wide variety of possibly hostile environments.[46] Although the design of such a network that meets these requirements may seem daunting, the designer of a wireless sensor network is not without tools. The stringent power and cost requirements come with more relaxed requirements in other areas.

1.2.6 Data Throughput

As already mentioned, wireless sensor networks have limited data throughput requirements when compared with Bluetooth (IEEE 802.15.1) and other WPANs and WLANs. For design purposes, the maximum desired data rate, when averaged over a one-hour period, may be set to be 512 b/s (64 bytes/s), although this figure is somewhat arbitrary. The typical data rate is expected to be significantly below this; perhaps 1 b/s or lower in some applications. Note that this is the data throughput, not the raw data rate as transmitted over the channel, which may be significantly higher.

This low required amount of data throughput implies that with any practical amount of protocol overhead (headers, addressing, etc.), the communications efficiency of the network will be very low – especially when compared against a network sending TCP/IP packets that may be 1500 bytes long. No matter what design is chosen, the efficiency will be very low, and the situation, therefore, may be viewed in a positive light: the protocol designer has the ability to design free of the consideration of communications efficiency, often a critical parameter in protocol design.

1.2.7 Message Latency

Wireless sensor networks have very liberal Quality of Service (QoS) requirements, because, in general, they do not support isochronous or synchronous communication, and have data throughput limitations that prohibit the transmission of real-time video and, in many applications, voice. The message latency requirement for wireless sensor networks is, therefore, very relaxed in comparison to that of other WPANs; in many applications, a latency of seconds or minutes is quite acceptable.

1.2.8 Mobility

Wireless sensor network applications, in general, do not require mobility. Because the network is therefore released from the burden of identifying open communication routes, wireless sensor networks suffer less control traffic overhead and may employ simpler routing methods than mobile ad hoc networks (e.g., MANET).

1.3 CONTRIBUTIONS OF THIS BOOK

The preceding constraints outline the requirements of a self-organizing, wireless, ad hoc communication network that trades lower data throughput and higher message latency, when compared with a conventional WPAN, for lower cost and lower power drain. This book describes the design of physical, data link, and network layers of communications protocols that meet these cost, power, and performance requirements. The physical layer design includes an efficient modulation method that combines good sensitivity and a high over-the-air data rate (for long battery life) with the possibility of low-cost digital implementation. The data link layer design employs a novel method by which low-cost devices may temporarily synchronize to exchange information, while maintaining low duty cycle and, therefore, power-efficient operation. Also reported are simulation results of a wireless network employing the novel data link layer design, including message throughput, message delay, effective node duty cycle, and channel collision performance.

Many factors affecting the design of practical wireless sensor network nodes are discussed at length, including the partitioning of node functions into integrated circuits, low-power system design techniques, the selection of the proper power source, and the interaction between antenna selection and product design. In addition, design techniques are presented to improve the electromagnetic compatibility of the wireless sensor network node and to reduce the likelihood of damage due to electrostatic discharge. Finally, a discussion on IEEE standards relevant to wireless sensor networks is presented, including the IEEE 802.15.4 Low-

Rate Wireless Personal Area Network standard, and the IEEE 1451.5 Wireless Sensor standard.

1.4 ORGANIZATION OF THIS BOOK

This text focuses on communication protocols suitable for wireless sensor networks, and the architecture of systems needed to implement them. Existing work in the field of digital communication networks, including a study of early wireless radiotelegraphic message-handling networks, is surveyed in Chapter 2. Chapter 3 describes a wireless sensor network physical layer designed for long battery life and low-cost implementation. A novel logical link control layer protocol, the Mediation Device protocol, is presented in Chapter 4; the protocol enables communication between low-cost devices with low duty cycle operation. Chapter 5 describes a wireless sensor network layer design and a discrete-event simulation tool for it that includes the behavior of the Mediation Device protocol; simulation results, including network message throughput, message latency, device duty cycle, and packet collisions, are included. Chapter 6 considers system implementation issues, including system partitioning and interfaces. Chapter 7 presents the complicated problem of power management for wireless sensor network nodes, including the need to match the power requirements of source and load. Antennas and how they affect overall network performance are discussed in Chapter 8, as are the advantages of describing node performance in terms of message error rate rather than bit error rate. Design techniques for electromagnetic compatibility (EMC) are presented in Chapter 9. These techniques reduce the likelihood of interference to and from other devices, and also to and from circuits inside the network node itself. The design of small, battery-powered products with good protection from electrostatic discharge (ESD) events is presented in Chapter 10. Chapter 11 outlines standards relevant to wireless sensor networks, and Chapter 12 concludes and identifies possible directions for future research.

References

1. This section is an extension of: Jose A. Gutierrez et al. IEEE 802.15.4: A developing standard for low-power, low-cost wireless personal area networks, *IEEE Network*, v. 15, n. 5, September/October 2001, pp. 12–19.
2. Institute of Electrical and Electronics Engineers, Inc., IEEE Std 802.11-1999(ISO/IEC 8802-11:1999), IEEE Standard for Information Technology — Telecommunications and Information Exchange between Systems — Local and Metropolitan Area Networks — Specific Requirements — Part 11: Wireless LAN Medium Access Control (MAC) and Physical Layer (PHY) Specifications. New York: IEEE Press, 1999.
3. Bob O'Hara and Al Petrick, *The IEEE 802.11 Handbook: A Designer's Companion*. New York: IEEE Press, 1999.
4. Chris Heegard et al., High-performance wireless ethernet, *IEEE Commun.*, v. 39, n. 11, November 2001, pp. 64–73.

5. Bill McFarland et al., The 5-UP protocol for unified multiservice wireless networks, *IEEE Commun.,* v. 39, n. 11, November 2001, pp. 74–80.

6. Jim Lansford and Paramvir Bahl, The design and implementation of HomeRF: a radio frequency wireless networking standard for the connected home, *Proc. IEEE,* v. 88, n. 10, October 2000, pp. 1662–1676.

7. Institute of Electrical and Electronics Engineers, Inc., IEEE Std 802.15.1-2002, IEEE Standard for Information technology — Telecommunications and Information Exchange between Systems — Local and Metropolitan Area Networks — Specific Requirements — Part 15.1: Wireless Medium Access Control (MAC) and Physical Layer (PHY) Specifications for Wireless Personal Area Networks (WPANs). New York: IEEE Press, 2002.

8. Jaap C. Haartsen, The Bluetooth radio system, *IEEE Pers. Commun.,* v. 7, n. 1, February 2000, pp. 28–36.

9. Jaap C. Haartsen and Sven Mattisson, Bluetooth – A new low-power radio interface providing short-range connectivity, *Proc. IEEE,* v. 88, n. 10, October 2000, pp. 1651–1661.

10. Jeyhan Karaoğuz, High-rate wireless personal area networks, *IEEE Commun.,* v. 39, n. 12, December 2001, pp. 96–102.

11. D. Estrin et al., Instrumenting the world with wireless sensor networks, *Proc. IEEE Intl. Conf. Acoustics, Speech, and Signal Processing,* 2001, v. 4, pp. 2033–2036.

12. Robert Fricke et al., *Wireless Sensor Review Final Report,* United States Air Force Research Laboratory Report AFRL-HE-WP-TR-2001-0167. Springfield, VA: National Technical Information Service, March 2001.

13. Ed Callaway et al., Home networking with IEEE 802.15.4: a developing standard for low-rate wireless personal area networks, *IEEE Commun. Mag.,* v. 40, n. 8, August 2002, pp. 70–77.

14. Emil M. Petriu et al., Sensor-based information appliances, *IEEE Instrumentation & Measurement Magazine,* v. 3, n. 4, December 2000, pp. 31–35.

15. Donald A. Norman, *The Invisible Computer.* Cambridge, MA: MIT Press, 1998.

16. Cf. research projects of the Toys of Tomorrow special interest group of the MIT Media Lab (http://toys.media.mit.edu/).

17. Gregory D. Abowd et al., Cyberguide: a mobile context-aware tour guide, *Wireless Networks,* v. 3, 1997, pp. 421–433.

18. Abhaya Asthana, Mark Cravatts, and Paul Krzyzanowski, An indoor wireless system for personalized shopping assistance, *Proc., Wksp. Mobile Computing Sys. and Applications,* 1994, pp. 69–74.

19. Abhaya Asthana and Paul Krzyzanowski, A small domain communications system for personalized shopping assistance, *IEEE Intl. Conf. Personal Wireless Communications,* 1994, pp. 199–203.

20. R. G. Swank, *Implementation Guidance for Industrial-Level Security Systems Using Radio Frequency Alarm Links,* Westinghouse Hanford Company Technical Security Document WHC-SD-SEC-DGS-002. Springfield, VA: National Technical Information Service, July 15, 1996.

21. R. Lacoss and R. Walton, Strawman design for a DSN to detect and track low flying aircraft, *Proc. Distributed Sensor Nets Conf.,* Carnegie-Mellon Univ., Pittsburgh, PA, December 1978, pp. 41–52.

22. Mike Horton et al., Deployment ready multimode micropower wireless sensor networks for intrusion detection, classification, and tracking. Sensors and Command, Control, Communications, and Intelligence (C3I) Technologies for Homeland Defense and Law Enforcement, Edward M. Carapezza, Ed., *Proc. SPIE,* v. 4708, 2002, pp. 290–295.

23. Mark Hewish, Little brother is watching you: unattended ground sensors, *Jane's Int. Defense Review,* v. 34, n. 6, June 2001, pp. 46–52.

24. Ryan Y. Miyamoto and Tatsuo Itoh, Retrodirective arrays for wireless communications, *IEEE Microwave Mag.,* v. 3, n. 1, March 2002, pp. 71–79.

25. Ryan Y. Miyamoto, Yongxi Qian, and Tatsuo Itoh, A reconfigurable active retrodirective/direct conversion receiver array for wireless sensor systems, *IEEE MTT-S International Microwave Symp. Digest,* v. 2, 2001, pp. 1119–1122.

26. Ryan Y. Miyamoto et al., An adaptive multi-functional array for wireless sensor systems, *IEEE MTT-S Int. Microwave Symp. Digest,* v. 2, 2002, pp. 1369–1372.

27. K. Yao et al., "Beamforming performance of a randomly distributed sensor array system," *Proc., IEEE Wksp. Design and Implementation of Signal Processing Systems,* 1997, pp. 438–447.

28. National Research Council, Division on Engineering and Physical Sciences, Computer Science and Telecommunications Board, Committee on Networked Systems of Embedded Computers, *Embedded, Everywhere: A Research Agenda for Networked Systems of Embedded Computers.* Washington, D.C.: National Academy Press, 2001.

29. J. Lee Schoeneman, Authenticated tracking and monitoring system (ATMS) tracking shipments from an Australian uranium mine. Presented at the *39th Inst. Nuclear Materials Management Annual Meeting,* Technical Report DE98007251. Springfield, VA: National Technical Information Service. 1998.

30. J. Lee Schoeneman, Heidi Anne Smartt, and Dennis Hofer, WIPP transparency project — container tracking and monitoring demonstration using the authenticated tracking and monitoring system (ATMS), Presented at the Waste Management Conference (WM2k), 2000, Tucson, AZ. http://www.osti.gov.

31. C. L. Britton Jr. et al., MEMS sensors and wireless telemetry for distributed systems, Smart Materials and Structures 1998, Smart Electronics and MEMS, Vijay K. Varadan et al., Eds., *Proc. SPIE,* v. 3328, pp. 112–123.

32. Bonnie Berkowitz, Technology catches up to runners, *Washington Post,* April 20, 2001, sec. E, p. 1.

33. Juha Pärkkä et al., A wireless wellness monitor for personal weight management, *Proc. IEEE EMBS Intl. Conf. on Information Technology Applications in Biomedicine,* 2000, pp. 83–88.

34. Olga Boric-Lubecke and Victor M. Lubecke, Wireless house calls: using communications technology for health care monitoring, *IEEE Microwave Mag.,* v. 3, n. 3, September 2002, pp. 43–48.

35. Yuh-Shyong Yang, Ude Lu, and Ben C. P. Hu, Prescription chips, *IEEE Circuits Devices Mag.,* v. 18, n. 5, September 2002, pp. 8–16.

36. http://wireless.fcc.gov.

37. Florian Michahelles and Bernt Schiele, Better rescue through sensors, presented at the *First Int. Wkshp. on Ubiquitous Computing for Cognitive Aids,* 2002, at UbiCom 2002, Göteborg (Gothenburg), Sweden, 2002, http://www.vision.ethz.ch.

38. Thad Starner, Human-powered wearable computing, *IBM Sys. J.,* v. 35, n. 3 & 4, 1996, pp. 618–629.

39. Anantha Chandrakasan et al., "Design considerations for distributed microsensor systems," *Proc. IEEE Custom Integrated Circuits Conf.,* May 1999, pp. 279–286.

40. Richard C. Braley, Ian C. Gifford, and Robert F. Heile, Wireless personal area networks: An overview of the IEEE P802.15 working group, *ACM Mobile Computing Commun. Review,* v. 4, n. 1, January 2000, pp. 26–34.

41. Institute of Electrical and Electronics Engineers, Inc., IEEE Std 802.15.4-2003, IEEE Standard for Information Technology — Telecommunications and Information Exchange between Systems — Local and Metropolitan Area Networks — Specific Requirements — Part 15.4: Wireless Medium Access Control (MAC) and Physical Layer (PHY) Specifications for Low Rate Wireless Personal Area Networks (WPANs). New York: IEEE Press, 2003.

42. Roger B. Marks, Ian C. Gifford, and Bob O'Hara, Standards in IEEE 802 unleash the wireless Internet, *IEEE Microwave*, v. 2, n. 2, June 2001, pp. 46–56.
43. Hubert Zimmermann, OSI reference model — the ISO model of architecture for open systems interconnection, *IEEE Trans. Commun.*, v. COM-28, n. 4, April 1980, pp. 425–432.
44. Dimitri Bertsekas and Robert Gallager, *Data Networks,* 2nd ed. Upper Saddle River, NJ: Prentice Hall, 1992, p. 19.
45. Institute of Electrical and Electronics Engineers, Inc., IEEE Std 802.2-1998 (ISO/IEC 8802-2: 1998), IEEE Standard for Information Technology — Telecommunications and Information Exchange between Systems — Local and Metropolitan Area Networks — Specific Requirements — Part 2: Logical Link Control. New York: IEEE Press, 1998.
46. Ian F. Akyildiz, Weilian Su, Yogesh Sankarasubramaniam, and Erdal Cayirci, A survey on sensor networks, *IEEE Commun. Mag.*, v. 40, n. 8, August 2002, pp. 102–114.

Chapter 2
The Development of Wireless Sensor Networks

2.1 EARLY WIRELESS NETWORKS

Wireless communication networks have a long history. In 1921, for example, the U.S. Army Signal Corps received authorization to establish the War Department Radio Net,[1] a nationwide radiotelegraphic network that, by 1925, numbered 164 stations, stretched to Alaska, and was called "the largest and most comprehensive radio net of its kind in the world today."[2] It carried more than 3.8 million words of message traffic its first year and, by 1933, more than 26 million words were carried annually.[3] Another early example is the Army-Amateur Radio System (AARS), established in 1925 by the Signal Corps and the American Radio Relay League (ARRL), an organization of amateur radio operators, to pass wireless message traffic between Army, National Guard, and Reserve units nationwide.[4-6] In this system, volunteer amateur radio operators, employing Morse code, were organized into "tactical nets," which were, in turn, organized in a hierarchy based on the Army command structure. At that time, the Army was organized into nine geographical "Corps Areas" within the continental United States. Within each Corps Area, amateur radio nets were organized with stations representing each military unit concerned:

1. A Corps Area Radio Net, comprising the headquarters of each of its Organized Reserve Divisions, the Governor's office in each state within its area, and a Corps Area Headquarters station acting [as a] Net Control Station.
2. A Division Radio Net for each of the Organized Reserve Divisions with Brigade, Regimental and such other nets as are necessary to properly provide radio communication for the units of the Organized Reserves.
3. A radio net for the National Guard of each to be called the Governor's Radio Net and which will comprise all of the units of the National Guard of that state, grouped into Brigade, Regimental and

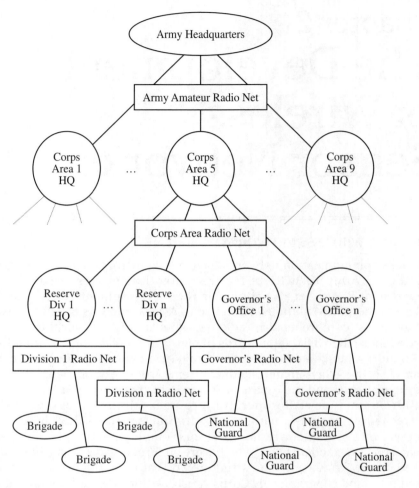

Exhibit 1. The Army Amateur Radio System (1925)

such other nets as are necessary to properly provide radio communication for all of the units of the National Guard.[7] (Published with the permission of the ARRL. Copyright 1925, ARRL.)

The Corps Area Headquarters will be connected in an Army Amateur Radio Net with an Army Headquarters Station located at The Signal School, Fort Monmouth, New Jersey.[8]

This describes a tree network (Exhibit 1), in which the sender addressed each message with the hierarchical destination address. Messages generated at the lower levels were passed up the tree to a station that had the destination below it in the tree; at that point, the message was routed back down the tree to its destination. With the exception of the leaves of the tree, each member of the tree acted as a Net Control Station (NCS) of a first "tactical net" consisting of itself and stations immediately below it in the hierarchy, and as a member of a second tactical net controlled by the station directly above it. (Being at the bottom of the hierarchy, the leaves were only net members, and did not function as NCSs.) To pass messages, a net would meet at a prearranged time on a prearranged frequency band. When the net was in session, messages would be passed between the NCS and other member stations, at the direction of the NCS.[9] In modern terminology, each station acted as a master of a star network at one time, and as a slave of another network, encompassing a larger physical area, at another time. The members of the tree closer to the root (i.e., higher in the military hierarchy) were generally better-equipped stations capable of greater range.[10]

Because of much favorable publicity following the 1928 Florida hurricane, when the AARS was the only link to West Palm Beach, the AARS was reorganized in 1929 to more closely align it with the Red Cross.[11,12] The tree network concept must have been viewed as a success, however, for the new organization included an additional layer of hierarchy. Now, in addition to the root-level Army Amateur Radio Net (composed of one station in each Corps Area, controlled by the Fort Monmouth station), each Corps Area contained:

- A Corps Area Amateur Radio Net, comprising one station in the capital of each of its states. A corps area amateur station or a selected civilian amateur station will act as Net Control Station.
- State Amateur Radio Nets, based on the division of each state into approximately five geographical areas. The stations will normally be located in the principal city or town of each geographical area, or near the center of the area. The state capital station will act as Net Control Station.
- District Amateur Radio Nets, each comprising approximately five stations so distributed as to best accomplish [the purposes of the AARS]. The geographical area station referred to previously in 2. [i.e., the previous bullet] will act as Net Control Station.
- Local Amateur Radio Nets, comprising all amateurs in the local areas for which the respective substations of the District Net may act as Net Control Station. Local Nets will operate on schedules prepared by the net control station and approved by the corps area signal officer.[13] (Published with the permission of the ARRL. Copyright 1929, ARRL.)

The new regulations also contained a "master traffic schedule," defining the time the net at each level of hierarchy, for each Corps Area, was to be active. In this way, the system achieved nationwide coordination between its hierarchical layers;[14] the schedule was "designed to permit a message filed anywhere in the United States to flow through the nets as organized and to reach its approximate destination the same night."[15]

By all accounts, the AARS was a very efficient communication network, especially after the 1929 reorganization, producing communication links for the Army,[16] satisfaction for the operators,[17] and emergency communications during natural disasters for the nation.[18,19] During the period from July 1, 1936 to June 30, 1937, the root Army Headquarters Station handled 22,458 messages (counting an origination as 1, a delivery as 1, and a relay as 1); the entire AARS handled 504,330 messages in the same period.[20] At that time, there were 1151 active AARS members,[21] a number that rose to 2400 by December 1941.[22]

On January 27, 1938, the system was tested by mailing 14 messages to selected AARS members, with instructions to file as soon as possible for transmission to the Army Headquarters Station in Washington, D.C. (the station had been moved from Fort Monmouth in 1930[23]). Twelve of the messages were received; the winning elapsed time was 12 minutes, from St. Augustine, Florida, via a two-hop route. Seven of the 12 received messages arrived in 1 hour, 15 minutes or less; the longest elapsed time was 22 hours, 37 minutes, from New Orleans, Louisiana, via a three-hop route. A four-hop route from Stafford, Arizona, took only 2 hours, 16 minutes.[24] The network was routinely tested by flooding from the root, with return acknowledgment requested.[25]

Interestingly, a second, all-amateur (i.e., one with no Army involvement) communication network existed contemporaneously with the AARS and, in fact, predated it by many years. Message relaying by amateur radio in the United States was a popular activity even prior to 1917, when the United States entered World War I and operation ceased for the duration. Due to the primitive nature of the equipment available at the time, most stations had a very limited range; ad hoc relaying began as a pragmatic means of range extension. Gradually, it came to be recognized that messages could be relayed more effectively if the stations involved were to become organized and messages were relayed in a more structured way. Several regional relaying organizations appeared, quickly followed by national organizations. An example of the former was the Central Radio Association, formed in 1911, which was relaying messages over hundreds of miles in the Midwest by 1914.[26] Examples of the latter include the ARRL and the Radio League of America (RLA), founded in 1915 by Hugo Gernsback.[27]

On February 22, 1916, the RLA organized the multicast transmission of a message from the Rock Island (Illinois) Arsenal to the governors of every state in the union, President Wilson, and the heads of many city governments. A total of 34 of the 48 governors received the message, as did the President, 129 city mayors, 6 town commissioners, and 2 constables.[28] Although this was less than perfect, the fact that the majority of the messages were received, and that message latency was low (the message reached the Atlantic, Pacific, and Gulf coasts in an hour or less), led most observers to view the test as a success.[29,30] This test, however, was a "stunt," in that the relay schedules were established explicitly for the transmission of this message; there was no continuing network organization.

During 1916, the ARRL began the development of what would be identified today as a true transcontinental communications network. Also using Morse code, this all-amateur network employed what became known as the "Five-Point" system.[31] Each amateur station kept four schedules — one each with a station to its North, South, East, and West — plus a fifth schedule to a station in the nearest large city. This formed a grid network of peer-to-peer communication links covering the nation; these links became known as "trunk lines." The trunk lines system was first proposed by Hiram Percy Maxim, co-founder of the ARRL, in 1916,[32] but it had its origins in a regional network Maxim formed as early as 1914.[33,34]

On January 27, 1917, the first three transcontinental wireless test messages were sent on the trunk lines system, from Los Angeles, California, to Hartford, Connecticut, via a four-hop route. (Some of the participating stations had also participated in the 1916 RLA test.) On February 6, 1917, a test message was sent from New York to California, and a reply message was received, in 80 minutes. This message exchange also used a four-hop route.[35]

A form of location routing was used in the trunk lines system, in that messages were passed along the grid to the station in the direction of the destination.[36] Viewed retrospectively, the major weaknesses of the trunk line network were that each operator was required to maintain five schedules per day, a grueling long-term commitment that limited participation,[37] and accommodate schedules that were not globally synchronized across the network. The former weakness, analogous to the power consumption concerns of modern wireless communication systems, resulted in operator burnout over time; the latter often resulted in greatly delayed message traffic (high message latency) during ordinary (i.e., nontesting) operation.

The trunk lines system was restarted after World War I, and it grew rapidly through the 1920s and 1930s. In 1941, all amateur transmitting ceased for the duration of World War II. After the war, the Army elected not to restart the AARS; however, the amateur trunk lines system was restarted. The performance of the system, though, was unsatisfactory: Although prior to the war, the range of the typical amateur station communication link was limited to 200–300 miles,[38] the improved performance of postwar radio equipment meant that long distance, and even transcontinental, links between two stations could be established at almost any time. "As a result, the trunk lines were slowly turning into large wide area nets instead of the relay lines that they were in the early days."[39] In addition, the location routing method of relaying a message to the station in the direction of the destination broke down because the relay chains could be so easily bypassed.

To solve these problems, and thereby improve the efficiency of the system, George Hart of the ARRL headquarters staff proposed a radical change in 1949.[40–43] The new plan, dubbed the ARRL National Traffic Plan and, later, the ARRL National Traffic System (NTS), was based on the prewar AARS tree network, but superimposed it on the ARRL organization structure.[44] At the time, the ARRL was geographically organized into 73 sections, largely along state boundaries. Hart used these as the leaves of the tree (Exhibit 2):

1. The plan called for local section nets in each of the ARRL sections. Then, traffic would be relayed between sections on regional nets.
2. Traffic between regions would be exchanged on area nets, and originally traffic between areas was handled by special area net liaison stations.
3. The original plan called for 4 area nets and 13 regional nets. Regional nets would be established in each of the four areas. The area nets were to be established on the basis of time zones: Eastern, Central, Mountain, and Pacific.
4. The Plan called for an integrated time schedule that would move traffic within sections, between section, between regions and between area nets, all following a smooth time pattern. Traffic would move from east to west in one night and from west to east in a two-day cycle.[45] (Published with the permission of the ARRL. Copyright 1988, ARRL.)

This integrated time schedule was reminiscent of the 1929 modified AARS system:

> Section and Region nets met twice each evening, with Area nets meeting once. At 7:00 p.m. local time, Section nets would convene, followed by the Region nets at 7:45 and the Area net at 8:30. The Region nets and subsequently the Section nets reconvened [at 9:15 and 10:00 p.m., respectively[46,47]] and traffic gleaned from the Area net was cleared.[48] (Published with the permission of the ARRL. Copyright 1999, ARRL.)

Other improvements reflect recognition of the workload placed on the NCSs in the AARS network, which had to both control their nets and relay all messages through them. In the NTS, the NCS assignment was not fixed, but was allowed to rotate among net members. Hart also established the con-

(Local Nets)

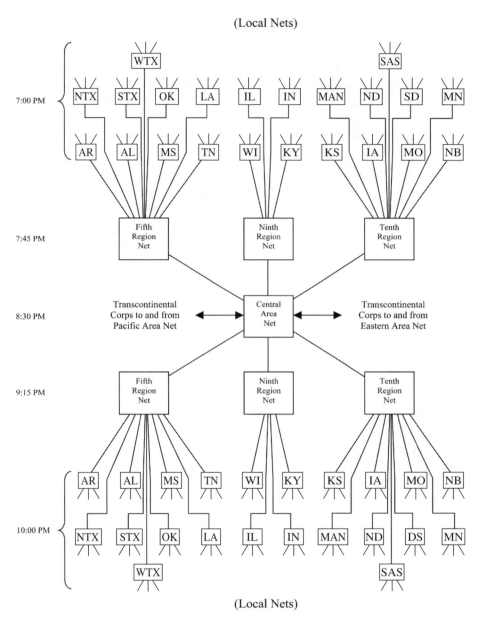

Exhibit 2. Organization of the National Traffic System, Central Area (Adapted from American Radio Relay League, Public Service Communications Manual, FSD-235, p. 28. Published with the permission of the ARRL. Copyright 1996, ARRL.)

cept of designated representatives of the Section and Region nets.[49] When a Section net, for example, was in session, messages for other Sections were passed not to the NCS, as done previously, but to a designated representative of the Section, who would then attend the Region net instead of the NCS.[50] This had two significant benefits: it spread some of the NCS workload among the other members of the net (analogous to the modern concept of power-aware routing metrics[51]), and it enabled the separation of the control and data passing functions of the net when it was in session.

Separating the control and data functions into separate logical channels was a critical advance. In the AARS approach, because all traffic to other nets had to pass through the NCS, all other net activity had to wait while this was accomplished (on the NCS channel). In the NTS system, net members needing to send messages to, or receive them from, other nets were instructed by the NCS to meet the net representative station on a different frequency (channel), pass the message traffic there, and then return to the NCS frequency. This direct member–member (or slave–slave) communication, always under the control of the NCS (master), greatly increased the possible data throughput of the net, because data transfers could occur simultaneously with net maintenance functions, such as member association and dissociation, performed by the NCS. This capability was enabled by the same key performance improvements in postwar radio technology, frequency stability, and receiver sensitivity and selectivity, which had led to the disruption of the earlier trunk lines system.[52]

As in the AARS system, when a net was in session, "the NCS is in supreme command and no station transmits unless directed to do so by him. Net stations communicate with each other only to transact business authorized by the NCS."[53] Although the net was in session, if not otherwise occupied, the NCS would regularly make short transmissions (beacons), including the name of the net and his or her identity (i.e., call sign). These beacons indicated that the NCS was ready to receive more stations into the net or respond to transmission requests from other net members in the short pause, or contention period, that followed. Medium access control was handled by a simple Request To Send/Clear To Send (RTS/CTS) system. When a net member desired to transmit, it would first listen on the net channel during the contention period following the NCS beacon, to ensure the channel was idle, then make a very short RTS transmission — either the call sign of the NCS[54] or a single letter (usually the first letter of the member's call sign suffix).[55,56] The NCS, if accepting the transmission request, performed a CTS reply by transmitting either the letters BK, if its call sign had been sent (literally, "break"),[57] or a repeat of the single letter sent.[58,59] The exact protocol used varied between nets.[60]

Messages handled by the NTS were transmitted in a standardized format. In addition to the address, message body, and signature fields, the message included a "preamble" (header) with several fields:[61]

- Message number (usually serialized by the originating station, resetting at the beginning of the year)
- Precedence (Routine, Welfare, Priority, or Emergency)
- Handling instructions (optional) (e.g., return receipt requested)
- Call sign of the station originating the message
- Check (number of words in the message, excluding the address and signature)
- Place of origin of the message (a city name)
- Message filing time (optional)
- Date (month and day) the message was filed

The NTS also included a method of multicasting messages, called "book" messages. This was a very flexible message format in which the standard header was modified to start with "book of," followed by the number of recipients, then the precedence and the rest of the header fields. After the header was sent, the common part of the message (usually the text) was sent, followed by a serialized list of the different parts of the message (usually the addresses).[62]

In later years, the Mountain Area net suffered from low activity and was folded into the Pacific Area net, leaving the nation with only three Area nets. Communication between them took place by a peer-to-peer network called the Transcontinental Corps (TCC), representatives of which attended each Area net session. Because the Area nets met at 8:30 p.m., local time, there was an hour between the start of the Eastern and Central Area nets (and two hours between the Eastern and Pacific Area nets) that was used by the TCC to send messages westward. Messages traveling eastward from the Pacific Area net were delayed until the following day.[63] This wireless communication network, with minor modifications, has been in operation for more than 50 years;[64] due to its low power ad hoc design (and longevity), it has been proposed as the basis for a wireless sensor network architecture.[65]

2.2 WIRELESS DATA NETWORKS

2.2.1 The ALOHA System

The ALOHA system[66] is generally recognized as being the first successful wireless data communication network, and was certainly the first data network to employ a random channel access protocol. Now generally remembered by its eponymous medium access control method,[67] the ALOHA system provided a 24 kBaud interactive link between a University of Hawaii mainframe computer and computer users on four of the Hawaiian Islands, using two radio frequency (RF) channels (one for transmission, one for reception) in the 400 MHz band. Although sometimes considered a "packet" radio system, a message was not broken up into multiple packets — each packet corresponded to a single message. Packets had a fixed length of 640 payload bits (80 bytes), plus 32 identification and control information bits and 32 parity bits (for a total of 704 bits in each packet).[68]

2.2.2 The PRNET System

Based at least partially on the success of the ALOHA system, plus the developing technology of the Advanced Research Projects Agency (ARPA-Net),[69] the U.S. Defense Advanced Research Projects Agency (DARPA) began a series of packet radio development programs, beginning with the DARPA Packet Radio Network (PRNET) project in 1972. The PRNET was a 12.8 Mc/s direct sequence spread spectrum system, operating at 1800 MHz, that provided a network of 138 entities (any combination of packet radios and attached devices, such as host computers) with fully distributed network management. The data rate was 100 or 400 kb/s, switchable based on link conditions; interleaving, convolutional coding — at three rates, 7/8, 3/4, and 1/2 — and a 32-bit cyclic redundancy checksum (CRC) were used.[70] Beacons were transmitted by each packet radio every 7.5 seconds to announce its existence and inform its neighbors of its understanding of the network topology. The beacons listed the number of hops from the transmitting radio to each packet radio in the network, information obtained by monitoring the beacons of its neighbors. This information was used for routing purposes.[71]

The PRNET system was not without its weaknesses. These weaknesses included the limitation to relatively small networks, the size and power consumption of PRNET nodes, and a susceptibility to certain forms of electronic attack, and were addressed in a follow-on DARPA project, the Survivable Radio Networks (SURAN) program, initiated in 1983.[72]

2.2.3 Amateur Packet Radio Networks

The first amateur packet radio communications controller, called a Terminal Node Controller (TNC), was built before 1980.[73] Following a change in Federal Communications Commission (FCC) rules to permit it, the first U.S. amateur data packet repeater station was built in 1980. Operating on a single channel at 146 MHz, it transmitted an identification beacon once every five minutes, using 1200 Baud frequency shift keying (2-FSK); the rest of the time was a contention period during which it would retransmit any error-free packet received in a simplex fashion.[74] Packet collision avoidance was performed by a Carrier Sense Multiple Access (CSMA) method as follows:

... the TNC performs a carrier-sense check (to see if anyone is using the channel). Just to reduce the possibility of two TNC boards hearing nothing and bursting packets at exactly the same time, a variable time delay is built in. Because the time delay at each TNC (user) is changing, repeated collisions between the same pair of users should not occur.[75] (Published with the permission of the ARRL. Copyright 1981, ARRL.)

Source routing was employed, but no provision was made for more than two-hop (i.e., one repeat) communication. However, the intent was to develop multi-hop capability to enable an ad hoc nationwide computer network to augment the amateur NTS.[76] This network was to be at 144 or 220 MHz;[77] while transcontinental,[78] and even intercontinental,[79] digital communication was possible at shortwave frequencies (14 and 28 MHz, respectively), difficulties with multipath propagation at these frequencies led to very low data throughput with the simple physical and link layers used (audio FSK modulation and automatic repeat request (ARQ) with no interleaving or forward error correction). This nationwide ad hoc packet communication network was eventually established, but only with the use of low-rate shortwave links between regions;[80] because it lacked a unified protocol above the link layer,[81] routing was done manually at the source node, which limited network utility.

Packet network gateways to the nascent Internet,[82] the development of several physical layers for shortwave packet communication,[83] and success using satellite-based packet networks[84] notwithstanding, probably the major advance due to amateur packet networks was the development of the Multiple Access with Collision Avoidance (MACA) channel access protocol.[85] This protocol combined the existing CSMA protocol with the RTS/CTS transmission request sequence in Apple's Localtalk network (and used manually in the NTS), to solve the "hidden terminal" and "exposed terminal" problems then extant (see Chapter 4, Section 4.2.2). MACA has since been extended and refined in a number of ways (e.g., to improve the fairness of its backoff algorithm).[86]

2.2.4 Wireless Local Area Networks (WLANs)

In 1990, the Institute of Electrical and Electronics Engineers (IEEE) 802 LAN/MAN Standards Committee (LMSC) established the 802.11 Working Group to create a WLAN standard. The first standard was released in 1997, and enabled data rates of 1 and 2 Mb/s. The standard has been revised and amended several times since its initial publication, and now includes five physical layers:[87]

- Infrared at 1 and, optionally, 2 Mb/s
- Frequency hopping spread spectrum at 1 and, optionally, 2 Mb/s at 2.4 GHz
- Direct sequence spread spectrum from 1 to 11 Mb/s at 2.4 GHz[88]

- Orthogonal Frequency Division Multiplex (OFDM) up to 54 Mb/s at 5 GHz[89,90]
- A choice between DSSS and OFDM up to 54 Mb/s at 2.4 GHz.

The 802.11 standard specifies the carrier sense multiple access with collision avoidance (CSMA/CA) channel access method, which is a refinement of the MACA protocol that employs a binary exponential random backoff mechanism.[91]

Both an ad hoc peer-to-peer connectivity and an infrastructure-based network requiring fixed access points (APs) acting as gateways between the wired (IEEE 802.3, i.e., Ethernet) and wireless networks, are specified in the standard.

2.2.5 Wireless Personal Area Networks (WPANs)

The development of WPANs began in 1997, with the formation of the Home RF Working Group,[92] and in 1998, with the formation of the Bluetooth Special Interest Group.[93] Revision 1.0 of both specifications was released in 1999 and, while the Home RF Working Group disbanded in January 2003, development of Bluetooth continues today. Both of these WPANs operate in the 2.4 GHz ISM band, employing frequency hopping spread spectrum. HomeRF had a raw data rate of 800 kb/s using 2-level frequency-shift keying (2-FSK), and optionally 1.6 Mb/s using 4-level FSK (4-FSK); a data throughput on a lightly loaded system of near 1 Mb/s was claimed, using 4-FSK.[94] Bluetooth has a raw data rate of 1 Mb/s, also using 2-FSK, and claims a maximum data throughput of 721 kb/s.[95]

HomeRF systems used the Shared Wireless Access Protocol-Cordless Access (SWAP-CA) protocol, which incorporates features of the Digital European Cordless Telephony (DECT) standard from the European Telecommunication Standards Institute (ETSI) for isochronous traffic,[96] and the IEEE 802.11 WLAN standard for conventional data transfer. SWAP-CA, therefore, employed a superframe in which a polled time division multiple access (TDMA) medium access control (MAC) method was used, for isochronous traffic, interrupted by a contention period in which the CSMA/CA MAC method was used, for asynchronous traffic. Both an ad hoc peer-to-peer and a star network, (called a "managed network") were provided; in the peer-to-peer network, only the CSMA/CA MAC was used — the entire superframe was a contention period.[97] The standard considered four types of devices — connection points (gateways), isochronous nodes (voice-centric), asynchronous nodes (data-centric), and combined isochronous and asynchronous nodes — thereby enabling heterogeneous networks.[98]

In contrast, Bluetooth employs a frequency hopping/time division duplex (FH/TDD) access method. Bluetooth's ad hoc networks, called piconets, are restricted to the star topology, with a single master and up to seven slaves; however, all devices are physically identical (i.e., the network is homogeneous). The channel is divided into 625 μs slots, during which a network node may alternately transmit or receive a single packet; the channel frequency is changed for each slot. Slaves may only transmit upon being polled by the master; there is no contention period for channel access.[99,100] There is no provision in the Bluetooth specification itself for the creation of networks larger than a single piconet (so-called "scatternets"); however, this is an active area of research (e.g., Salonidis et al.[101]).

2.3 WIRELESS SENSOR AND RELATED NETWORKS

Similar to the development of other wireless personal communications systems,[102] the present-day development of wireless sensor networks has many roots, going back at least to the 1978 DARPA-sponsored Distributed Sensor Nets Workshop at Carnegie-Mellon University in Pittsburgh, Pennsylvania (e.g., Lacoss and Walton[103]). Interest due to military surveillance systems[104] led to work on the communication and computation trade-offs of sensor networks, including their use in a ubiquitous computing environment.[105–110] (Lyytinen and Yoo[111] distinguish between ubiquitous computing, which they define as requiring a high level of user mobility, and pervasive computing, requiring a low level of user mobility — a definition that meshes better with wireless sensor networks.) This interest rose with the DARPA low-power wireless integrated microsensors (LWIM) project of the mid-1990s[112] and continued with the launch of the SensIT project[113] in 1998, which focuses on wireless, ad hoc networks for large distributed military sensor systems. A total of 29 research projects, from 25 institutions, were funded under this project. A short list of the more prominent development efforts (not all funded through SensIT) follows.

2.3.1 WINS[114]

The University of California at Los Angeles, often working in collaboration with the Rockwell Science Center, has had a Wireless Integrated Network Sensors (WINS) project since 1993. It has now been commercialized with the founding of the Sensoria Corporation (San Diego, California) in 1998.[115] The program covers almost every aspect of wireless sensor network design, from microelectromechanical system (MEMS) sensor and transceiver integration at the circuit level,[116] signal processing architectures,[117] and network protocol design,[118] to the study of fundamental principles of sensing and detection theory.[119] The group envisions that WINS will "provide distributed network and Internet access to sensors, controls, and processors deeply embedded in equipment, facilities, and the environment."[120]

The WINS communication protocol data link layer is based on a TDMA structure; separate slots are negotiated between each pair of nodes at network initiation;[121,122] the physical layer employs RF spread spectrum techniques for jamming resistance.[123]

2.3.2 PicoRadio[124]

Jan M. Rabaey of the University of California at Berkeley started the PicoRadio program in 1999 to support "the assembly of an ad hoc wireless network of self-contained mesoscale, low-cost, low-energy sensor and monitor nodes."[125] The physical layer proposed for the PicoRadio network is also direct sequence spread spectrum; the MAC protocol proposed "combines the best of spread spectrum techniques and Carrier Sense Multiple Access (CSMA)."[126] A node would randomly select a channel (e.g., a code or time slot) and check for activity. If the channel were active, the node would select another channel from the remaining available channels, until an idle channel was found and the message sent. If an idle channel was not found, the node would back off, setting a random timeout timer for each channel. It would then use the channel with the first expired timer and clear the timers of the other channels.[127]

Note that the PicoRadio program at Berkeley is distinct from its perhaps better-known "Smart Dust" program,[128] in which MEMS-based motes "could be small enough to remain suspended in air, buoyed by air currents, sensing and communicating for hours or days on end."[129] The Smart Dust physical layer would be based on passive optical transmission, employing a MEMS corner reflector to modulate its reflection of an incident optical signal.[130,131]

2.3.3 μAMPS[132]

The μAMPS program, led by Principal Investigator Anantha Chandrakasan at the Massachusetts Institute of Technology (Cambridge, Massachusetts), is focused on the development of a complete system for wireless sensor networks, emphasizing the need for low power operation. Their work has led to the development of a sensor network communication protocol called Low Energy Adaptive Clustering Hierarchy (LEACH).[133] LEACH features a node-clustering algorithm that randomizes the assignment of the high-power-consuming cluster head function to multiple nodes in the network. This spreading of the cluster head function to multiple self-elected nodes lengthens the lifetime of the network.

2.3.4 Terminodes,[134] MANET,[135] and Other Mobile Ad Hoc Networks

A field related to wireless sensor networks is the study of mobile ad hoc networks; the Terminodes project[136] and the Mobile Ad Hoc Networks (MANET) Working Group of the Internet Engineering Task Force (IETF)[137] are two examples. Although these programs are concerned with ad hoc networks of nodes with low power consumption, the main issue addressed by them is the routing problem created in an ad hoc network when mobility of network nodes is added. This problem does not arise in wireless sensor networks, the nodes of which are assumed substantially stationary. Further, mobile ad hoc networks are usually assumed to be composed of Internet Protocol (IP)-addressable nodes;[138] the ability to handle Transmission Control Protocol/Internet Protocol (TCP/IP) is generally considered to be beyond the capability of wireless sensor network nodes due to cost (e.g., required buffer memory size) and power concerns.

2.3.5 Underwater Acoustic and Deep Space Networks

A wireless sensor network protocol has design features in common with protocols for a wide range of network types, from underwater acoustic networks[139] to deep space radio networks. The need in oceanography for underwater acoustic networks, which share the low power consumption, low data throughput, large physical area network coverage, and high message latency tolerance characteristics of wireless sensor networks,[140] led to the development of several systems by the early 1990s.[141] State-of-the-art underwater acoustic networks employ phase shift keying (PSK) in the physical layer, a MACA-derived protocol for medium access control, and multi-hop routing techniques[142] — all features that would be familiar to the designer of a wireless sensor network protocol. Similarly, the strict power consumption constraints, ad hoc network architecture, and tolerance of message latency requirements common in deep space communication networks[143] also are common with wireless sensor networks.[144]

2.4 CONCLUSION

In the past 85 years, wireless data networks have gone from manually operated transcontinental radiotelegraphic networks to fully automatic local and personal area networks employing spread spectrum techniques. The methods of MAC, and network organization and operation developed for early radiotelegraphic networks were often independently reinvented for use in the computer communication networks that arose from the development of packet-switched systems in the 1960s.

Although public funding of packet-switched systems development (mostly for defense applications) has continued from the 1960s to the present, commercial development of WLANs beginning in 1990 accelerated research in wireless packet data systems for a wide variety of applications, including WPANs and wireless sensor networks. The SensIT DARPA program supported 29 research programs in the field of wireless sensor networks, and the first commercial interest in wireless sensor network systems is now appearing.

References

1. Dulaney Terrett, *The Signal Corps: The Emergency.* World War II 50th Anniversary Commemorative Edition, CMH Pub 10-16-1. Washington, D.C.: Center of Military History, United States Army. 1994. p. 49.
2. Rebecca Robbins Raines, *Getting the Message Through: A Branch History of the U.S. Army Signal Corps.* Washington, D.C.: Center of Military History, United States Army. CMH Pub 30-17. 1996. p. 224.
3. Terrett, p. 49.
4. American Radio Relay League, "The army links up with the amateur," *QST,* v.9, n. 10, October 1925, pp. 22–24.
5. Raines, p. 224.
6. Terrett, p. 54.
7. American Radio Relay League, ibid.
8. Ibid.
9. George Hart, personal communication, January 7, 2002.
10. Ibid.
11. Terrett, p. 55.
12. Clinton B. DeSoto, *Two Hundred Meters and Down: The Story of Amateur Radio.* West Hartford, CT: The American Radio Relay League. 1936. p. 164.
13. American Radio Relay League, The Army-Amateur Radio System is revised, *QST,* v. 13, n. 3, March 1929, pp. 21–25.
14. American Radio Relay League, Changes in the regulations for the Army-Amateur Radio System, *QST,* v. 14, n. 2, February 1930, pp. I–III (insert).
15. American Radio Relay League, The Army-Amateur Radio System is revised, ibid.
16. Raines, p. 235.
17. American Radio Relay League, The Army-Amateur Radio System, *QST,* v. 14, n. 5, May 1930, p. 58.
18. Terrett, ibid.
19. American Radio Relay League, The editor's mill, *QST,* v. 20, n. 5, May 1936, pp. 7–8.
20. American Radio Relay League, Army-Amateur Radio System activities, *QST,* v. 21, n. 9, September 1937, pp. 39, 105.
21. Ibid.
22. Terrett, ibid.
23. American Radio Relay League, Army-Amateur Radio System, 1930–31, *QST,* v. 14, n. 10, October 1930, pp. V–VI (insert).
24. American Radio Relay League, Army-Amateur Radio System activities, *QST,* v. 22, n. 4, April 1938, p . 44–45.
25. American Radio Relay League, Army-Amateur Radio System, 1930–31, ibid.
26. DeSoto, p. 39.
27. The Radio League of America, *The Electrical Experimenter,* December 1915, p. 381.
28. William H. Kirwan, The Washington's birthday amateur radio relay, *The Electrical Experimenter,* May 1916, p. 24.
29. DeSoto, pp. 46–48.
30. Schumacher, p. 66.
31. George Hart, A look at traffic handling prior to World War II, *QST,* v. 82, n. 6, June 1998, p. 73.
32. Hiram Percy Maxim, Practical relaying, *QST,* v. 1, n. 3, February 1916, pp.19–22.
33. Alice Clink Schumacher, *Hiram Percy Maxim.* Cortez, CO: Electric Radio Press. 1998. pp. 64–66.
34. DeSoto, pp. 38–41.
35. DeSoto, p. 49.
36. George Hart, personal communication, January 7, 2002.

37. Due to the volume of traffic carried, the life of an operator at the root of the AARS hierarchy was perhaps not noticeably better. George Hart, Random recollections of an old ham, *QCWA J.,* v. LI, n. 4, Winter 2002, pp. 40–45. Available from the Quarter Century Wireless Association, Inc., Eugene, OR.

38. George Hart, A look at traffic handling prior to World War II, *QST,* v. 82, n. 6, June 1998, p. 73.

39. Jean A. Gmelin, The History of the National Traffic System - Part 1, *QST,* v. 72, n. 11, November 1988, p. 69.

40. George Hart, New National Traffic Plan, *QST,* v. 33, n. 9, September 1949, pp. 50–51, 96, 98.

41. George Hart, A brief history and perspective of the ARRL National Traffic System, *QST,* v. 79, n. 8, August 1995, p. 91.

42. Gmelin, ibid.

43. Jean A. Gmelin, The History of the National Traffic System — Part 2, *QST,* v. 73, n. 4, April 1989, pp. 66–67.

44. American Radio Relay League, *Public Service Communications Manual, FSD-235.* Newington, CT: American Radio Relay League. 1996. Section II.

45. Jean A. Gmelin, The History of the National Traffic System — Part 1, ibid.

46. Robert J. Halprin, Hip packet, *QST,* v. 65, n. 4, April, 1981, p. 91.

47. George Hart, Message handling, in *The ARRL Operating Manual,* 2nd ed., Robert Halprin, Ed. Newington, CT: American Radio Relay League. 1985. Chapter 4.

48. Rick Palm, Golden anniversary: a look at fifty years of the National Traffic System, *QST,* v. 83, n. 9, September 1999, pp. 50–53.

49. Ibid.

50. Hart, ibid.

51. Suresh Singh, Mike Woo, and C. S. Raghavendra, Power-aware routing in mobile ad hoc networks, *Proc. MOBICOM,* 1998, pp. 181–190.

52. Ed Callaway, Sam Hsu, and Ravi Shankar, On the design of early radiotelegraphic wireless networks, Technical Report, TR-CSE-02-02, Department of Computer Science and Engineering, Florida Atlantic University, Boca Raton, FL, February 2002.

53. William Walker, How a C.W. traffic net operates, *QST,* v. 36, n. 4, April 1952, pp. 48–49, 128, 130.

54. Ibid.

55. George Hart, Message handling, ibid.

56. George Hart, personal communication, January 7, 2002.

57. Walker, ibid.

58. George Hart, Message handling, ibid.

59. George Hart, personal communication, January 7, 2002.

60. George Hart, Message handling, ibid.

61. Ibid.

62. Ibid.

63. American Radio Relay League, *Public Service Communications Manual, FSD-235.* Section II, ibid.

64. Palm, ibid.

65. Ed Callaway, Sam Hsu, and Ravi Shankar, JAN: A communications model for wireless sensor networks, Technical Report, TR-CSE-02-12, Department of Computer Science and Engineering, Florida Atlantic University, Boca Raton, FL, April 2002.

66. Norman Abramson, The ALOHA System — Another alternative for computer communications, *Proc. AFIPS Fall Joint Comput. Conf.,* v. 37, 1970, pp. 281–285.

67. Norman Abramson, Multiple access in wireless digital networks, *Proc. IEEE,* v. 82, n. 9, September 1994, pp. 1360–1370.

68. Abramson, The ALOHA System, ibid.

69. Vinton G. Cerf, Packet communication technology, in *Protocols and Techniques for Data Communication Networks,* Franklin F. Kuo, Ed., Englewood Cliffs, NJ: Prentice Hall. 1981.
70. John Jubin and Janet D. Tornow, The DARPA packet radio network protocols, *Proc. IEEE,* v. 75, n. 1, January 1987, pp. 21–32.
71. C.-K. Toh, *Ad Hoc Mobile Wireless Networks: Protocols and Systems.* Upper Saddle River, NJ: Prentice Hall. 2002. pp. 19–21.
72. Charles E. Perkins, Ed., *Ad Hoc Networking.* Boston: Addison-Wesley. 2001. pp. 35–36.
73. Philip R. Karn, Harold E. Price, and Robert J. Diersing, Packet radio in the amateur service, *IEEE J. Selected Areas in Commun.,* v. SAC-3, n. 3, May 1985, pp. 431–439.
74. Hank Magnuski, First packet repeater operational in U.S., *QST,* v. 65, n. 4, April 1981, p. 27.
75. David W. Borden and Paul L. Rinaldo, The making of an amateur packet-radio network, *QST,* v. 65, n. 10, October 1981, pp. 28–30.
76. Robert J. Halprin, Hip packet, *QST,* v. 65, n. 4, April, 1981, p. 91.
77. Borden and Rinaldo, ibid.
78. Ed Kalin, First transcontinental packet-radio QSO, *QST,* v. 66, n. 4, April 1982, p. 32.
79. American Radio Relay League, Intercontinental packet radio a reality, *QST,* v. 67, n. 8, August 1983, p. 15.
80. Harold Price, A closer look at packet radio, *QST,* v. 69, n. 8, August 1985, pp. 17–20.
81. Stan Horzepa, Packet-radio networking leaps forward, *QST,* v. 71, n. 5, May 1987, p. 66.
82. Patty Winter, Internet gateways expand the world of packet radio, *QST,* v. 77, n. 12, December 1993, p. 105.
83. Tim Riley et al., A comparison of HF digital protocols, *QST,* v. 80, n. 7, July 1996, pp. 35–39.
84. Philip R. Karn, Harold E. Price, and Robert J. Diersing, Packet radio in the amateur service, *IEEE J. Selected Areas in Commun.,* v. SAC-3, n. 3, May 1985, pp. 431–439.
85. Phil Karn, MACA — A new channel access method for packet radio, *ARRL/CRRL Amateur Radio 9th Computer Networking Conf.,* 1990, pp. 134–140.
86. Vaduvur Bharghavan, Alan Demers, Scott Shenker, and Lixia Zhang, MACAW, *ACM SIGCOMM Computer Commun. Review, Proc. Conf. on Communications Architectures, Protocols and Applications,* v. 24, n. 4, October 1994, pp. 212–225.
87. Roger B. Marks, Ian C. Gifford, and Bob O'Hara, Standards in IEEE 802 unleash the wireless Internet, *IEEE Microwave,* v. 2, n. 2, June 2001, pp. 46–56.
88. Chris Heegard et al., High-performance wireless ethernet, *IEEE Commun.,* v. 39, n. 11, November 2001, pp. 64–73.
89. Richard van Nee, A new OFDM standard for high rate wireless LAN in the 5 GHz band, *Proc. IEEE Veh. Tech. Conf.,* 1999, v. 1, pp. 258–262.
90. Bill McFarland et al., The 5-UP protocol for unified multiservice wireless networks, *IEEE Commun.,* v. 39, n. 11, November 2001, pp. 74–80.
91. Bob O'Hara and Al Petrick, *The IEEE 802.11 Handbook: A Designer's Companion.* New York: IEEE Press. 1999.
92. Jim Lansford and Paramvir Bahl, The design and implementation of HomeRF: a radio frequency wireless networking standard for the connected home, *Proc. IEEE,* v. 88, n. 10, October 2000, pp. 1662–1676.
93. Jaap C. Haartsen, The Bluetooth radio system, *IEEE Pers. Commun.,* v. 7, n. 1, February 2000, pp. 28–36.
94. Lansford and Bahl, ibid.
95. Jaap C. Haartsen et al., Bluetooth: Vision, goals, and architecture, *ACM Mobile Computing and Communications Review,* v. 2, n. 4, October 1998, pp. 38–45.
96. That is, traffic generated at regular intervals for which low latency is important (e.g., interactive voice communication).
97. Lansford and Bahl, ibid.

98. Kevin J. Negus et al., HomeRF and SWAP: Wireless networking for the connected home, *ACM Mobile Computing and Communications Review,* v. 2, n. 4, October 1998, pp. 28–37.

99. Haartsen et al., ibid.

100. Chatschik Bisdikian, An overview of the Bluetooth wireless technology, *IEEE Commun.,* v. 39, n. 12, December 2001, pp. 86–94.

101. Theodoros Salonidis et al., Distributed topology construction of Bluetooth personal area networks, *Proc. 20th Annu. Joint Conf. IEEE Computer and Commun. Societies,* 2001, v. 3, pp. 1577–1586.

102. Donald C. Cox, Wireless personal communications: what is it? *IEEE Personal Commun. Mag.,* v. 2, n. 2, April 1995, pp. 20–35.

103. R. Lacoss and R. Walton, Strawman design for a DSN to detect and track low flying aircraft, *Proc. Distributed Sensor Nets Conf.,* Carnegie-Mellon Univ., Pittsburgh, PA, December 1978, pp. 41–52.

104. Mark Hewish, Little brother is watching you: Unattended ground sensors, *Jane's Int. Defense Review,* v. 34, n. 6, June 2001, pp. 46–52.

105. Y.-C. Cheng and T. G. Robertazzi, Communication and computation trade-offs for a network of intelligent sensors, *Proc. Computer Networking Symp.,* 1988, pp. 152–161.

106. Y.-C. Cheng and T. G. Robertazzi, Distributed computation with communication delay (distributed intelligent sensor networks), *IEEE Trans. Aerospace and Electronic Sys.,* v. 24, n. 6, November 1988, pp. 700–712.

107. Mark Weiser, The computer for the 21st century, *Scientific American,* v. 265, n. 9, September 1991, pp. 94–104.

108. Mark Weiser, Some computer science issues in ubiquitous computing, *Commun. ACM,* v. 36, n. 7, July 1993, pp. 75–84.

109. Harold Abelson et al., Amorphous computing, *Commun. ACM,* v. 43, n. 5, May 2000, pp. 74–82.

110. National Research Council, Division on Engineering and Physical Sciences, Computer Science and Telecommunications Board, Committee on Networked Systems of Embedded Computers, *Embedded, Everywhere: A Research Agenda for Networked Systems of Embedded Computers.* Washington, D.C.: National Academy Press. 2001.

111. Kalle Lyytinen and Youngjin Yoo, Issues and challenges in ubiquitous computing, *Commun. ACM,* v. 45, n. 12, December 2002, pp. 62–65.

112. K. Bult et al., Low power systems for wireless microsensors, *Proc. Int. Symp. Low Power Electronics and Design,* 1996, pp. 17–21.

113. http://dtsn.darpa.mil.

114. http://www.janet.ucla.edu.

115. Gregory J. Pottie. and W. J. Kaiser, Wireless integrated network sensors, *Commun. ACM,* v. 43, n. 5, May 2000, pp. 51–58.

116. G. Asada et al., Wireless Integrated Network Sensors: Low power systems on a chip, *Proc. 24th European Solid-State Circuits Conf.,* 1998, pp. 9–16.

117. M. J. Dong, K. G. Yung, and W. J. Kaiser, Low power signal processing architectures for network microsensors, *Proc, Int. Symp. Low Power Electronics and Design,* 1997, pp. 173–177.

118. Loren P. Clare, Gregory J. Pottie, and Jonathan R. Agre, Self-organizing distributed sensor networks, *Proc. SPIE Conf. Unattended Ground Sensor Technologies and Applications,* 1999, v. 3713, pp. 229–237.

119. Pottie and Kaiser, ibid.

120. Ibid.

121. Katayoun Sohrabi and Gregory J. Pottie, Performance of a novel self-organization protocol for wireless ad-hoc sensor networks, *Proc. IEEE Veh. Tech. Conf.,* 1999, pp. 1222–1226.

122. Katayoun Sohrabi et al., A self organizing wireless sensor network, *Proc. 37th Annu. Allerton Conf. on Communication, Control, and Computing,* 1999, pp. 1201–1210.
123. Gregory J. Pottie, Wireless sensor networks, *Information Theory Workshop,* 1998, pp. 139–140.
124. http://bwrc.eecs.berkeley.edu.
125. Jan M. Rabaey et al., PicoRadio supports ad hoc ultra-low power wireless networking, *IEEE Computer,* v. 33, n. 7, July 2000, pp. 42–48.
126. Lizhi Charlie Zhong et al., An ultra-low power and distributed access protocol for broadband wireless sensor networks, presented at *Networld+Interop: IEEE Broadband Wireless Summit,* Las Vegas, 2001. http://bwrc.eecs.berkeley.edu.
127. Ibid.
128. http://robotics.eecs.berkeley.edu.
129. J. M. Kahn, R. H. Katz, and K. S. J. Pister, Next century challenges: mobile networking for Smart Dust, *Proc. Fifth Annual ACM/IEEE Int. Conf. on Mobile Computing and Networking,* 1999, pp. 271–278.
130. Brett A. Warneke et al., Smart dust: communicating with a cubic-millimeter computer, *Computer,* v. 34, n. 1, January 2001, pp. 44–51.
131. Brett A. Warneke et al., An autonomous 16 mm^3 solar-powered node for distributed wireless sensor networks, *Proc. IEEE Sensors,* 2002, v. 2, pp. 1510–1515.
132. http://www-mtl.mit.edu.
133. W. R. Heinzelman, A. Chandrakasan, and H. Balakrishnan, Energy-efficient communication protocol for wireless microsensor networks, *Proc. 33rd Annu. Hawaii Int. Conf. on System Sciences,* 2000, pp. 3005–3014.
134. http://www.terminodes.org.
135. http://www.ietf.org.
136. Jean-Pierre Hubaux et al., Toward self-organized mobile ad hoc networks: the Terminodes project, *IEEE Commun.,* v. 39, n. 1, January 2001, pp. 118–124.
137. M. Scott Corson, Joseph P. Macker, and Gregory H. Cirincione, Internet-based ad hoc networking, *IEEE Internet Computing,* v. 3, n. 4, July–August 1999, pp. 63–70.
138. Ibid.
139. Bertrand Baud, A communication protocol for acoustic ad hoc networks of autonomous underwater vehicles, Master's Thesis, 2001, Florida Atlantic University, Boca Raton, FL.
140. Ethem M. Sozer, Milica Stojanovic, and John G. Proakis, Underwater acoustic networks, *IEEE J. Oceanic Eng.,* v. 25, n. 1, January 2000, pp. 72–83.
141. John G. Proakis et al., Shallow water acoustic networks, *IEEE Commun.,* v. 39, n. 11, November 2001, pp. 114–119.
142. Ibid.
143. Clare, Pottie, and Agre, ibid.
144. J. Agre and L. Clare, An integrated architecture for cooperative sensing networks, *Computer,* v. 33, n. 5, May 2000, pp. 106–108.

Chapter 3
The Physical Layer

3.1 INTRODUCTION

This chapter describes a physical layer for wireless sensor networks. A unique Direct Sequence Spread Spectrum (DSSS) modulation method is described and evaluated that enables a high data rate, which is desirable to minimize total transceiver active time and, therefore, maximize battery life, while minimizing transceiver complexity. Performance is simulated in Additive White Gaussian Noise (AWGN) and a Bluetooth™ interference environment.

A communication protocol physical layer

> ... provides mechanical, electrical, functional, and procedural character-istics to establish, maintain, and release physical connections (e.g., data circuits) between data link entities.[1]

Wireless sensor networks are designed to handle very low data through-put (as low as a few bits/day), exchanging lower throughput and higher message latency for longer node battery life, lower cost, and self-organiza-tion. Target applications for these networks are inventory management, industrial control and monitoring, security, intelligent agriculture, and con-sumer products such as wireless keyboards and personal computer (PC)-enhanced toys. The networks often have stringent cost and battery life goals; nodes in an Institute of Electrical and Electronics Engineers (IEEE) 802.15.4 Low-Rate Wireless Personal Area Network (LR-WPAN), for exam-ple, must have multiyear operation from a 750 mAh AAA cell while being considerably simpler (having fewer logic gates, smaller analog circuitry, and less memory) than a comparable Bluetooth device. In addition, the system must be capable of universal (worldwide) unlicensed operation, which greatly limits possible frequency band, modulation, and other phys-ical layer alternatives.

Because the major design metrics of a wireless sensor network are its cost and power consumption, the physical layer design of a wireless sen-sor network is critical to its success.[2] Unfortunate choices of modulation, frequency band of operation, and coding can produce a network that, although functional, does not meet the cost or battery life goals needed for market acceptance. Although one-way networks have been proposed (in which sensors transmit to a central receiver) to achieve low cost and low power consumption,[3] this work will consider only two-way networks, in which all network nodes may both transmit and receive. Further, the networks will be capable of multi-hop operation to reduce overall network power consumption.[4,5]

3.2 SOME PHYSICAL LAYER EXAMPLES

Although wireless sensor networks may be conceived to communicate via many different mechanisms, including both electromagnetic (both radio frequency (RF) and infrared) and audio means, this chapter will focus on an RF physical layer. A survey of some relevant RF physical layers includes the following subsections.

3.2.1 BLUETOOTH

Bluetooth, a WPAN, operates in the 2.4 GHz Industrial, Scientific, and Medical (ISM) band. It employs binary Gaussian Frequency Shift Keying (GFSK), with a symbol rate of 1 MBaud. At the time of its development, devices in the 2.4 GHz ISM band were required by regulation in many countries to employ spread spectrum techniques. Bluetooth performs frequency hopping spread spectrum, randomly hopping across 79, 1-MHz channels (in the United States) at a rate of 1600 hops per second.[6,7]

Although Bluetooth is often suggested for sensor network applications, the Bluetooth physical layer is not very suitable for wireless sensor networks. Frequency hopping spread spectrum makes network discovery and association difficult because the network and prospective member node are asynchronous and the search must be done on a packet rate time scale. This causes a power consumption problem as the prospective member searches the band for the network, which may, in fact, not be present, and is especially problematic for low duty rate systems. Further, frequency hopping systems employ relatively narrowband modulation schemes (1 MHz wide in the case of Bluetooth); this means that integrated channel filtering will be difficult and expensive due to the large component (usually capacitor) values required to achieve the low-frequency filter corner. In addition, the large capacitor values associated with low-frequency circuits must be charged to a bias voltage during circuit warm-up, prior to operation; this lengthened warm-up period negatively affects node duty cycle. Further, the narrow channel separation makes the phase noise requirements of signal sources in both transmitter and receiver more difficult.

3.2.2 IEEE 802.11b

The IEEE 802.11b WLAN standard also specifies operation in the 2.4 GHz ISM band. Although similar, regulations for the 2.4 GHz band are not the same around the world:

The standard provides for 14 overlapping channels of 22 MHz between 2.4 and 2.5 GHz. Not all channels are usable in all regulatory areas; e.g., only channels 1 through 11 may be used in the United States. Channel centers are spaced 5 MHz apart.[8]

As noted in Chapter 2, the original 802.11 standard specified three physical layer options at 1 Mb/s (with an optional 2 Mb/s) — infrared as well as both frequency hopping and direct sequence spread spectrum in the 2.4 GHz band. The direct sequence layer, operating at a chip rate of 11 Mc/s, employed differential binary phase shift keying (DBPSK) at 1 Mb/s, and differential quadrature phase shift keying (DQPSK) at 2 Mb/s; 802.11b extends the direct sequence layer to 5.5 and 11 Mb/s by a technique called complementary code keying (CCK). In CCK, the chip rate and modulation type remain the same as for the 2 Mb/s rate — 1 Mc/s and DQPSK, respectively — but additional information is placed in each symbol by selecting different spreading codes. For the 5.5 Mb/s rate, four possible spreading codes can be used, so two bits of information can be placed in the DQPSK modulation, and two bits of information can be placed by choice of spreading code, for a total of four bits per transmitted symbol. For the 11 Mb/s rate, 64 spreading codes are possible, so a total of 8 bits (1 byte) may be transmitted per symbol. The standard provides for an alternative, optional technique, packet binary convolutional coding (PBCC), which produces the same 5.5 and 11 Mb/s data rate as CCK but with improved sensitivity, at the cost of increased receiver complexity.[9]

The original 1- and 2-MHz direct sequence 802.11 physical layer is a possible candidate for a wireless sensor network; its hardware requirements are simple, the data rate is high enough, and the use of direct sequence means that it avoids the problems of frequency hopping systems. However, the use of an 11-Mc/s chip rate is somewhat onerous for a low power device; it would be desirable to operate circuits associated with the pseudonoise (PN) chip sequence at a lower rate. The cost and power consumption of the extended 802.11b systems, on the other hand, are far beyond that feasible for wireless sensor networks; for example, the large number of correlators needed to identify and decode the CCK chip sequences would be far too large, and dissipate far too much power, for the receiver of a wireless sensor node.[10]

3.2.3 Wireless Sensor Networks

Many studies of wireless sensor networks have not defined the physical layer and have used existing wireless components for their first demonstration networks. However, some investigators have considered the physical layer.

3.2.3.1 PicoRadio. Rabaey et al.[11] find ultrawide band systems attractive for the physical layer of wireless sensor networks because ease of integration (i.e., low cost) is of higher priority than bandwidth efficiency. An interesting feature of the PicoRadio physical layer is the proposed use of "wake-up radios," namely:

> *To support sleep mode, a wakeup radio is used in the physical layer, which consumes less than 1 μW running at full duty cycle. It wakes up the main radio and informs it of the channel it should tune to. Because the destination ID is modulated into the wakeup signal, only the destination node will be waken [sic] up.[12]*

This concept had also been considered in the Piconet program of the University of Cambridge (England) and the Olivetti and Oracle Research Laboratory,[13] Cambridge, England, since renamed the Prototype Embedded Network (PEN) program.[14–16] (The PEN program was discontinued in 2002.) The wakeup receiver provides low average power consumption without the strict internode timing synchronization that would otherwise be necessary because only short pulses of the high power consumption ultrawide band receiver would be possible.[17]

3.2.3.2 WINS. The WINS system employs spread spectrum signaling, in either the 900 MHz or 2.4 GHz ISM bands. The physical layer design is "directed to CMOS circuit technology to permit low cost fabrication along with the additional WINS components."[18]

3.2.3.3 μAMPS. Wang et al.[19] and Shih et al.[20] note the importance of multilevel signaling to minimize the time a network node is actively communicating and, therefore, maximize the time it is sleeping (and thereby conserving power). Further, they identify the problem of start-up energy, which is the energy taken by a wireless transceiver during its transition between the sleep and active modes, and note that the energy consumed during start-up may be greater than that consumed during active operation. This occurs because the transmission time may be very short, but the start-up time is limited by physical time constants inherent in the physical layer implementation.

3.3 A PRACTICAL PHYSICAL LAYER FOR WIRELESS SENSOR NETWORKS

This section concerns the design of a physical layer suitable for wireless sensor networks. Because the primary metrics of wireless sensor protocol design are cost and battery life, it is instructional to consider each in detail.

3.3.1 Cost

The physical layer's effect on network cost is largely due to the cost of hardware — the cost of chip(s) and external parts. Integration typically reduces cost because fewer discrete parts must be purchased, handled, and placed in the final product; in the limit, one may consider a single-chip design with only an antenna and a battery as external parts (the integration of antennas[21] and batteries[22–24] will be substantially more difficult, although not impossible). Aside from the antenna and battery, the most difficult component to integrate may be the quartz crystal frequency reference. A promising alternative to the discrete quartz crystal is the integrated microelectromechanical system (MEMS) resonator.[25] One consideration in the design of the physical layer is the required reference frequency stability. Although quartz crystal resonators routinely achieve less than five ppm (pages per minute) frequency variation over consumer temperature ranges, they are a mature technology; first- or second-generation MEMS resonators may not be capable of such stability, and the physical layer design should not require it.

Another consideration for low-cost integrated devices is the relative amounts of analog and digital circuitry required. As is well known, digital circuits dimensionally shrink with lithographic improvements in integrated circuit processing; analog circuits, however, generally do not shrink with process improvements. Not only are the dimensions of analog transistors controlled by factors other than process capability (such as current density), they also often employ passive components, the dimensions of which are controlled by physical factors such as capacitor dielectric constant. Therefore, as process technology advances, digital circuits become smaller and, therefore, cheaper, while the size of analog circuits remains relatively unchanged. Further, as integrated circuit technology advances and minimum feature size is reduced, the cost of die area actually increases. This penalizes analog circuits, which become more expensive with process improvements.

The lowest cost implementation, therefore, is either (1) a nearly all-analog approach in a large-dimension, relatively old process with a low cost per die area, or (2) a nearly all-digital approach in a small-dimension, state-of-the-art process with a low cost per digital gate. The deciding factor is that the desired system-on-a-chip (SoC) design must include RF circuits, as well. These likely cannot be made in a large-dimension process, due to the lower gain-bandwidth product of the transistors. The nearly all-digital approach is, therefore, the proper approach.

Note that the digital approach is the proper approach even if, at a given point in integrated circuit process development, the total circuit area of the digital-centric approach is larger than an alternative analog-centric approach. With process improvements over time, the cost of the digital approach will monotonically drop, while the cost of the analog approach will, at best, remain constant (if one stays with an aging process) or may even increase (if the integrated circuit (IC) vendor no longer supports the old process and the design must be moved to a more modern one).

One of the largest circuits in many transceivers is the channel filtering of the receiver. The channel filters, if analog, are typically large because the filters require relatively large capacitors to obtain sufficient strong signal (dynamic range) performance;[26] even if the main channel filtering is done with a digital filter, after an analog-to-digital converter, the analog-to-digital converter requires an antialias filter before it. The size of the analog channel filter is inversely proportional to the filter's corner frequency; if possible, the physical layer design should maximize the required channel filtering corner frequency (i.e., have a wide bandwidth) to minimize the die area, and therefore the cost, of the resulting IC. In addition, the amount of channel selectivity required in the physical layer should be as little as possible, consistent with proper network operation, so that as few poles of filtering as possible may be specified.

Another factor in network cost that is affected by the physical layer is the need to maximize market size and, therefore, achieve cost reductions due to volume production. This means that the physical layer should be compatible with the regulations in as many countries worldwide as possible. The first decision is the frequency band in which to operate. Clearly, it is impractical to license individual wireless sensor network nodes, or even the actual networks, just due to their expected ubiquity and low maintenance and cost goals. This then requires the identification of a band, available nearly worldwide, that allows unlicensed operation and has sufficient bandwidth. The bands that meet these requirements are the unlicensed ISM bands.

An additional requirement is the ability to integrate, with available or soon-to-be-available technology, a radio transceiver for the chosen band in the network node SoC.[27,28] Although operation as high as 60 GHz has been proposed for some consumer products,[29] and future integration of a 60-GHz transceiver with SoC-compatible technologies (e.g., silicon CMOS) does not seem out of the question,[30] at present 60-GHz operation requires "exotic" materials (e.g., compound semiconductors, such as gallium arsenide) that make a SoC design impractical due to cost and current drain considerations. Node operation at relatively low frequencies, below 1 GHz, suffers from the poor efficiency of electrically small antennas because the physical size of network nodes is limited.[31] Choice of the proper ISM band is, therefore, a compromise between the cost and power drain associated with high-frequency operation, and the physical size and antenna efficiency limitations associated with low-frequency operation. Above 1 GHz and below 60 GHz, three ISM bands are used: the 2.4-GHz band (2.400–2.4835 GHz in the United States); the 5.8-GHz band (5.725–5.875 GHz in the United States);[32] and the 24-GHz band (24.000–24.250 GHz in the United States). Of these, the 2.4-GHz band was chosen as having the best combination of antenna efficiency and power consumption.

Note that the 2.4-GHz band is not unoccupied; similar to all ISM bands, it is populated with many services. The IEEE 802.11b (Wi-Fi™) WLAN and the Bluetooth WPAN, among a host of other services, are in the 2.4-GHz band. The study of how these services coexist is, therefore, of significant interest, and must be considered in the evaluation of the proposed wireless sensor network physical layer.[33,34] As discussed in this chapter and in Section 4.3.4 of Chapter 4, a complicating factor in unlicensed bands (beyond the simple reduction of data throughput caused by packet collisions and retransmissions) is that different services often employ different channel access mechanisms; this situation can result in unfair channel access to one service over another. These problems can be mitigated to a degree by a physical layer design that is resistant to the interference environment (e.g., by using spread spectrum techniques to provide processing gain).

An alternative that has recently become available is the use of the 3.1–10.6 GHz ultrawideband (UWB) frequency band, defined by the Federal Communications Commission (FCC).[35] The use of UWB techniques instead of conventional carrier-based modulation schemes can offer improved location determination performance (down to the centimeter range), can support very high network node densities, and can have a very low probability of detection — all critical factors for some wireless sensor network applications. Although much of the world may follow the lead of the FCC and establish similar regulations (perhaps after a period of cautious observation of the new technology), at present, such a physical layer would be limited to the United States.

3.3.2 Power

Two aspects of the power problem must be considered in the physical layer design: the behavior of the power source and the power consumption of the system.

3.3.2.1 Power Source. Because the average power consumption of a wireless sensor network node is expected to be very low, on the order of 50 μW or less, unconventional power sources become plausible alternatives. These include those based on scavenging from ambient sources of energy, such as solar, RF, and mechanical vibration sources.[36,37] However, despite their usefulness in particular applications, each of these sources has limitations — solar-powered devices must be lit, and vibration-powered devices, of course, must have vibration. To maintain generality, this chapter considers the more conventional battery.

Although many battery technology alternatives are available, and many criteria exist for the selection of a battery technology for a specific application,[38] the design of a protocol physical layer can take advantage of a critical behavior of batteries that seems to be technology independent. A battery undergoing pulsed discharge has a greater capacity than an identical battery undergoing a constant discharge. Further, the lower the duty cycle of the pulsed discharge, the greater the battery capacity. This charge recovery effect has been demonstrated experimentally in manganese–zinc, alkaline–manganese,[39] and lead–acid[40] cells, modeled and demonstrated experimentally in manganese–zinc[41] and lithium–ion[42] cells, and modeled in zinc–zinc oxide cells;[42] it has even been modeled in more esoteric battery technologies, such as the lithium–aluminum, iron sulfide battery.[43]

Use of the charge recovery effect has been proposed as a method of extending the battery life of portable communication devices;[45–48] its use in the physical layer of wireless sensor networks, which have low duty cycle, bursty communication, is a clear opportunity. In practice, taking advantage of the charge recovery effect means that the protocol should activate high-power dissipation components (e.g., the transceiver) in infrequent bursts (resulting in a low duty cycle of operation). Also, the bursts should be separated from each other by the maximum extent possible, thereby giving the cell the maximum possible time to recover before the next discharge pulse.

3.3.2.2 Power Consumption. Because the active power consumption of the transceiver in a wireless sensor network node is much greater than its standby power consumption, the node must operate its transceiver in a low duty cycle mode to get low average power consumption. One is then placed in the happy position in which the operating condition of the node that provides the lowest average power drain is exactly the condition that maximizes the available battery capacity.

A typical 2.4-GHz CMOS transceiver of modest specification and power output (e.g., 0 dBm) will have a power drain of about 32 mW while transmitting and about 38 mW while receiving (taking the values from Appendix C[49] and adding 5 mW for microcomputer power consumption). For the purpose of a duty cycle calculation in a peer-to-peer network, one may make the simplifying approximation that the power drains of both transmitter and receiver are the same (an average of 35 mW), and the further simplifying assumption that the communication links are symmetrical (that is to say, equal time is spent transmitting and receiving). To meet a 1-year battery life goal with such a transceiver under these assumptions, its duty cycle must be extremely low: if we assume a battery life goal of one year (8760 hours) from two, 750-mAh AAA cells in series, an average current drain of

$$I_{avg} = 750 \text{ mAh}/8760 \text{ h} = 86 \text{ }\mu\text{A} \tag{1}$$

is required. If the node transceiver is supplied from these batteries through a 1.8 V linear voltage regulator, in which the output current equals the input current, an average power drain

$$P_{avg} = 1.8 \text{ V} * 86 \text{ } \mu\text{A} = 154.8 \text{ } \mu\text{W} \tag{2}$$

is required. For this system, the average current drain is

$$I_{avg} = T_{on} * I_{on} + (1 - T_{on}) * I_{stby} \tag{3}$$

where

T_{on} = Fraction of time either receiver or transmitter is on

I_{on} = Current drain from the battery when either the receiver or transmitter is on

I_{stby} = Current drain from the battery when both transmitter and receiver are off

From this equation, and estimates of the current consumption of practical hardware, the maximum acceptable T_{on} may be determined as a function of the maximum device average battery current. For example, if I_{on} = 19.5 mA, I_{stby} = 30 μA, and I_{avg} = 86 μA, T_{on} = 0.0029, or 0.29 percent. This low value implies that, despite the low data throughput of wireless sensor networks, a relatively high data rate is required so that an active device may finish communication quickly and return to sleep, to produce the low duty cycle needed to meet the system battery life requirements. This must be achieved, of course, without increasing I_{on} excessively.

Active power consumption of a transceiver is more closely tied to the symbol rate than to the raw data rate. Although it is desirable to maximize the raw data rate to minimize power consumption (by minimizing T_{on}), it is, therefore, also desirable to minimize the symbol rate, to minimize I_{on}. The simultaneous requirements of maximizing data rate and minimizing symbol rate leads one to the conclusion that multilevel or *M-ary* signaling (i.e., sending more than one bit per symbol) should be employed in the physical layer of wireless sensor networks.[50,51] The use of *M-ary* signaling, however, cannot be done indiscriminately; for example, if simple multiple phase *M-ary* signaling were used, the result would be a reduction of detection sensitivity that, to recover link margin, would require an increase in transmitted power that could result in no net power consumption savings. To avoid this, some type of orthogonal *M-ary* signaling should be employed.[52]

The value of T_{on} includes any warm-up time of the communication circuits, during which no communication is possible, but large amounts of current may be drawn from the battery. If the actual communication time is sufficiently small, the primary factor determining battery life may, in fact, be warmup current drain.[53,54] It is, therefore, important to minimize the length of the warmup time; the physical layer protocol may ease the

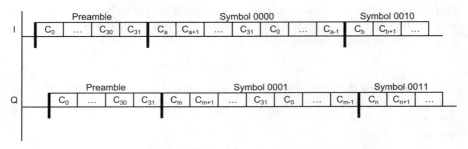

Exhibit 1.   Code Position Modulation

task of the hardware designer by considering channel frequencies and spacing that allow a high synthesizer reference frequency to be used (for fast lock time) and reducing or eliminating requirements for narrow bandwidth filtering (for example, in channel selectivity filters). This gives wide bandwidth systems, such as direct sequence spread spectrum systems, an advantage over narrowband systems for low duty cycle networks.[55] To minimize I_{on}, however, it is important to minimize the chip rate of the direct sequence system, just as the symbol rate should be minimized.

An additional advantage of direct sequence systems is that their implementations have a high digital circuit content, and relatively low analog circuit content. As noted earlier, this is desirable for cost reasons due to the economics of integrated circuit fabrication.

Although the rules have recently been modified in the United States,[56-58] in many other parts of the world, operation in the 2.4-GHz band requires some form of spread spectrum modulation if a significant amount of transmitted power is to be employed. (Lower-power classes of operation that do not require spreading are used in most countries.) Further, the European regulations allow a higher transmit power, which may be desired in certain applications if the raw data rate is at least 250 kb/s.[59] For these reasons, plus its cost and low power advantages, the proposed physical layer employs direct sequence spread spectrum modulation, with a raw data rate of 250 kb/s.

The modulation scheme used in this work is called differential code (or pulse) position modulation; it employs two 32-chip (augmented) random pseudonoise (PN) sequences, one each on the I and Q channels, transmitted with offset-quadrature phase shift keying (QPSK). Half-sinusoidal pulse shaping of the chips is employed; the resulting modulation then has a constant envelope[60] and is, therefore, suitable for inexpensive nonlinear power amplifiers. The PN sequences used are maximal-length sequences (m-sequences). The I channel uses the PN sequence with the characteristic polynomial 45 (octal), and the Q channel uses the PN sequence with the characteristic polynomial 75 (octal).[61] The symbol rate is 31.25 kSymbols /s. As illustrated in Exhibit 1, information is placed on each channel in each symbol by cyclically shifting the PN sequence of each channel by one of 16 possible shift values.[62,63] The information is placed on each channel differentially (i.e., the information is the difference in shift values between that of the present symbol and the one immediately preceding it on the same channel). Because one of M = 16 shift values (each containing k = 4 bits of information) are placed on each of the I and Q channels during one symbol time, a total of eight bits (one byte) are transmitted per symbol. (Because 32 chips are in the PN sequence, it is possible to set M = 32 and send k = 5 bits in each symbol, but this was not done for reasons of complexity; it is easier to parse eight-bit bytes into four bits than five, and having fewer shift values simplifies receiver implementation.) Because the packet lengths will be short (< 100 bytes), symbol synchronization is achieved during a PHY packet preamble and fixed during data transmission.

Code position modulation is a form of orthogonal coding in which the symbols transmitted on each of the I and Q channels are selected from a set of orthogonal codewords, in this case each 32 chips long. For equally likely, equal-energy orthogonal signals, the probability of bit error $P_B(k)$ in a single I or Q channel can be bounded by[64]

$$P_B(k) \leq (2^{k-1})Q\left(\sqrt{\frac{kE_b}{N_0}}\right) \tag{4}$$

where

$$Q(x) = \frac{1}{\sqrt{2\pi}} \int_x^\infty \exp\left(-\frac{u^2}{2}\right) du \tag{5}$$

is the complementary error function, k is the number of information bits sent in each symbol, E_b is the received signal energy/bit, and N_0 is the thermal noise power spectral density.[65] Because data is encoded differentially in the proposed modulation method, however, symbol errors occur in pairs; this doubles the bit error rate (BER):

$$P_B(k) \leq (2^k)Q\left(\sqrt{\frac{kE_b}{N_0}}\right) \tag{6}$$

Orthogonal coding enjoys increased, instead of decreased, detection sensitivity over the uncoded binary waveform; that is, the value of E_b/N_0

required for a given bit error probability is lower for orthogonal coding than for the uncoded binary waveform. The cost is an increase in bandwidth; however, bandwidth efficiency is not a critical performance metric for wireless sensor networks, due to their low offered data throughput.

Code position modulation, by employing a single PN sequence per channel, offers significant cost savings over a conventional multilevel direct sequence system, by simplifying PN code generation. The conventional system requires that separate PN sequences be generated for each possible transmitted sequence and compared with the received sequence. Code position modulation allows the PN code to be generated once, compared with the received sequence, and then cyclically shifted and compared again. In many integrated circuit implementations, the generation and storage of PN sequences can represent a large portion of the die area, and, therefore, the cost, of the receiver. For the proposed system in which 16 possible sequences per channel may be sent, the reduction in die area enabled by generating one sequence per symbol, instead of 16, can be significant; because shifting a PN sequence is a lower-power activity than generating a PN sequence, power consumption is also lower.

The orthogonal signaling technique has been applied to high-speed physical layers (e.g., IEEE 802.11b[66]), but its application to wireless sensor networks is believed to be a novel approach.

3.4 SIMULATIONS AND RESULTS

This physical layer was simulated in version 4.5 of the Signal Processing Worksystem (SPW) from Cadence Design Systems, San Jose, California, to determine the detector sensitivity (E_b/N_0), and susceptibility to interference from Bluetooth transmissions, which may also be present in the 2.4-GHz band. SPW is described in Appendix A.

3.4.1 Simulations

As illustrated in Exhibit 2, the BER simulation is both simple and flexible. Both I and Q channels are generated and detected independently, so that studies of possible differences in performance between them (due to autocorrelation differences between PN sequences, for example) may be made. This results in some slowdown of simulations, but because the packet size in these systems is so small (< 800 bits), there is little need for BER information below BER = 10^{-5}. This means that the long simulations needed (to achieve a statistically significant number of bit errors) for lower BERs are unnecessary, so simulation speed is not of critical importance.

The modulator block (Exhibit 3) produces the desired shifted PN sequence, based on the incoming data, from the 16 possible choices. After combining the I and Q channels, the transmitted signal is sent over an

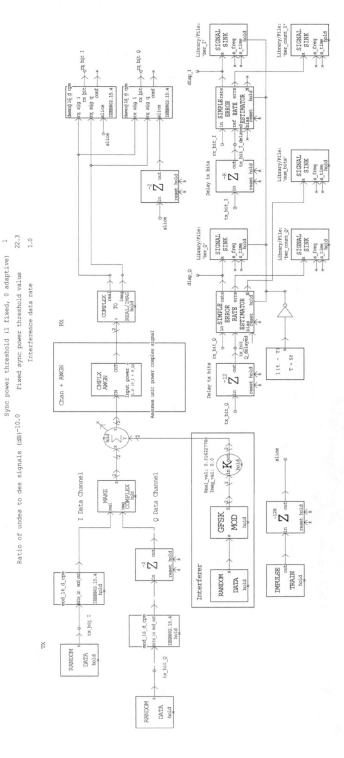

Exhibit 2.   Code Position Modulation Simulation in SPW

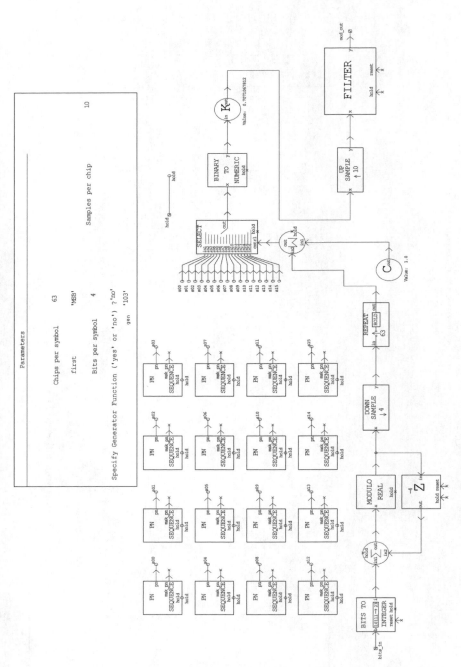

Exhibit 3. SPW Simulation of CPM Modulator

AWGN channel block representing receiver noise. This block is a large file of random noise, used so that repeatable results could be obtained even when modifications to the system were made.

The range of a wireless sensor network communication link is usually limited by design to 10 meters or less, so the one-way flight delay of the RF signal, even if via an indirect (e.g., reflected) path, is 30 nanoseconds or less. Because the duration of a chip is one microsecond, the effects of multipath interference were not considered significant and so were not simulated.

The demodulator block (Exhibit 4) is a correlator to each of the 16 possible transmitted sequences on each channel, matched to the half-sine-shaped chips. Because the purpose of this study was to investigate the performance of the proposed physical layer during data transmission, instead of the performance of potential symbol synchronization algorithms, the symbol synchronization was fixed (i.e., "hard-wired" from transmitter to receiver).

To investigate the effects of potential interference from Bluetooth devices, which will also occupy the 2.4-GHz ISM band, a Bluetooth interfering scenario was simulated (Exhibit 5).

The Gaussian minimum shift keying (GMSK) modulator block in the SPW communication library was modified to reduce its modulation index from 0.5 to 0.315, the nominal Bluetooth value. An adjustable gain block was placed after the GFSK modulator, to control the ratio of undesired to desired signal level. The GFSK modulator was fed with 1-Mb/s random data, and no attempt was made to offset or randomize the phases of the 1-Mb/s Bluetooth data bits and the 1-Mc/s direct sequence chips. Similarly, no attempt was made to model any higher-protocol stack levels (e.g., frame structure) of either Bluetooth or the sensor network, so neither packet error rate (PER) nor message error rate (MER) was simulated.

3.4.2 Results

The BER versus E_b/N_0 performance of the proposed modulation method, as simulated by SPW, is given in Exhibit 6, and is compared with the theoretical prediction of Equation 5 for $k = 1$ and $k = 4$. The SPW Grouping parameter script was run to link the number of samples simulated with the E_c/N_0 parameter (and, therefore, the expected number of bit errors), so that each data point was simulated just long enough to obtain 100 bit errors. This minimized simulation time while ensuring statistically valid results. The E_b/N_0 ratio, required in this simulation to achieve 1 percent BER using the proposed modulation method, is approximately 5.2 dB. This is within 1.25 dB of the theoretical value calculated using Equation 5 with $k = 4$, and is approxi-

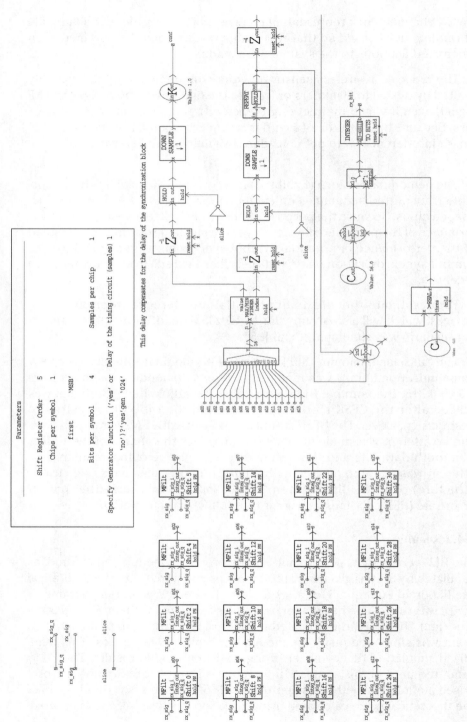

Exhibit 4. SPW Simulation of CPM Demodulator

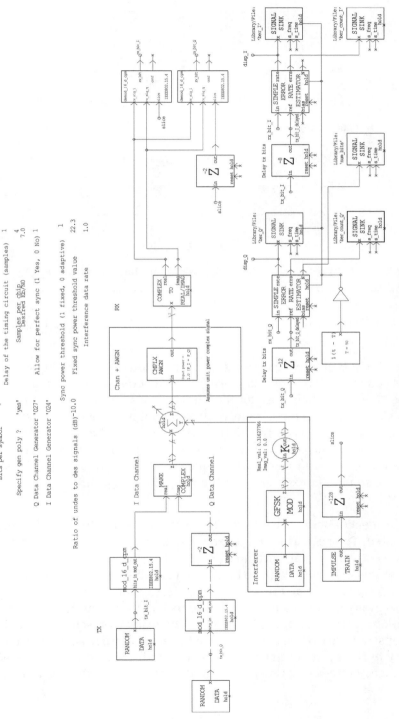

Exhibit 5. SPW Simulation of Bluetooth Interference Scenario

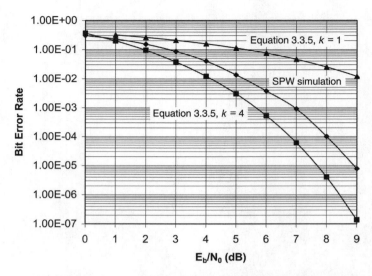

Exhibit 6.   BER Performance of Code Position Modulation versus E_b/N_0

mately 4 dB better than that obtained with $k = 1$ (i.e., binary orthogonal signaling).

The 1.25-dB difference between the simulation results and Equation 5 is largely due to the nonideal autocorrelation and cross-correlation properties of the PN sequences chosen for the I and Q codes. The correlation between a code and, itself shifted in time, is close to, but not exactly, zero; similarly, the correlation between the 45 (octal) code on the I channel and the 75 (octal) code on the Q channel is not zero. This means that the signals used are not precisely orthogonal, as assumed by the theory; this trade of sensitivity was made in exchange for the simple circuit implementation that results.

The effect of Bluetooth interference on the sensor network BER was also simulated. The scenarios simulated were interference to a relatively weak sensor network signal, with an E_b/N_0 level of 4 dB, a somewhat stronger sensor network signal, with an E_b/N_0 level of 7 dB, and a very strong sensor network signal, with an E_b/N_0 level of 50 dB. The amount of Bluetooth interference was parameterized and allowed to vary in amplitude relative to the desired signal level, from −10 to +15 dB. The results, presented in Exhibit 7, demonstrate clearly the effect of spread-spectrum processing gain, and the ability of the system to survive, albeit at an increased BER, even in the presence of interfering Bluetooth signals stronger than the desired sensor network signal.

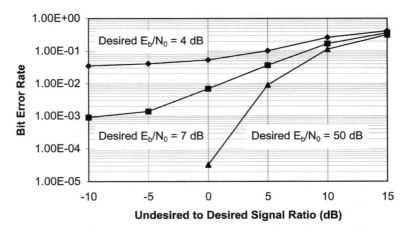

Exhibit 7. BER Performance of Code Position Modulation in the Presence of Bluetooth Interference

3.5 CONCLUSION

Features required to meet the cost and power consumption requirements of a wireless sensor network physical layer were reviewed, including the need to support high levels of integration and low duty cycle operation. Code position modulation, a type of modulation not previously proposed for wireless sensor networks, was shown to be compatible with these features. This modulation method enables a relatively high data rate, which is needed to reduce device duty cycle and thus improve battery life. This method is also compatible with a high level of digital, as opposed to analog, integration and, therefore, can enable low-cost implementations. The BER versus E_b/N_0 performance of code position modulation was simulated with SPW, and good agreement was found between the simulation result and that predicted by analytical theory. Because the network will operate in the 2.4-GHz band, and may experience interference from Bluetooth devices, a second simulation was performed to evaluate the BER in the presence of varying levels of Bluetooth interference. The expected resistance to narrowband signals caused by the processing gain of the spread spectrum modulation was observed in the simulation.

References

1. Hubert Zimmermann, OSI reference model — the ISO model of architecture for Open Systems Interconnection, *IEEE Trans. Commun.*, v. COM-28, n. 4, April 1980, pp. 425–432.
2. A. A. Abidi, Gregory J. Pottie, and William J. Kaiser, Power-conscious design of wireless circuits and systems, *Proc. IEEE*, v. 88, n. 10, October 2000, pp. 1528–1545.

3. Carl Huben, The Poisson network, *Circuit Cellar*, n. 113, December 1999, pp. 12–19.

4. I. Stojmenovic and Xu Lin, Power-aware localized routing in wireless networks, *IEEE Trans. Parallel and Distributed Syst.*, v. 12, n. 11, November 2001, pp. 1122–1133.

5. Priscilla Chen, Bob O'Dea, and Ed Callaway, Energy efficient system design with optimum transmission range for wireless ad hoc networks, *Proc. IEEE Intl. Conf. Commun.*, 2002, v. 2, pp. 945–952.

6. Jaap C. Haartsen and Sven Mattisson, Bluetooth — a new low-power radio interface providing short-range connectivity, *Proc. IEEE*, v. 88, n. 10, October 2000, pp. 1651–1661.

7. Tom Siep, *An IEEE Guide: How to Find What You Need in the Bluetooth Spec.* New York: IEEE Press. 2001.

8. Roger B. Marks, Ian C. Gifford, and Bob O'Hara, Standards in IEEE 802 unleash the wireless Internet, *IEEE Microwave*, v. 2, n. 2, June 2001, pp. 46–56.

9. Chris Heegard et al., High-performance wireless ethernet, *IEEE Commun.*, v. 39, n. 11, November 2001, pp. 64–73.

10. Brent A. Myers et al., Design considerations for minimal-power wireless spread spectrum circuits and systems, *Proc. IEEE*, v. 88, n. 10, October 2000, pp. 1598–1612.

11. Jan M. Rabaey et al., PicoRadios for wireless sensor networks: The next challenge in ultra-low power design, *IEEE Int. Solid-State Circuits Conf. Digest of Technical Papers*, 2002, v. 1, pp. 200—202; v. 2, pp. 156—157, 444–445.

12. Lizhi Charlie Zhong et al., An ultra-low power and distributed access protocol for broadband wireless sensor networks, *Networld+Interop: IEEE Broadband Wireless Summit*, Las Vegas, 2001. http://bwrc.eecs.berkeley.edu.

13. Frazer Bennett et al., Piconet: embedded mobile networking, *IEEE Pers. Commun.*, v. 4, n. 5, October 1997, pp. 8–15.

14. http://www.uk.research.att.com.

15. Gray Girling et al., The design and implementation of a low power ad hoc protocol stack, *Proc. IEEE Wireless Communication and Networking Conf.*, September 2000, pp. 282–288.

16. Gray Girling et al., The PEN low power protocol stack, *Proc. 9th IEEE In. Conf. on Computer Communication and Networks*, v. 3, October 2000, pp. 1521–1529.

17. Rabaey et al., ibid.

18. G. Asada et al., Wireless integrated network sensors: low power systems on a chip, *Proc. 24th European Solid-State Circuits Conf.*, 1998, pp. 9–16.

19. Andrew Y. Wang et al., Energy efficient modulation and MAC for asymmetric RF microsensor systems, *IEEE Intl. Symp. Low Power Electronics and Design*, 2001, pp. 106–111.

20. Eugene Shih et al., Physical layer driven protocol and algorithm design for energy-efficient wireless sensor networks, *Proc. MOBICOM*, 2001, pp. 272–287.

21. K. Kim and K. K. O, Characteristics of integrated dipole antennas on bulk, SOI, and SOS substrates for wireless communication, *Proc. IITC*, 1998, pp. 21–23.

22. J. B. Bates, D. Lubben, and N. J. Dudney, Thin-film Li-LiMn$_2$O$_4$ batteries, *IEEE AESS Syst. Mag.*, v. 10, n. 4, April 1995, pp. 30–32.

23. W. C. West et al., *Thin Film Li Ion Microbatteries for NASA Applications.* NTIS order number N20000109952. Springfield, VA: National Technical Information Service. 1999.

24. Nelson (Hou Man) Chong et al., Lithium batteries for powering sensor arrays, in M. A. Ryan et al., Eds., *Power Sources for the New Millennium.* Pennington, NJ: Electrochemical Society proceedings volume PV2000-22, October 2000, pp. 266–271.

25. Clark T.-C. Nguyen, Micromechanical circuits for communication transceivers, *Proc. IEEE Bipolar/BiCMOS Circuits and Technology Meeting*, 2000, pp. 142–149.

26. Yannis Tsividis, Continuous-time filters in telecommunications chips, *IEEE Commun.*, v. 39, n. 4, April 2001, pp. 132–137.

27. Peter G. M. Baltus and Ronald Dekker, Optimizing RF front ends for low power, *Proc. IEEE*, v. 88, n. 10, October 2000, pp. 1546–1559.
28. Mehmet Soyuer et al., Low-power multi-GHz and multi-Gb/s SiGe BiCMOS circuits, *Proc. IEEE*, v. 88, n. 10, October 2000, pp. 1572–1582.
29. Peter Smulders, Exploiting the 60 GHz band for local wireless multimedia access: prospects and future directions, *IEEE Commun.*, v. 40, n. 1, January 2002, pp. 140–147.
30. Thomas H. Lee and Simon Wong, CMOS RF integrated circuits at 5 GHz and beyond, *Proc. IEEE*, v. 88, n. 10, October 2000, pp. 1560–1571.
31. However, regulatory requirements for certain applications, such as the Medical Implant Communications Service for implantable devices, may require operation as low as 402 MHz. U.S. Code of Federal Regulations, 47 C.F.R. §§95.601–95.673. Washington, D.C.: U.S. Government Printing Office. 2001.
32. Pete Fowler, 5 GHz goes the distance for home networking, *IEEE Microwave Mag.*, v. 3, n. 3, September 2002, pp. 49–55.
33. Jon M. Peha, Wireless communications and coexistence for smart environments, *IEEE Pers. Commun.*, v. 7, n. 5, October 2000, pp. 66–68.
34. Ivan Howitt, WLAN and WPAN coexistence in UL band, *IEEE Trans. Veh. Tech.*, v. 50, n. 4, July 2001, pp. 1114–1124.
35. U.S. Federal Communications Commission, *First Report and Order,* FCC 02-48, 14 February 2002. Washington, D.C.: U.S. Government Printing Office.
36. S. Meninger et al., Vibration-to-electric energy conversion, *Proc. Int. Symp. Low Power Electronics and Design,* 1999, pp. 48–53.
37. S. Meninger et al., Vibration-to-electric energy conversion, *IEEE Trans. Very Large Scale Integration (VLSI) Systems,* v. 9, n. 1, February 2001, pp. 64–76.
38. J. F. Freiman, Portable computer power sources, *Proc. Ninth Annu. Battery Conf. on Applications and Advances,* 1994, pp. 152–158.
39. S. Okazaki, S. Takahashi, and S. Higuchi, Influence of rest time in an intermittent discharge capacity test on the resulting performance of manganese-zinc and alkaline-manganese dry batteries, *Prog. in Batteries & Solar Cells,* 1987, v. 6, pp. 106–109.
40. Rodney M. LaFollette, Design and performance of high specific power, pulsed discharge, bipolar lead acid batteries, *10th Annu. Battery Conf. on Applications and Advances,* 1995, pp. 43–47.
41. E. J. Podlaha and H. Y. Chen, Modeling of cylindrical alkaline cells: VI variable discharge conditions, *J. Electrochem. Soc.,* v. 141, n. 1, January 1994, pp. 28–35.
42. Thomas F. Fuller, Marc Doyle, and John Newman, Relaxation phenomena in lithium-ion-insertion Cells, *J. Electrochem. Soc.,* v. 141, n. 4, April 1994, pp. 982–990.
43. Mark J. Isaacson, Frank R. McLarnon, and Elton J. Cairns, Current density and ZnO precipitation-dissolution distributions in Zn-ZnO porous electrodes and their effect on material redistribution: a two-dimensional mathematical model, *J. Electrochem. Soc.,* v. 137, n. 7, July 1990, pp. 2014–2021.
44. Richard Pollard and John Newman, Mathematical modeling of the lithium-aluminum, iron sulfide battery II: the influence of relaxation time on the charging characteristics, *J. Electrochem. Soc.,* v. 128, n. 3, March 1981, pp. 503–507.
45. C. F. Chiasserini and R. R. Rao, A model for battery pulsed discharge with recovery effect, *Proc. Wireless Commun. Networking Conf.,* 1999, v. 2, pp. 636–639.
46. C. F. Chiasserini and R. R. Rao, Routing protocols to maximize battery efficiency, *Proc. MILCOM,* 2000, v. 1, pp. 496–500.
47. C. F. Chiasserini and R. R. Rao, Stochastic battery discharge in portable communication devices, *15th Annual Battery Conf. on Applications and Advances,* 2000, pp. 27–32.
48. C. F. Chiasserini and R. R. Rao, Improving battery performance by using traffic shaping techniques, *IEEE J. Selected Areas in Commun.,* v. 19, n. 7, July 2001, pp. 1385–1394.
49. Supplied by Motorola.
50. Shih et al., ibid.

51. Wang et al., ibid.
52. Bernard Sklar, *Digital Communications: Fundamentals and Applications.* Englewood Cliffs, NJ: PTR Prentice Hall. 1988.
53. Shih et al., ibid.
54. Wang et al., ibid.
55. Jose A., Gutierrez et al., IEEE 802.15.4: A developing standard for low-power, low-cost wireless personal area networks, *IEEE Network,* v. 15, n. 5, September/October 2001, pp. 12–19.
56. U.S. Code of Federal Regulations, 47 C.F.R. § 15.247. Washington, D.C.: U.S. Government Printing Office. 2001.
57. U.S. Federal Communications Commission, *Further Notice of Proposed Rulemaking and Order,* FCC 01-158, 11 May 2001. Washington, D.C.: U.S. Government Printing Office.
58. U.S. Federal Communications Commission, *Second Report and Order,* FCC 02-151, 30 May 2002. Washington, D.C.: U.S. Government Printing Office.
59. European Telecommunication Standards Institute, Electromagnetic Compatibility and Radio Spectrum Matters (ERM); Wideband Transmission Systems; Data Transmission Equipment Operating in the 2,4 GHz ISM Band and Using Spread Spectrum Modulation Techniques. Document EN 300-328. Sophia-Antipolis, France: European Telecommunication Standards Institute. 2001.
60. Leon W. Couch II, *Digital and Analog Communication Systems,* 5th ed. Upper Saddle River, NJ: Prentice Hall. 1997.
61. The characteristic, or generating, polynomial is a compact way of identifying a particular PN sequence. For a detailed discussion, see Jhong Sam Lee and Leonard E. Miller, *CDMA Systems Engineering Handbook.* Boston: Artech House. 1998.
62. Isao Okazaki and Takaaki Hasegawa, Spread spectrum pulse position modulation — a simple approach for Shannon's limit, *Proc. ICCS/ISITA,* 1992, v. 1, pp. 300–304.
63. Isao Okazaki and Takaaki Hasegawa, Spread spectrum pulse position modulation and its asynchronous CDMA performance — a simple approach for Shannon's limit, *Proc. IEEE Sec. Intl. Symp. Spread Spectrum Techniques and Applications,* 1992, pp. 325–328.
64. Bernard Sklar, *Digital Communications: Fundamentals and Applications.* Englewood Cliffs, NJ: PTR Prentice Hall. 1988.
65. The equality

$$P_B(k) = \frac{M}{2(M-1)} \sum_{m=1}^{M-1} (-1)^{m+1} \frac{1}{m+1} C_m^{M-1} \exp\left(-\frac{mk}{m+1}\left(\frac{E_b}{N_0}\right)\right)$$

may be derived from Marvin K. Simon, Sami M. Hinedi, and William C. Lindsey, *Digital Communication Techniques: Signal Design and Detection.* Englewood Cliffs, NJ: PTR Prentice Hall. 1995. p. 325.
66. Bob O'Hara and Al Petrick, *The IEEE 802.11 Handbook: A Designer's Companion.* New York: IEEE Press. 1999.

Chapter 4
The Data Link Layer

4.1 INTRODUCTION[1]

Recent technology advances have made it conceivable to place wireless sensor networks in many applications. The wireless nodes of these networks will be comprised of a transducer (sensor or actuator), communications circuitry, and behavior logic. These nodes will be embedded in ceiling tiles, in factory equipment, in farmland, and on the battlefield, and will locate things, sense danger, and control the environment with minimal human effort. Crucial to the success of these ubiquitous networks is the availability of small, lightweight, low-cost nodes. Even more important, the nodes must consume ultra-low power to eliminate frequent battery replacement.[2]

To reduce the power consumption of these devices, both the operation and standby power consumption must be minimized; also, as described in Chapter 3, because the operation power consumption is greater than the standby power consumption, the communication time or time the communication circuits are active must be minimized. That is, the communication duty cycle must be reduced to a minimum.[3] For an asynchronous, multi-hop network (one in which no single node can serve as a global, centralized time reference), however, an ultra-low duty cycle makes it difficult for the wireless devices to discover and synchronize with each other, because they are so rarely active.

Low-cost crystal, ceramic, and on-chip Micro Electro-Mechanical System (MEMS) resonators[4] are regarded as key enabling technologies for ultra-low-cost wireless devices. The issue with these technologies is their relatively poor frequency stability, which compounds the device synchronization problem for ad hoc networks employing conventional time-division duplex (TDD) or time-division multiple access (TDMA) methods. Therefore, if the TDD beacon (or frame) period is nominally T_{beacon} seconds, and the timebase stability of each node is ε ppm, a node must begin every receiving period at least $2\varepsilon T_{beacon}$ seconds prior to the nominal reception time to ensure reception of the beacon (because both transmitter and receiver time bases may vary). Further, in the worst-case scenario reception of the beacon may not begin until $2\varepsilon T_{beacon}$ seconds after the nominal reception time, so the receiver could stay active until that time, waiting for the beacon to start, plus the time needed to receive the beacon itself. If the length of the beacon itself is defined to be a communications time T_c, and a symmetrical distribution of time base offsets around the nominal value (i.e., no systematic offsets among the nodes) is assumed, the average time T_o the receiver must be active per beacon period is

$$T_o = \frac{T_c + (2\varepsilon T_{beacon} + 2\varepsilon T_{beacon} + T_c)}{2} = 2\varepsilon T_{beacon} + T_c. \tag{1}$$

The lowest possible average duty cycle of a receiving node in a conventional TDD system is, therefore,

$$\frac{2\varepsilon T_{beacon} + T_c}{T_{beacon}} = 2\varepsilon + \frac{T_c}{T_{beacon}}. \tag{2}$$

Clearly, no matter how short the beacon length T_c is made, nor how long the beacon period T_{beacon} is made, the duty cycle is limited by the attainable time base stability, ε. Recognizing that the time variation between sequential beacons is probably highly correlated, it is possible for the receiving node to reduce the device duty cycle by tracking the relative received beacon time from beacon to beacon. Therefore, a reception window smaller than $2\varepsilon T_{beacon} + T_c$ is employed. For a (possibly dense) ad hoc peer-to-peer network in which nodes are expected to receive beacons from many different nodes, however, the memory overhead associated with assigning a window time and size to each node in range becomes difficult. One must also consider the case of a node leaving the network unexpectedly; remaining network nodes in range may conclude that their windows must be widened to "find" the missing node. The resulting increase in power consumption may cause a chain reaction of other nodes leaving the network unexpectedly, due to exhaustion of primary power sources, leading to network failure. This scenario is especially threatening for systems employing energy-scavenging techniques that may have only small temporary energy caches (e.g., capacitors).

In this chapter, a novel data link layer protocol is introduced. It enables low duty cycle devices to communicate with each other easily, and it does not require a high-accuracy synchronization reference, thereby overcoming the issue of poor time base stability.

4.2 MEDIUM ACCESS CONTROL TECHNIQUES

The medium access problem for wireless networks has received a great deal of attention in recent years; Chandra, Gummalla, and Limb[5] is a good review of protocols proposed in the literature.

4.2.1 ALOHA

The ALOHA system[6] is generally described as the first wireless computer communication system employing random access. The ALOHA network uses a star topology, consisting of a central computer and several remote user sites, and employs two physical channels: one for outbound transmissions and one for inbound transmissions. The remote sites accessed the central computer on the inbound channel in a completely asynchronous manner. Packet collisions, when they occurred, were handled by retransmissions after a random backoff time. If the offered traffic is assumed to be of a Poisson distribution, it can be shown that the channel throughput rate is Ge^{-2G}, where G is the mean offered traffic rate (in packets per unit time); ALOHA thus has a maximum throughput of $1/(2e) \approx 0.184$.[7]

This relatively low value has lead to criticism of ALOHA for its poor channel efficiency, and a search for protocols, such as slotted ALOHA,[8] having higher throughput. However, this statement is usually made in the context of an infinite channel signal-to-noise power ratio. Abramson, considering the more realistic case of finite power ratios, notes that

> [W]e see that the channel efficiency of an ALOHA channel approaches one for the important case of small values of throughput and small values of the signal-to-noise power ratio. In other words, under these conditions it is not possible to find a multiple access protocol which has a higher capacity for a given value of average power and a given bandwidth.[9]

Wireless sensor networks, of course, have small values of throughput, and small values of the signal-to-noise power ratio to conserve battery life. The ALOHA channel access scheme thus appears relevant to the present work; however, ALOHA requires a star network with a master node that is receiving at all times, something incompatible with a power-sensitive, multi-hop wireless sensor network.

4.2.2 Carrier Sense Multiple Access (CSMA)

The CSMA channel access protocols are a large family of protocols that attempt to improve the relatively poor channel efficiency of ALOHA access by requiring nodes to sense (i.e., receive on) the channel prior to packet transmission to check for channel occupancy.

> The various protocols … differ by the action (pertaining to packet transmission) that a terminal [i.e., network node] takes after sensing the channel. However, in all cases, when a terminal learns that its transmission was unsuccessful, it reschedules the transmission of the packet according to a randomly distributed retransmission delay.[10]

The simplest form of CSMA is nonpersistent CSMA. In this form, if the channel is sensed to be idle, the packet is transmitted. If it is sensed that the channel is busy, the node reschedules the transmission to occur at

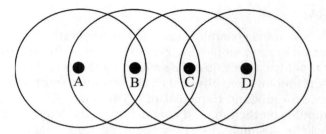

Exhibit 1.   The Hidden and Exposed Terminal Problems

some later, randomly determined time, when the channel is sensed again and the process repeated.

It was recognized, however, that the channel may be unoccupied during the random backoff time of nonpersistent CSMA, and that channel capacity could be improved further by detecting such idle time and transmitting immediately, instead of waiting for the previously determined random time period to elapse. However, should two nodes desiring to transmit employ this "persistent CSMA" strategy, their packet transmissions would certainly collide once they both detected the idle channel. To avoid this scenario, p-persistent CSMA was proposed, in which nodes transmit upon idle channel detection only with a probability, p. With a probability $(1 - p)$, they delay their transmission. The optimum value of p is a function of the offered traffic rate.[11]

CSMA access protocols suffer from the "hidden terminal" problem. As shown in Exhibit 1, nodes A and C are in range of node B, but are not in range of each other. If node A is transmitting to node B, and node C desires to transmit to node B, node C may sense the channel, find it idle, and transmit, causing interference at node B with node A's transmission. CSMA access protocols also suffer from the "exposed terminal" problem. That is, if node B is transmitting to node A, and node C desires to transmit to node D, node C may sense the channel, find it occupied by B, and delay its transmission, even though a transmission by C would not cause interference at node A (and B cannot cause interference at node D).

Both the hidden and exposed terminal problems reduce channel capacity and are a serious problem for services, such as Wireless Local Area Networks (WLANs), where channel capacity is important.[12] The first solution proposed was the use of "busy tones,"[13] wherein a receiving node simultaneously transmits a signal on a second channel, indicating that it is receiving a packet. Nodes desiring to transmit first sense the second channel for the presence of busy tones; if none are detected, the node may transmit.

The busy tone solution, however, suffers from some practical difficulties, mostly centered on the requirement that each node be capable of simultaneous reception and transmission. This duplex operation greatly adds to the cost of nodes, increases their active power drain, and requires additional channel bandwidth (the busy tone channel). Although work continues on busy tone methods (e.g., Haas et al.[14]), these problems led to a search for alternatives.

One of the first proposals was Multiple Access with Collision Avoidance (MACA), which added a request-to-send-clear-to-send (RTS-CTS) exchange between transmitting nodes and intended receiving nodes to the Carrier Sense Multiple Access (CSMA) protocol.[15] Nodes intending to transmit first sent a short RTS packet to the intended recipient. If the node received a CTS packet in reply, it would begin transmission of the data packet. This proposal led to many variations, including CSMA with collision avoidance, CSMA/CA, used in the Institute of Electrical and Electronics Engineers (IEEE) 802.11 WLAN standard;[16] MACAW, developed at the Xerox Palo Alto Research Center in Palo Alto, California; [17] and Floor Acquisition Multiple Access (FAMA).[18] All of these employ a form of RTS-CTS handshake that reduces, but does not eliminate, the possibility of packet collisions.

A difficulty in applying CSMA techniques to wireless sensor networks is the amount of time nodes are required to sense the channel before transmitting. In an ad hoc, multi-hop network, there may be no global time reference; this requires both candidate transmitting and (especially) potential receiving nodes to sense the channel for long periods, degrading their battery life. This is an especially serious problem for dense networks of high order, in which individual network nodes may have to listen for their neighbors at different times. When the number of neighbors grows, the amount of time available for sleeping becomes very small, and duty cycle suffers. This is the same problem faced by the human wireless telegraphers employing the "5-point" system in the 1920s (Chapter 2). (With five different schedules to keep each day, operating in the network became a grueling task.) This problem may be overcome by the use of a global time reference; however, the establishment and maintenance of a global time reference in an ad hoc, multi-hop network of arbitrary physical geometry is a nontrivial task.

4.2.3 Polling

An alternative to CSMA for wireless channel access is polling. In the polling algorithm, network nodes may transmit on the channel only after receiving permission to do so from a master node in the network. Messages originating at nonmaster (i.e., slave) network nodes are handled by requiring the master node to regularly and individually poll all network nodes, asking if they have a desire to transmit. Each slave device replies in turn to this query. A node indicating a desire to transmit is then given permission to do so by the master. In this way, a single entity, the master, controls access to the channel.

Polling offers some advantages over CSMA. First, channel access timing may be deterministic and not suffer from the random delays inherent in the CSMA algorithm. This can be an important advantage for applications requiring low message latency and applications sensitive to jitter in message arrival times, such as multimedia (e.g., high-definition television [HDTV]) applications. Further, because channel access is controlled by a single entity, the polling policy of the master node may be easily adjusted to provide different levels of access to the channel to each slave node. In this way, each node may be given the quality of service (QoS) needed for the application it is serving; nodes with high data throughput or low message latency requirements may be polled more often than other nodes. In addition, because it is controlled by a central authority, channel access can also be guaranteed to be fair. Finally, the hidden terminal problem, inherent in CSMA, is avoided.

With that said, the polling algorithm has some features that make it unsuitable for wireless sensor networks:

- First among these is the load on the master device. Because it must constantly transmit and receive, it is incapable of the low duty cycles commensurate with long battery life. Slave nodes must also bear the load of constantly receiving polls of the master intended for other slaves, and occasionally replying to the master when they are addressed that, no, they have no messages to send.
- Second, as the order of (i.e., the number of devices in) the network grows, the amount of network time spent polling network nodes grows accordingly. It is difficult to envision a polling algorithm suitable for the several hundred to several thousand nodes expected to be present in a large wireless sensor network.
- Finally, the conventional polling channel access method requires a single-hop network. Nodes must all be in range of the master to gain access to the channel.

Many wireless sensor networks, however, are most useful if they can be implemented as multi-hop networks and can extend their physical presence farther than the range of any single node. Proposals have been made, however, for polling channel access systems in multi-hop networks;[19] these rely on the large ratio of channel capacity to data throughput common in wireless sensor networks to avoid an unacceptably large number of channel collisions (due to the number of nodes in range of one another).

The most well-known application of the polling method of channel access in Wireless Personal Area Networks (WPANs) is in Bluetooth™. Bluetooth, a single-hop network with a maximum of seven slaves and the requirement to transmit synchronous data (e.g., digitized voice), is particularly suited for the polling channel access method. The selection of an appropriate polling policy (i.e., when to poll which slaves) is an active and interesting area of queuing research (e.g., Capone et al.[20]). Even Bluetooth, however, can benefit from reduced power consumption; Bluetooth has incorporated three lower duty cycle/power-reduction modes in which the slave does not have to actively participate in the polling process. In order of decreasing duty cycle, these are the HOLD, SNIFF, and PARK modes.[21]

In HOLD mode, a slave is put to sleep for a fixed period; it remains synchronous with the network and, upon expiration of the HOLD period, it returns to active network participation. It is a single event; to reenter the sleep mode, HOLD mode must be reentered. In the SNIFF mode, the slave can repeatedly skip certain master-transmission time slots, remaining asleep to save power. When it awakes, it becomes active for a few master-transmission time slots, then returns to sleep. The slave and master agree on the slots the slave will receive beforehand. In PARK mode, the slave is again put to sleep, but for longer periods, interspersed by reception of a special PARK beacon to maintain synchronization with the network. During this sleep time, the node's logical address in the network may be reassigned; the slave may only regain access to the channel after obtaining permission to do so from the master. Permission is requested in a reply to a PARK beacon.

Management of these multiple power-reduction modes in an optimal way is a nontrivial task in the Bluetooth system; even determining the optimum set of parameters for the SNIFF mode alone (i.e., how long the active period should be, how long the interval between active periods should be, etc.) is an interesting problem.[22]

4.2.4 Access Techniques in Wireless Sensor Networks

Wireless sensor networks, due to their ad hoc network design and long battery life needs, impose difficult requirements on a medium access protocol. Techniques studied in some of the major wireless sensor network research programs now under way are briefly described next.

4.2.4.1 WINS. The WINS group at the University of California, Los Angeles, has studied a polling, relative-TDMA approach. The TDMA frame in this multi-hop network is asynchronous between nodes, because there is no global time reference (i.e., beacon) directly received by all network nodes.[23–25] Nodes negotiate a mutually acceptable slot between them. This is reminiscent of the "5-point" system used in the amateur trunk lines grid network of the 1920s, in that network nodes schedule a unique time slot for each communication link. It also suffers from a similar problem; in a dense network this leads to many active slots, randomly placed within the TDMA frame, and a high node duty cycle leading to poor battery life. It is easy to implement and suitable for a wide variety of message types, however. As in all polling schemes, message latency and arrival time jitter can be controlled, making the system suitable for multimedia and other applications requiring isochronous data.

4.2.4.2 PicoRadio. The PicoRadio program at the Berkeley Wireless Research Center (Berkeley, California) has proposed a multichannel access protocol that assigns ~30 code division multiple access (CDMA) codes such that, for every node in the network, each of its neighbors has a distinct (different) code.[26] Because the CDMA spreading codes are assumed to be orthogonal, packet collisions are avoided. All nodes are asynchronous; no centralized time reference enables power-saving sleep modes, so every node must have a receiver active at all times. To retain good battery life, the use of an ultra-low-power "wake-up radio" is proposed. Such a radio would have an active power consumption of 1 µW; its sole purpose is to listen to the channel at all times, waking the main receiver when a "wakeup beacon" was decoded.[27,28]

The viability of the PicoRadio approach for low-power networks depends on the viability of the wakeup radio. The wakeup radio is proposed to be a simple radio frequency (RF) amplifier, followed by a filter and a simple energy detector.[29]

4.2.4.3 Others. The Piconet project[30] employs Singly Persistent Data Sensing for Multiple Access (1-DSMA). To establish that the channel is busy, a node must decode a valid packet preamble. Although this is superior to simple CSMA in an unlicensed band, because a node will transmit over interference from other services, it requires a long channel sensing period because the preamble must be detected rather than a simple channel energy measurement made. Because the receiver will, therefore, be active longer, this represents a loss of power efficiency.

Haas has proposed the use of a polling scheme employing two transceivers per node.[31,32] Each node continuously polls its neighbors for traffic, in an asynchronous manner. The difficulty in this approach is that, because each node is polled individually, a node spends much time actively transmitting and receiving; in a dense network with many neighboring nodes, it is not possible to achieve the low duty cycle needed for long battery life.

The preceding access protocols all have weaknesses when employed in the wireless sensor network application, particularly when very low duty cycle operation is desired in a network of large order. One solution to this problem is a relatively new protocol, the Mediation Device protocol, which is designed to enable even dense, multi-hop networks to enjoy low duty cycle operation.

4.3 THE MEDIATION DEVICE (MD)

4.3.1 The MD Protocol

The time-average power consumption of a network node can be written as

$$P = \alpha \cdot P_o + (1 - \alpha) \cdot P_s \qquad (3)$$

where

    P = time-average power consumed, W
    α = duty cycle of operation
    P_o = power consumed while in active operation, W
    P_s = power consumed while in standby mode, W

To reduce the time-average power consumption, both the operation power and the standby power should be reduced. For a given technology and set of applications supported by the network, however, a practical limitation exists for both the operation power and the standby power. For most applications, the operation power is much larger than the standby power:

$$P_o \gg P_s \qquad (4)$$

Under this assumption, we can see that by reducing the duty cycle, low power consumption levels and associated long battery lifetimes can be reached. For example, for a node with 10-mW operation power and 10-μW standby power, if the duty cycle is 0.1 percent, then the time-average power drain is about 19.99 μW. If the node is supplied by a 750 mAh AAA battery, linearly regulated to 1 V, it will have a battery life of more than 37,000 hours, or more than 4 years.

For such an aggressive low duty cycle system, it is very difficult for nodes to discover and synchronize with each other; all nodes in the network are inactive 99.9 percent of the time the network is in operation. To keep the power consumption to a minimum, and still achieve reliable communication, a mediation device (MD) is introduced. The MD acts as a mediator between wireless devices within the network, and is capable of recording and playing back control message related information, as illustrated in Exhibit 2.

In normal network operation, each node in the network transmits a short (< 1 ms) query beacon, listing its node identity (address) while stating that it has no message traffic for any other node and is available to receive traffic. Following each beacon, the node listens for a short period

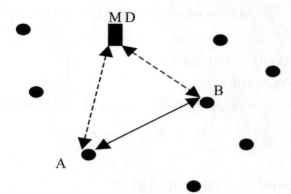

Exhibit 2. Mediation Device Operation

---

to receive any replies. The beacons are sent at a fixed, periodic rate defined for the network (e.g., every 2 seconds); however, because the time bases for the nodes are independent and generated from inexpensive components, beacons may slowly drift in time, with respect to each other, in an uncontrolled fashion. This creates an unslotted ALOHA multiple access mechanism. If a node generates a message for another node, it indicates this by stopping query beacon transmission and instead transmitting a request-to-send, or RTS, beacon, containing its address and the address of the message destination. The RTS beacon exactly replaces the query beacon; it is sent periodically at the same rate and, like the query beacon, is followed by a brief receiving period.

In the simplest case, and for pedagogical purposes, the MD may be assumed to be mains powered, with its receiver continuously active. It records (or, equivalently, calculates from an internal timebase) the relative time differences (offsets) between the received beacons of each node in range. Because the data throughput of wireless sensor networks is expected to be low, the majority of the beacons the MD receives will be query beacons.

Assuming node A needs to send a message to node B, a message transfer would proceed as follows:

- Step 1: Node A begins to transmit a series of RTS beacons and waits for an RTS reply message. Node B is asynchronous with node A, so it will not be able to directly receive node A's RTS beacon. The MD will record node A's RTS beacon, which includes the identity of both node A and node B.
- Step 2: When node B wakes up, node B transmits a query beacon. The MD records the time offset between the beacon schedule of node A

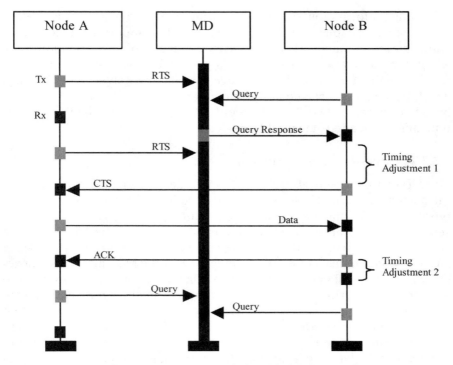

Exhibit 3. Timing Schedule for the Mediation Device Protocol

and node B (i.e., the time difference between the RTS beacon of node A and the query beacon of node B). Upon reception of node B's query beacon, the MD transmits a query response message to node B, indicating the message available at node A, and including the time offset between the two nodes.

• Step 3: After node B receives the query response message from the MD, node B adjusts its timing and synchronizes with node A, which is still transmitting RTS beacons. Node B now transmits a CTS packet to node A during the receiving period of node A, and the two nodes may start transferring the message.

• Step 4: After acknowledgment of message receipt, node A returns to periodic transmission of query beacon packets. To avoid collisions with node A, node B readjusts its timing to its original state and also returns to periodic transmission of query beacon packets.

A detailed timing schedule procedure for this protocol is illustrated in Exhibit 3. Intuitively, the MD is seen to function in a manner analogous to that of a telephone answering machine; callers may call, and machine owners may listen to their machines, in an asynchronous manner, but the scheduling information left in the recorded message (e.g., "Please call at noon today.") enables both caller and owner to temporarily synchronize and achieve communication. This synchronization of network nodes only when communication is necessary is known as "dynamic synchronization." Note that the particular sequence of beacons received at the MD (i.e., which beacon was received first) is unimportant; should device B's beacon have been received first, the MD would still be able to calculate the proper time difference and transmit it to device B.

4.3.2 The Distributed MD Protocol

The MD can be a dedicated device, as just described; however, ensuring the placement of an MD within range of all network nodes only implies the type of system administration and planning that self-organizing wireless sensor networks attempt to avoid. Further, requiring the MD to receive constantly is not compatible with the concept of a low-power network. An improvement on this situation can be realized if the MD function is distributed throughout all nodes in the network. In this scheme, each node in the network temporarily operates as an MD at a time chosen by an independent random process (i.e., a node begins operation as an MD at a random time, independent of other nodes). To ensure all beacons within range are received, each node receives for a period of time equal to one beacon period, plus the length of a beacon transmission, and then makes transmissions required of an MD to other network nodes when the nodes are receiving. After this procedure, the node returns to normal (i.e., non-MD) operation. Because each device functions as an MD only rarely, the average communication duty cycle of each device can remain very low, and the overall network to remain a low-power, low-cost asynchronous network.

Assuming node A needs to send a message to node B, a message transfer under the distributed MD protocol would proceed as follows:

- Step 1: Node A begins to transmit a series of RTS beacons and waits for an RTS reply message. Nodes B and C are asynchronous with node A, and so will not be able to directly receive node A's RTS beacon. However, at this time, node C changes to "MD mode," as the result of a random process, and will record node A's RTS beacon, which includes the identities of both node A and node B.
- Step 2: When node B wakes up, node B transmits a query beacon. Node C, still operating in MD mode, records the time offset between the beacon schedule of node A and node B (i.e., the time difference between the RTS beacon of node A and the query beacon of node B). Upon reception of node B's query beacon, node C transmits a query response message to node B, indicating the message available at node A and including the time offset between the two nodes.

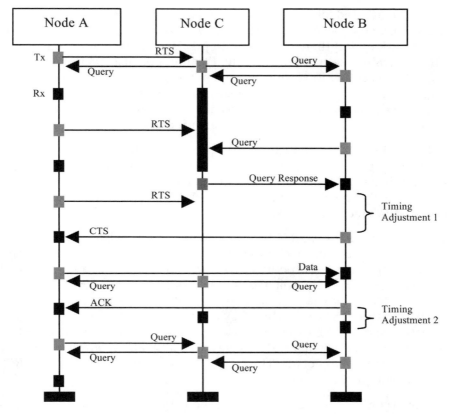

Exhibit 4. Timing Schedule for the Distributed Mediation Device Protocol

- Step 3: After node B receives the query response message from node C, node B adjusts its timing and synchronizes with node A, which is still transmitting RTS beacons. Node B now transmits a CTS packet to node A during the receiving period of node A, and the two nodes may start transferring the message. Node C, having completed its MD function, returns to normal operation and begins transmitting regular query beacons.
- Step 4: After acknowledgment of message receipt, node A returns to periodic transmission of query beacon packets. To avoid collisions with node A, node B readjusts its timing to its original state and also returns to periodic transmission of query beacon packets.

A detailed timing schedule procedure for this protocol, the "distributed Mediation Device protocol," is illustrated in Exhibit 4.

The distributed MD protocol overcomes many weaknesses of the original MD protocol. Because the MD function is randomly distributed among all network nodes, there is no need to precisely locate dedicated MDs to protect against network portioning. Further, because each device only rarely becomes an MD, its duty cycle is not unduly affected. However, message latency may be affected, because a node must wait for a nearby node to become a mediation device to transmit a message. Message latency of the distributed MD protocol is analyzed in Section 4.4.

One issue with the distributed MD protocol is the possibility that two nodes may enter the MD mode simultaneously. If these nodes are both within range of nodes A and B, the MD nodes will simultaneously respond to node A's RTS beacon, resulting in a packet collision at node B and a failure of node synchronization.

This issue is addressed in two ways. When an MD finishes its MD reception period, it transmits a short beacon announcing this. Any other MD listening is now aware of the first MD, and can either (1) go to sleep and assume that the first MD will handle any available message transfers or (2) remain listening and eavesdrop on the activities of the first MD. If the first MD handles all RTS beacons heard by the second MD, the second MD takes no action and returns to sleep; however, the second MD may hear an RTS beacon from a node outside the range of the first MD, upon which the first MD (of course) does not act. The second MD may then respond to that RTS beacon without fear of interference from the first MD.

The second method of avoiding MD collisions is to randomize the MD period of each node within a range. A range of ± 25 percent about a nominal value is an appropriate value; after each MD activity, the node selects a random time period within this range to wait before entering MD mode again. This prohibits nodes from synchronizing their MD activities and repeatedly interfering with each other.

4.3.3 "Emergency" Mode

In some applications, only the minimum possible message latency is permissible. Applications such as security systems require very low message latency, yet still require very long battery life. This may be an issue with the distributed MD protocol, especially if the beacon period is long (for low duty cycle and good battery life) and only a few network nodes are within range. A solution to this problem is the use of the distributed MD protocol, while disabling the random process used to induce nodes to act as MDs. Instead, nodes begin "MD mode" only when a message is generated (or received for forwarding). During the MD receiving period, the node receives the beacon of the desired node (if present); the node then contacts the desired node and transmits the message. This mode has the advantage of minimal latency; however, it adds the duty cycle overhead of receiving for half of the beacon period (on average) for each transmitted message. If messages are frequent, this can seriously affect battery life. For some applications, such as security systems, messages are infrequent, and this may not be an issue. Also, note that, as described in Section 4.4, if a significant number of nodes are within range, the latency of the distributed MD protocol can be acceptably low.

4.3.4 Channel Access

The MD protocols described so far do not consider channel access per se; they rely on the ALOHA strategy (due to the low data throughput/channel capacity ratio) to minimize packet collisions within the network. This would be quite suitable if the network was a licensed service and no other services were present in the channel.

Implementation of wireless sensor networks is expected to be mainly (if not exclusively) in unlicensed radio bands, however. The selection of a suitable channel access technique for a wireless device operating in an unlicensed radio band, such as an Industrial, Scientific, and Medical (ISM) band, is a complicated one, especially if coexistence with other services, such as WLANs, is required. In an unlicensed band, the usual assumption made in licensed bands that any signal present must be from the same network, or at most a different network in the same service, does not apply; any signals, from sources as varied as microwave ovens and WLANs, may be present. In such interference-limited cases, if a Carrier Sense Multiple Access with Collision Avoidance (CSMA-CA) strategy is employed, based solely on detected energy in the channel, the "fairness doctrine" for channel access often fails. Data throughput may suffer as the device repeatedly backs off in the presence of interference from other services. Some of these services will not employ CSMA-CA in return to ensure channel access at a later time (e.g., microwave ovens). In the MD protocols, channel status assessments are not made prior to beacon transmissions because beacon timing is of critical importance to network operation. Because the duty cycle is so low, this causes little interference to other coexisting services. One alternative to this approach would be to employ a simple channel energy detection CSMA-CA algorithm with the backoff quanta set to be a beacon period (i.e., devices back off only in multiples of the beacon period).

For channel status assessment in unlicensed bands, it is generally necessary not just to detect energy in the channel prior to transmission, but also to identify (as far as possible) the signal present. This must be done quickly, for the channel status (i.e., presence of interference) may change rapidly; in addition, channel status assessment costs precious battery life. Algorithms meeting these requirements are an area for future research.

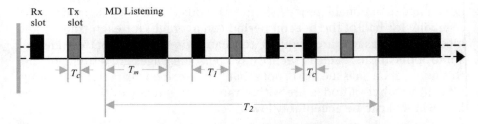

Exhibit 5.   A Typical Timing Schedule for a Node Employing the Distributed Mediation Device Protocol

---

4.4 SYSTEM ANALYSIS AND SIMULATION

In the following discussion, the node duty cycle and message latency of a network employing the distributed MD protocol are evaluated. For simplicity, it is also assumed that all the devices are within communication range of each other.

4.4.1 Duty Cycle

Exhibit 5 illustrates a typical timing schedule for a node employing the distributed MD algorithm.

In Exhibit 5, T_c is the communication (Tx or Rx) time duration for each communication slot, T_m is the MD listening time duration, T_1 is the repeating period of the communication (Tx or Rx) slots and T_2 is the average repeating period of the MD mode.

It is assumed that

$$\begin{cases} T_1 \gg T_c \\ T_2 \gg T_m \end{cases} \tag{5}$$

In the period of T_2, the total number of communication slots can be written as

$$n = \frac{T_2 - T_m}{T_1} \tag{6}$$

The total operation time t_o

$$t_o = T_m + n \cdot T_c \tag{7}$$

The duty cycle α can be written as

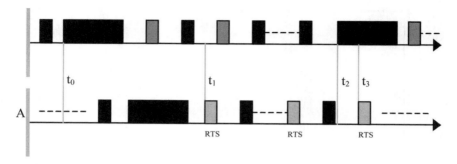

Exhibit 6. A Typical Timing Schedule for Two Devices

$$\alpha = \frac{t_o}{T_2} \tag{8}$$

Putting Equations 6 and 7 into Equation 8,

$$\alpha = \frac{T_m}{T_2} + \frac{T_c}{T_1} - \frac{T_m}{T_2} \cdot \frac{T_c}{T_1} \tag{9}$$

By considering Equation 5, the third term of the preceding equation can be ignored. Finally, the duty cycle α can be written as

$$\alpha \cong \frac{T_m}{T_2} + \frac{T_c}{T_1} \tag{10}$$

For example, if $T_c = 1$ millisecond, $T_1 = 1$ second, $T_m = 2$ seconds, and $T_2 = 1000$ seconds, then the overall duty cycle can be as low as 0.3 percent.

4.4.2 Latency

One concern for this protocol is message latency; specifically, the time a node must wait between first transmitting an RTS beacon and having it heard by an MD. (The average time between MD reception of an RTS beacon and transmission of data is, from inspection of Exhibit 3, $4T_1$.) An analysis of latency is as follows.

As illustrated in Exhibit 6, at time t_1, Node A transmits an RTS message. No MD is available at this time, so node A needs to repeat the RTS message until time t_3 when an MD becomes available. To guarantee that the RTS will not miss the MD period at time t_3, the duration of the MD listening time has a lower bound. Clearly, if

$$T_m >= 2 \cdot T_1 + T_c \tag{11}$$

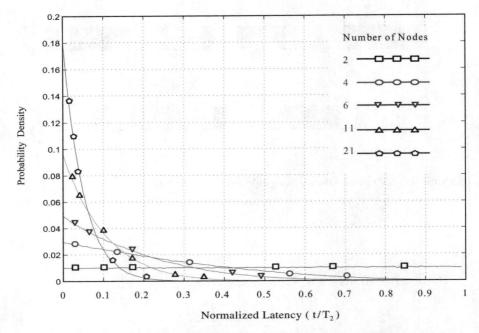

Exhibit 7. Normalized Latency Probability Density Function for Networks of Several Orders

the RTS will not miss the MD listening period. The longest latency, therefore, is about T_2.

The smallest network has only two nodes, one source node and one destination node. In this case, the probability density function of the latency can be written as

$$p(T_{latency} = t) = \begin{cases} \frac{1}{T_2} & 0 < t <= T_2 \\ 0 & Otherwise \end{cases} \quad (12)$$

The average latency is $T_2/2$.

For a larger network, at any given time there exists more than one node that may enter the MD mode; thus, the latency can be reduced. Exhibit 7 illustrates the simulated latency probability density function, normalized to T_2, for networks of order 2, 4, 6, 11, and 21. It is clear from Exhibit 7 that normalized message latency decreases significantly for relatively dense networks. Exhibit 8 illustrates the normalized average latency as a function of the number of nodes in range of a node generating a message. By increasing the number of nodes within the communication range, the latency can be dramatically reduced. This feature makes the distributed

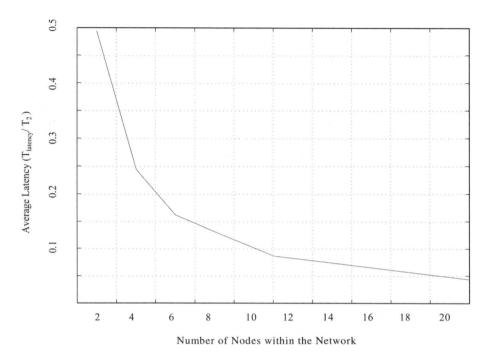

Number of Nodes within the Network

Exhibit 8. Normalized Average Latency versus Network Order

MD protocol especially suitable for wireless sensor networks, which are expected to be relatively dense in many applications.

4.5 CONCLUSION

The selection of a channel access method is a critical and difficult problem for wireless sensor networks, due to their need for low duty cycle operation in dense networks of high order. This has led to much creativity among investigators, and has resulted in many different proposed and implemented solutions. The effect of various media access control (MAC) protocols on battery power consumption has been reported in Chen et al.[33] and Woesner et al.[34]

ALOHA, although optimal for systems with low data throughput and low signal-to-noise ratio (SNR), is designed for star networks in which the master node is constantly receiving. Although some types of wireless sensor networks can support a constantly receiving master, it is generally incompatible with the long battery life required of them. Also, many applications of wireless sensor networks require a multi-hop network.

Conventional CSMA approaches require relatively high receive duty cycles, with frequent active periods, if the network nodes are not synchronized; conversely, if the network nodes must be globally synchronized, it becomes difficult to construct robust self-organizing ad hoc networks. CSMA methods also suffer from the hidden terminal problem. Solutions to the hidden terminal problem for CSMA systems include the use of busy tones (in which the receiving device transmits a constant signal on another channel while receiving) and the use of various RTS/CTS schemes (in which message-laden nodes send a short request-to-transmit message prior to transmission).

Polling methods place medium access control in the hands of a centralized network node, usually the master of a star network. Although very good for applications requiring low latency messages, it is typically difficult to achieve low duty cycle operation in a large network because all devices must be regularly listening to the master. It becomes even more difficult to design a multi-hop polling network, for the same reason.

This chapter described the MD protocol, which enables low duty cycle devices to communicate easily and does not require a high accuracy synchronization reference. In fact, by randomizing the times that different devices send query messages, a certain amount of synchronization reference instability is actually beneficial. The MD protocol also introduces the concept of "dynamic synchronization," which extends the random channel access feature of ALOHA to peer-to-peer networks, and adds the logical channel reservation of time division multiple access (TDMA) channel access methods. The distributed MD protocol does not require specialized devices and uses the expected high density of many wireless sensor networks to reduce message latency. By combining the use of lower-cost hardware with good battery life, the MD protocol improves the practicality of wireless sensor networks. This low duty cycle protocol is compatible with the IEEE 802.15.4 low rate WPAN standard.

References

1. This chapter is an extension of Qicai Shi and Edgar H. Callaway, An ultra-low power protocol for wireless networks, *Proc. 5th World Multi-Conference on Systemics, Cybernetics and Informatics,* 2001, v. IV, pp. 321–325.
2. Jan M. Rabaey, PicoRadio supports ad hoc ultra-low power wireless networking, *IEEE Computer,* v. 33, n. 7, July 2000, pp. 42–48.
3. Anantha Chandrakasan, Design considerations for distributed microsensor systems, *Proc. IEEE Custom Integrated Circuits Conf.,* May 1999, pp. 279–286.
4. Clark T.-C. Nguyen, Micromechanical circuits for communication transceivers, *Proc. IEEE Bipolar/BiCMOS Circuits and Technology Meeting,* 2000, pp. 142–149.
5. Ajay Chandra, V. Gummalla, and John O. Limb, Wireless medium access control protocols, *IEEE Commun. Surveys,* v. 3, n. 2, Second Quarter 2000, pp. 2–15.

6. Norman Abramson, The ALOHA System — Another alternative for computer communications, *Proc. AFIPS Fall Joint Comput. Conf.,* v. 37, 1970, pp. 281–285.
7. Ibid.
8. Lawrence G. Roberts, *ALOHA Packet System with and without Slots and Capture,* ARPANET Satellite System Note 8 (NIC Document 11290), ARPA Network Information Center, Stanford Research Institute, Menlo Park, CA, 26 June 1972. Reprinted in *Computer Communication Review,* v. 5, n. 2, April 1975, pp. 28–42.
9. Norman Abramson, Multiple access in wireless digital networks, *Proc. IEEE,* v. 82, n. 9, September 1994, pp. 1360–1370.
10. Leonard Kleinrock and Fouad A. Tobagi, Packet switching in radio channels: part I — carrier sense multiple-access modes and their throughput-delay characteristics, *IEEE Trans. Commun.,* v. COM-23, n. 12, December 1975, pp. 1400–1416.
11. Ibid.
12. Chane L. Fullmer and J. J. Garcia-Luna-Aceves, Solutions to hidden terminal problems in wireless networks, *ACM SIGCOMM Computer Communication Review, Proc. ACM SIGCOMM '97 Conf. on Applications, Technologies, Architectures, and Protocols for Computer Communication,* v. 27, n. 4, October 1997, pp. 39–49.
13. Fouad A. Tobagi and Leonard Kleinrock, Packet switching in radio channels: part II: the hidden terminal problem in carrier sense multiple access and the busy tone solution, *IEEE Trans. Commun.,* v. COM-23, n. 12, December 1975, pp. 1417–1433.
14. Zygmunt J. Haas, Jing Deng, and S. Tabrizi, Collision-free medium access control scheme for ad-hoc networks, *Proc. IEEE Military Communications Conference,* 1999, v. 1, pp. 276–280.
15. Phil Karn, MACA — a new channel access method for packet radio, *ARRL/CRRL Amateur Radio 9th Computer Networking Conf.,* 1990, pp. 134–140.
16. Roger B. Marks, Ian C. Gifford, and Bob O'Hara, Standards in IEEE 802 unleash the wireless Internet, *IEEE Microwave,* v. 2, n. 2, June 2001, pp. 46–56.
17. Vaduvur Bharghavan et al., MACAW, *ACM SIGCOMM Computer Communication Review, Proc. Conf. on Communications Architectures, Protocols and Applications,* v. 24, n. 4, October 1994, pp. 212–225.
18. Fullmer and Garcia-Luna-Aceves, ibid.
19. Loren P. Clare, Gregory J. Pottie, and Jonathan R. Agre, Self-organizing distributed sensor networks, *Proc. SPIE, Conf. Unattended Ground Sensor Technologies and Applications,* 1999, v. 3713, pp. 229–237.
20. Antonio Capone, Mario Gerla, and Rohit Kapoor, Efficient polling schemes for Bluetooth picocells, *Proc. IEEE Int. Conf. on Communications,* 2001, v. 7, pp. 1990–1994.
21. Jaap C. Haartsen and Sven Mattisson, Bluetooth — a new low-power radio interface providing short-range connectivity, *Proc. IEEE,* v. 88, n. 10, October 2000, pp. 1651–1661.
22. Ting-Yu Lin and Yu-Chee Tseng, An adaptive sniff scheduling scheme for power saving in Bluetooth, *IEEE Wireless Commun.,* v. 9, n. 6, December 2002, pp. 92–103.
23. Clare, Pottie, and Agre, ibid.
24. Katayoun Sohrabi, A self organizing wireless sensor network, *Proc. 37th Annu. Allerton Conf. on Communication, Control, and Computing,* 1999, pp. 1201–1210.
25. Katayoun Sohrabi Protocols for self-organization of a wireless sensor network, *IEEE Personal Commun.,* v. 7, n. 5, October 2000, pp. 16–27.
26. Chunlong Guo, Lizhi Charlie Zhong, and Jan M. Rabaey, Low power distributed MAC for ad hoc sensor radio networks, *Proc. IEEE Global Telecom. Conf.,* 2001, v. 5, pp. 2944–2948.
27. Lizhi Charlie Zhong, An ultra-low power and distributed access protocol for broadband wireless sensor networks, *Networld+Interop: IEEE Broadband Wireless Summit,* Las Vegas, 2001. http://bwrc.eecs.berkeley.edu.
28. Guo, Zhong, and Rabaey, ibid.

29. Jan M. Rabaey, PicoRadios for wireless sensor networks: the next challenge in ultra-low power design, *IEEE Int. Solid State Circuits Conf. Digest of Technical Papers,* 2002, v. 1, pp. 200—202; v. 2, pp. 156—157, 444–445.

30. Frazer Bennett et al., Piconet: Embedded mobile networking, *IEEE Personal Commun.,* v. 4, n. 5, October 1997, pp. 8–15.

31. Zygmunt J. Haas, A new routing protocol for the reconfigurable wireless networks, *Conference Record, IEEE 6th Int. Conf. on Universal Personal Communications,* 1997, v. 2, pp. 562–566.

32. Zygmunt J. Haas, On the performance of a medium access control scheme for the reconfigurable wireless networks, *Proc. IEEE Military Commun. Conf.,* 1997, v. 3, pp. 1558–1564.

33. Jyh-Cheng Chen, A comparison of MAC protocols for wireless local networks based on battery power consumption, *Proc. INFOCOM,* 1998, pp. 150–157.

34. Hagen Woesner, Power-saving mechanisms in emerging standards for wireless LANs: the MAC level perspective, *IEEE Trans. Personal Commun.,* v. 5, n. 3, June 1998, pp. 40–48.

Chapter 5
The Network Layer

5.1 INTRODUCTION

The International Standards Organization (ISO) model for Open Systems Interconnection (OSI) states that the network layer "provides functional and procedural means to exchange network service data units between two transport entities over a network connection. It provides transport entities with independence from routing and switching considerations."[1] The network layer controls the operation of the network and routes packets; it also addresses flow control.[2] Flow control is, by assumption, of little concern in most wireless sensor networks due to their very low offered data throughput and stated lack of quality of service [QoS] provisions. However, operation of, as well as routing in, a wireless sensor network represent significant challenges to the designer because of the need for low duty cycle operation of network nodes. This low duty cycle requirement limits the allowable overhead for synchronization, negotiation, coordination, and other network layer activities.[3]

In this chapter, a cluster network capable of multi-hop routing is developed and simulated. This network employs the distributed MD protocol. A discrete-event network simulator is employed to evaluate the throughput, message latency, and number of packet collisions of the network, and the network node duty cycle. The network is found to remain stable even with node duty cycles as low as 0.19 percent.

5.2 SOME NETWORK DESIGN EXAMPLES

The network layer of a wireless sensor network has two interrelated points of interest: the structure (topology) of the network itself and the algorithm used to route messages within it.

5.2.1 Structure

Because multi-hop wireless sensor networks are, by definition, self-organizing ad hoc networks without human administration, a discussion of their structure may seem of little value — the physical communication links between network nodes may take any topological form. It is productive, however, to discuss the logical structure of the network, due particularly to the problem of scalability. In a completely flat network without logical structure, all nodes must cooperate to control the network — identify link creation and loss, and node association and dissociation — at the cost of communication overhead, because nodes can only be directly aware of the network environment in their immediate vicinity. An example of this problem can be seen in the 1973 Defense Advanced Research Projects Agency (DARPA) packet radio network (PRNET) protocol.[4] In this protocol, each network node transmits a packet radio organization packet (PROP) every 7.5 seconds, stating its existence and transmitting a neighbor table that provides network topology information from its perspective. Information from received PROPs of nearby nodes is included in a node's own PROP, as is topology information (in the form of routing information) from received data packets.

It is clear that the transmission of the network's topology in toto, by every node every 7.5 seconds, cannot scale; for networks with 1000 or more nodes, this method of organization is wholly impractical. This problem was addressed in the linked cluster architecture (LCA),[5,6] in which

> [t]he network is organized into a set of node clusters, each node belonging to at least one cluster. Every cluster has its own cluster head which acts as a local controller for the nodes in that cluster.[7]

Gateway nodes provide communication between clusters. In the LCA, which employs TDMA, nodes periodically suspend regular data transmission to perform a distributed clustering algorithm during a control channel epoch. Each epoch is broken in to two frames. During its assigned slot in each frame, every node broadcasts a list of all network nodes it can receive; during all other slots in the frame it operates its receiver to detect the presence of (possibly new) neighbor nodes. At the end of the second frame, each node has a list of all nodes it can hear, and a report of its reception by each of its neighbors (i.e., a list of all bidirectional links). The nodes then perform an algorithm to choose cluster head node, ordinary node, or gateway node status.

Because each node only records and maintains information about its immediate environment, the LCA scales with network size; however, it is not without disadvantages when applied to wireless sensor networks. First, as a TDMA system, it relies on a global clock. Although this is not a drawback for its initial application, the Naval High Frequency Intra-Task Force (which has use of the Global Positioning System [GPS]), it is a problem for wireless sensor networks that have neither the cost nor the power budget for a GPS receiver (and may be indoors and, therefore, out of range of GPS, as well). A more fundamental problem is the detection of nonneighboring clusters. The LCA establishes a kind of "linked list" of clusters, in which each cluster is aware of itself and all its neighboring clusters. Without additional algorithms in place, however, no mechanism is available to advertise the existence of clusters outside a cluster's immediate neighborhood; network traffic generated in a given cluster cannot travel farther than a neighboring cluster.

This problem was addressed in the multicluster, mobile, multimedia radio network,[8] by employing the Bellman–Ford algorithm.[9-11] During the control phase, each node was required to broadcast its routing list, which identified the next node on the route to each destination and the number of hops. This solved the non-neighboring cluster problem, but did so by returning to the transmission of (possibly large) tables during the beacon period.

The problem of discovering non-neighboring clusters was solved in Hester,[12] by placing the clusters in a hierarchical tree, employing the cluster heads in a network backbone. Hierarchical routing is known to make efficient use of routing table space.[13] Hester was able to show that the resulting node complexity of the network was independent of the size of the network. However, the problem of optimal clustering arrangement was shown to be NP-hard to minimize network diameter; an $O(n^2)$ heuristic algorithm was used for clustering. The organization of clusters is similar to the organization of network nodes in Layer Net, a self-organizing network protocol that does not employ clustering.[14] While Hester employed cluster tree routing for messages destined for the tree root, however, it was assumed that messages sourced by the root were transmitted directly by the root to the member node (the root was assumed to have higher transmission power than other network nodes).

A concern in the use of clusters in wireless sensor networks is the larger-than-average power consumption of the nodes in the cluster head backbone, which, because they conduct a significant fraction of the message traffic, have higher duty cycles and, therefore, shorter battery life than other network nodes. This problem was addressed in the Low-Energy Adaptive Clustering Hierarchy (LEACH) protocol,[15] which randomly rotates the cluster head function among network nodes to more evenly distribute the energy load. Every "round," all network nodes select a random number between 0 and 1. If it is below a dynamic threshold (a function of the number of the current round, the number of rounds since the node has been a cluster head, and the desired percentage of cluster heads), the node elects itself to be a cluster head. It then announces this election, with a CSMA MAC protocol, to all other network nodes, which then associate themselves with cluster heads based on received signal strength. The nodes then announce their association decisions to their cluster heads, also via a CSMA MAC protocol.

Although it does address the cluster head power consumption problem, LEACH does not address possible network partitioning that may occur during random cluster generation if no cluster head is elected within range of a given node. A more significant problem is one of routing; one usually employs clusters to simplify the addressing and routing problem for large networks. It is unclear how network nodes maintain proper message route information through multiple cluster head changes.

5.2.2 Routing

The study of routing protocols in ad hoc wireless networks is an active area of research (e.g., Gao,[16] Ibriq[17]); Broch et al.,[18] Royer and Toh,[19] and Lee et al.[20] review the field and provide performance comparisons. Much of the attention, however, has been devoted to the problem of routing in networks wherein it is assumed that the nodes are mobile. Mobility greatly complicates the routing algorithm because the designer must balance the conflicting requirements of frequent path updates to account for node mobility, and low duty cycle to maximize battery life. (Perhaps the best example of this attention is the work of the Mobile Ad Hoc Networks (MANET) working group of the Internet Engineering Task Force (IETF), the goal of which is to write a standard for Internet-based mobile ad hoc networking.[21]) Protocols in this category include the destination-sequenced distance vector (DSDV) protocol,[22] the temporally ordered routing algorithm (TORA),[23] the Global State Routing (GSR) scheme,[24] the ad hoc on demand distance vector (AODV) protocol,[25] the fisheye state routing (FSR) and hierarchical state routing (HSR) protocols,[26,27] and the landmark ad hoc routing (LANMAR) protocol.[28]

Because it is assumed that the nodes in wireless sensor networks are quasi-stationary, however, routing protocols for them can avoid the overhead associated with mobility tracking and its associated link management, and reduce node power consumption accordingly. An example of this is the gradient routing (GRAd) protocol.[29] Nodes employing the GRAd protocol do not use a routing table per se. Instead, nodes maintain a cost table, listing the total cost (e.g., number of hops) that node requires to send a message to each destination. A node desiring to send a message simply broadcasts the message, along with the cost value it has for that destination (obtained from the cost table). Of all neighboring nodes receiving the broadcast, only those that can deliver the message at lower cost than the stated cost value will forward the message. All other nodes will simply ignore the message. In this way, the message slides down a "cost gradient" to the destination.[30,31]

On the positive side, the definition of cost used by GRAd is flexible; by incorporating power-conscious features, very low power operation should be possible. For example, power-aware routing metrics[32] may be used directly in the GRAd cost definition. The definition of power-aware metrics can include the node's source of power (e.g., battery or mains); for example, the routing scheme proposed in Ryu and Cho[33] for wireless home ad hoc networks, which prefers mains-powered relaying nodes, then results directly from the GRAd algorithm.

Although GRAd may be classified as an "on demand" routing method, because the originating node does not know in advance what route a given packet may take, it is also table-based — it requires a cost value for every potential destination in the network. For large networks, this can be a significant storage cost. Further, the cost to each destination must be discovered; this is performed initially by flooding the network. If the network is a sensor network with only one information sink (and, therefore, only one message destination for all nodes, save for the sink node), this is not a serious problem; if the network is required to support many destinations, however, the cost discovery process may itself be costly in terms of battery life.

A routing scheme employing connected dominating sets has also been proposed.[34,35] A set of network nodes is a dominating set of the network if all nodes in the network are in either the dominating set or neighbors of nodes in the set. Use of the dominating set for routing purposes is advantageous because it simplifies the routing problem to that of routing in the (smaller) dominating subnetwork; further, only nodes in the dominating subnetwork need to store routing information.

It can be shown that, to maximize the lifetime of the network, "traffic should be routed such that the energy consumption is balanced among the nodes in proportion to their energy reserves, instead of routing to minimize the absolute consumed power."[36] (This is the network metaphor of Oliver Wendell Holmes' "One-Hoss Shay."[37]) This principle is exploited in the battery energy efficient (BEE) protocol,[38] which attempts to balance the battery consumption among all network nodes. BEE assigns to each possible route a cost function that considers both node power consumption and battery behavior, including the charge recovery effect discussed in Chapter 3, Section 3.3; BEE then selects the one with the lowest cost. If the set of possible routes is large, this method becomes unfeasible; BEE then selects c routes at random from the set of possible routes (where c is a positive integer), computes the cost function value for each of the c routes, then selects the lowest cost route. Although this procedure may result in power-efficient routes, it requires a large table that includes multiple routes to each network node. In fact, Chiasserini et al.[39] suggest that c should be greater than four. Having a table storing five or more routes for each network node destination is a significant penalty for low-cost networks.

The problem of balancing energy use among network nodes is common to all networks in which message traffic is not uniformly distributed among nodes, including cluster-based networks, networks employing dominating-set-based routing, and networks employing a routing backbone or spine.[40,41] However, solutions to this problem do exist. The LEACH algorithm, as has been shown, addresses this issue for cluster networks. For networks employing dominating-set-based routing, Wu et al. have proposed a power-aware method to select the dominating set.[42]

Because wireless sensor networks have an ad hoc topology and are self-organizing, biological models of message transport are brought to mind. One model that has received attention is based on the communication paradigm of ants.[43-45] Ants communicate by means of pheromones deposited on the ground as they move. When an ant discovers food, for example, it returns to the nest, laying down a pheromone trail that other ants may follow to its source, the food. Pheromones diffuse and evaporate; diffusion widens the trail, while evaporation provides a time limit on its availability. Ants interpret a wide trail, which is diffused and expanded by the pheromones of many ants, as being more significant than a narrow one and, of course, no longer use an evaporated trail. The ants thereby establish a complicated organized behavior based solely on the relatively simple behavior of individuals without central control — the model of a distributed control system.

This communication model is appealing for wireless sensor networks because they, being self-organizing, lack a central control point. One model considers the ant nest as an information source (message origination) node, the food as an information sink (destination) node, and ants as messages. The ground between nodes is considered a collection of intermediate nodes. For the destination node to receive messages in this model, it must first transmit messages advertising its presence (i.e., an ant must first discover the food and return to the nest). This advertisement message is passed from node to node in a random, unicast way; each node receiving the message stores its source and the node from which it was received in a table (representing a pheromone) with a time stamp, then passes the message on to a random neighbor after incrementing a "number of hops taken" field in the message. Each node periodically "ages" its time stamp; old entries are deleted (evaporated).

When the message origination node transmits a data message, it sends it to a random neighbor, which checks its pheromone table to see if it has received an advertisement message from the destination. If not, it forwards the data message to a random neighbor. If it has received an advertisement message from the destination, it forwards the message to the neighbor from which it received the advertisement message. The data message is now "on the pheromone trail," and will be delivered to the destination node.

The model can be enhanced by the inclusion of "diffusion" effects to widen the pheromone trail (perhaps by having nodes broadcast their pheromone table entries to their neighbors), and to account for other factors such as finite network size. The biggest limitation to the ant model also has a biological metaphor: in the ant model, relatively few parcels of ground contain either food or nests — the majority of ground contains neither. Likewise, in the communication field, the model works well when only a few nodes are sourcing and sinking information. If one allows every node to do so, suddenly many advertisement messages are being transmitted — a condition that consumes a large amount of energy. One can imagine how poorly the ant communication system would work if every foraging ant discovered food on every parcel of ground.

Additional communication models, based on physical principles, can be used as models for information transport in wireless sensor networks. Directed diffusion[46] is a data-centric, instead of a network organization-centric, routing model:

> *Data generated by sensor nodes is named by attribute-value pairs. A node requests data by sending interests for named data. Data matching the interest is then "drawn" down towards that node.[47]*

This has similarities to the sensor protocols for information via negotiation (SPIN) protocols,[48] which describe their attributes as meta-data (i.e., the type of data a sensor can provide or the type of data desired at an information sink):

> *The protocol starts when a node obtains new data that it is willing to disseminate. It does this by sending an ADV [new data advertisement, containing meta-data] message to its neighbors, naming the new data (ADV stage). Upon receiving an ADV, the neighboring node checks to see whether it has already received or requested the advertised data. If not, it responds by sending an REQ [request for data] message for the missing data back to the sender (REQ stage). The protocol completes when the initiator of the protocol responds to the REQ with a DATA [data] message containing the missing data (DATA stage).[49]*

Both proposals model the information transport as a physical diffusion process, instead of the more conventional communication link. The diffusion model appears particularly appropriate for ubiquitous computing.[50,51]

5.3 A WIRELESS SENSOR NETWORK DESIGN EMPLOYING A CLUSTER TREE ARCHITECTURE[52]

The goal of this work was to develop a wireless sensor network capable of multi-hop routing and low-cost, low-power operation. A hierarchical tree of clusters, the cluster tree topology, was chosen for the network design due to its scalability and simple routing protocol. The cluster tree design also lends itself to a simple connection of the network to the Internet, via a gateway at the root. The design differs from that of Hester[53] in that, among other differences, routing from the root to other network nodes takes place along the tree and not in a single hop from a higher-power transmitter.

5.3.1 Network Design

As illustrated in Exhibit 1, the network design starts with the root, a "special" device called the Designated Device (DD).[54] The identity of the DD is the identity of the network (NetID); the DD is, by definition, also the cluster head of cluster 0. The DD has greater computing capability (e.g., storage capacity and processor speed) than an ordinary network node, but no greater transmitter power or receiver sensitivity. If desired, it may perform gateway duties between the wireless sensor network and the Internet via other wired or wireless communication protocols, such as Ethernet or Institute of Electronics and Electrical Engineers (IEEE) 802.11b. (For cost reasons, it is not desirable to include computing power in every network node sufficient to handle a Transmission Control Protocol/Internet Protocol [TCP/IP] stack.) The DD may be envisioned as a laptop computer, receiving data from the network, or a much smaller and dedicated microcontroller providing the gateway function.

The cluster tree network, as Exhibit 1 indicates, is a multi-hop network. As nodes join the network, the network node with which they communicate during the network association process is defined as the node's parent; the joining node becomes the child of the parent node. In Exhibit 1, node 3 in cluster 0 is the parent of nodes 10 and 22; nodes 10 and 22 are the children of node 3. The lines in Exhibit 1, therefore, indicate parent–child relationships, not communication links. Cluster heads are assigned node identification (NID) 0 in their clusters; the DD, as cluster head of cluster 0, is assigned both cluster identification (CID) 0 and NID 0.

By analogy to crystallography, self-organizing cluster tree networks may form in two ways. The first way can be considered the "monocrystalline" approach. In this algorithm, the network forms around the DD (the crystal seed). Nodes not in communication with the developing network may not communicate with each other. As nodes join the network, they are assigned functions, as either ordinary nodes or cluster heads. The hierarchy development is thus controlled by the DD. The second way is the "polycrystalline" approach. In this algorithm, nodes discover each other and form independent clusters. As the clusters grow, they become aware of each other and form linkages between each other. The hierarchical cluster tree is formed as the cluster containing the DD discovers other clusters. The DD thus has less control over the network structure because the clusters arrive already formed.

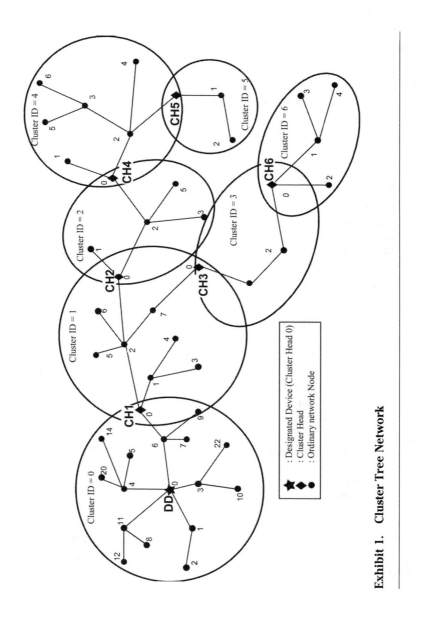

Exhibit 1.   Cluster Tree Network

As in crystallography, both types of growth have their advantages. The monocrystalline approach allows better control of the network as it is being grown. It also enables a certain type of network security; if, in a home automation application, for example, the DD is in the television set-top box, new nodes (from, for instance, a recently purchased light switch) may be programmed to respond only to that netID, while ignoring nodes from other networks such as those belonging to neighbors.

In the polycrystalline case, however, nodes are allowed to associate with, and form clusters with, all other nodes; netIDs only become apparent when a cluster contacts a cluster containing a DD. It, therefore, becomes impossible to limit the new node to communication with specific networks. In applications such as the home automation example, this can be a security problem.

To its defense, polycrystalline growth does have advantages. Some military applications, for example, require fast network formation, and security is obtained by other means (such as encoding in the physical layer). In these applications, polycrystalline growth is more appropriate. For this work, however, the monocrystalline approach was chosen.

5.3.2 Network Association

All nodes in the network periodically transmit beacons, as described in Chapter 4. These beacons contain the NetID, CID, and NID of the transmitting node, as well as its distance from its cluster head (in hops). Upon becoming active, a new node listens periodically for these transmitted beacons. When one is heard, the new node sends a CONNECTION REQUEST control message to join the network (Exhibit 2). If a beacon from more than one node is heard, the new node replies to the node closest to its cluster head; if a tie exists, it chooses the node with the lowest CID. If a tie still exists, it chooses the node with the lowest NID. The new node is now a candidate node. The prospective parent acknowledges the control message and sends a NODE ID REQUEST message to its cluster head, perhaps routed via intermediate nodes in the cluster.

If the cluster head can accept another node, it replies with a favorable NODE ID RESPONSE message that is relayed to the candidate node as a CONNECTION RESPONSE, giving it an NID in that cluster. If the cluster head cannot accept another node (Exhibit 3), it sends a CLUSTER ID REQUEST message to the DD, requesting the formation of a new cluster. The DD replies to the cluster head with a CLUSTER ID RESPONSE message, giving the candidate node a new CID (and an implied NID of 0). This response is then forwarded by the cluster head to the candidate node as a NODE ID RESPONSE message. After acknowledgment, the candidate node becomes a network member and begins transmitting beacons.

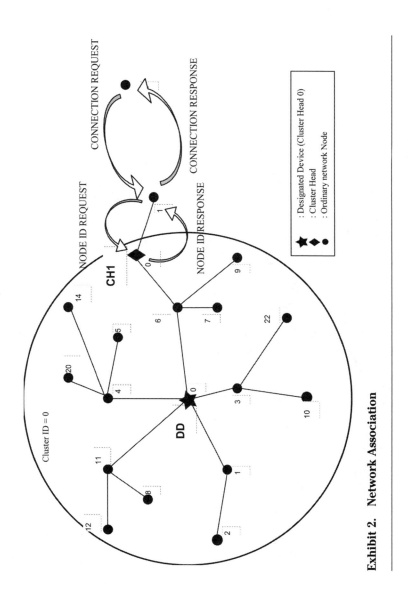

Exhibit 2.   Network Association

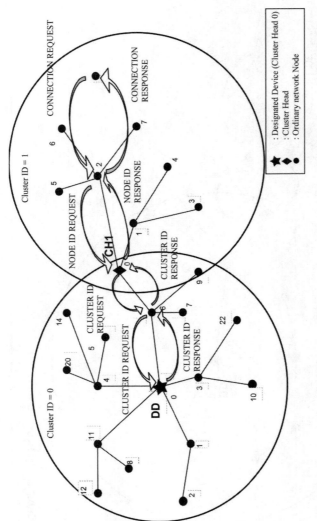

Exhibit 3.   Network Association, Forming New Cluster

Exhibit 4. Sample Routing Table for Node 7, Cluster 1

Destination		Next hop	
CID	**NID**	**CID**	**NID**
1	2 (parent)	1	2
1	4	1	4
2	0	2	0
3	0	3	0
2	—	2	0
3	—	3	0
6	—	3	0

The rule(s) used by the cluster head to determine whether or not it can accept another node must be heuristic; the minimum d-hop dominating set problem (i.e., the identification of the minimum set of nodes within d hops of all other nodes in the network) is NP-complete.[55]

5.3.3 Network Maintenance

Each network node keeps a routing table, initially containing just the CIDs and NIDs of its neighbors and the nodes it can hear while operating in MD mode (see Chapter 4). The table entry of its parent is identified.

Periodically, the DD sends a LINK STATE REQUEST message to all leaves of the hierarchical tree (i.e., all nodes that are not parents). Upon receiving the LINK STATE REQUEST message, a leaf generates a LINK STATE RESPONSE message. The LINK STATE RESPONSE message contains the leaf's CID and NID, plus the CIDs of all nodes it can hear that have CIDs differing from its own. The message is then sent to the cluster head. As this message passes through the network, relaying nodes "eavesdrop" on the contents of the message, identifying out-of-range "downstream" nodes and clusters. These downstream nodes and clusters are added to their routing tables, listing the node from which the LINK STATE RESPONSE message was received as the first relay node in the route. An example routing table for node 7 in cluster 1 of Exhibit 1 is presented in Exhibit 4. The relaying nodes then add their NIDs to the node list in the message and forward it "upstream" to the cluster head.

At the cluster head, the information in the LINK STATE RESPONSE message is placed in its own routing table. The cluster head then performs data fusion, generating a new LINK STATE RESPONSE message that contains its NID, but only the downstream CIDs — all downstream NID information is deleted. This data hiding reduces the size of the LINK STATE RESPONSE message without materially affecting routing efficiency. The new LINK STATE RESPONSE message is then forwarded to the DD; relaying nodes eavesdrop on the message on the way. Upon reception of all LINK STATE RESPONSE messages, the DD has an understanding of the network structure and, by incorporating the information into its own routing table, is able to route messages itself.

5.3.4 Routing

The routing algorithm of the proposed network is very simple and shown in Exhibit 5. If the CID of the destination differs from that of the relaying node, the node's routing table is searched to locate the CID. If the CID is found, the message is routed via the listed relay node. If the CID is not found, the message is routed to the node's parent.

If the CID of the destination is the same as that of the relaying node, the node's routing table is searched for a match of both CID and NID. If the destination found in the table is a neighbor, the neighbor is contacted directly and the data message delivered. If the destination is listed in the routing table because of eavesdropping on a LINK STATE RESPONSE message, the data message is sent downstream to the node that sent the LINK STATE RESPONSE message. If the destination is not listed in the routing table, the message is forwarded to the node's parent (upstream).

5.4 SIMULATIONS

Due to the unique nature of the distributed MD protocol, simulation of the proposed wireless sensor network with existing network simulators, such as Opnet[56] and ns-2,[57] is not very practical.[58] Instead, a discrete-event simulator was written in Visual C++ by Yan Huang, Ph.D., of the Florida Communication Research Laboratory, Plantation, a division of Motorola Labs. The simulator, WinneuRFon, was designed to simulate both the throughput (quantity of service) and message latency (quality of service) of a cluster tree network employing the distributed MD protocol. In particular, it provides the following message-centric outputs:

- Time and node of origination of each message
- Time between message origination and message delivery
- Number of nodes traversed
- Path taken by each message
- Time each message waited at each node

WinneuRFon also provides node-centric outputs, including the number of messages facilitated by a node in MD mode for each MD cycle, the number of messages generated and relayed by each node, and the operating duty cycle of each node, including any time spent in MD mode. The raw data is placed in a Microsoft Excel™-readable file for post-processing.

WinneuRFon is designed to model a wide range of physical network sizes, node densities, offered message rates, and beacon and MD periods. The location of the DD, as the root of the cluster tree, is deterministically

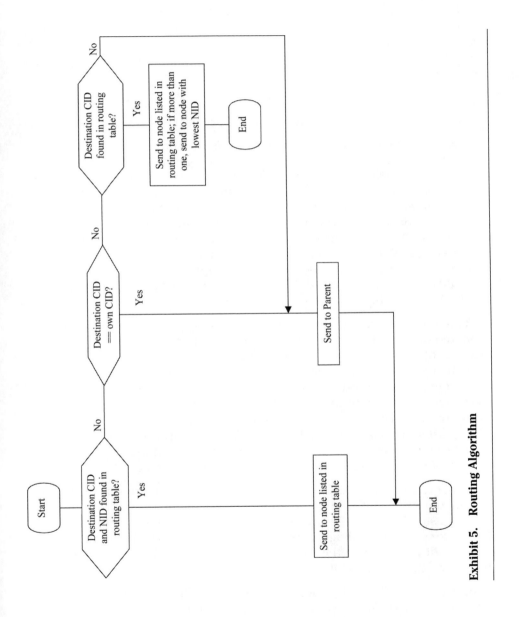

Exhibit 5. Routing Algorithm

placed within a rectangle of dimensions a × b; the other network nodes are randomly placed. WinneuRFon accepts as inputs:

- Physical array area, a × b (meters)
- Number of nodes in the area (randomly distributed in uniform distribution)
- Node beacon period (seconds)
- Length of node beacon (seconds)
- Range of MD period (seconds)
- Node range (meters)
- Coordinates of the DD within the array area
- Average number of messages offered to be sent per node per hour

Messages are generated at random times from random nodes. To best model the behavior of a network of sensors sending data to the Internet, all messages are short, identical in length (twice the node beacon length), and are destined for the DD. The phase of each node's beacon and its switches to MD mode are randomized; this models the effect of the low-cost time base of network nodes. WinneuRFon does not perform channel access protocols; however, it does detect packet collisions caused by simultaneous transmissions of nodes within range of a receiving node. When this happens, both packets are considered lost. A sample network generated by WinneuRFon is illustrated in Exhibit 6; in the figure, the lines represent parent–child relationships, not communication links.

WinneuRFon has an optional color graphics (animation) mode. In this mode, the physical layout of the generated network is displayed; a color code identifies the level of each node and the parent–child relationship. Nodes that enter MD mode turn black for the duration of MD mode; green lines stretch from the node in MD mode to all nodes from which it receives query beacons; and black lines stretch from the node in MD mode to all nodes from which it receives RTS beacons. Following MD activity, a red line indicates data message transmission between the appropriate RTS and query beacon nodes. WinneuRFon also has a "Graph Pace" input variable that can slow the animation display to a humanly comprehensible speed.

To ensure that their outputs are useful, simulation models should be verified and validated. The verification process evaluates the degree to which the simulator matches the intent of its author (that is, it attempts to identify, for example, coding errors in the software); the validation process evaluates how accurately the simulator matches the real-world phenomenon it purports to represent.[59]

Verification of WinneuRFon was simplified by performing simulations in its animation mode at a pace slow enough for a human expert to observe and record all simulated network behavior, and compare it with that expected. The resulting activity statistics were then computed by hand,

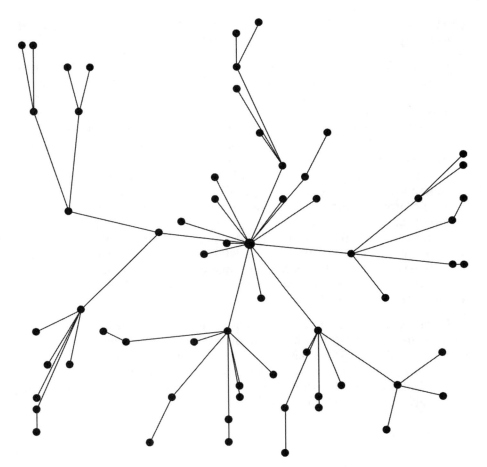

Exhibit 6. A Typical Network Generated by WinneuRFon

and compared to the Excel output files of WinneuRFon. Use of the well-characterized Excel for post-processing of data, instead of a custom-coded data processing module, reduced the possibility of data-processing errors. These verification simulations were performed over a wide range of input parameter combinations. The major branches of the simulation software were tested to ensure functional correctness as part of the structure-based software validation process. Parametric study of the output files was also performed to check for internal consistency; for example, the message latency increased, and data throughput decreased, as the MD period was increased. The simulated values of some output parameters were compared with analytical values where analytical models could be identified; for example, message latency was simulated in a single-hop network (i.e., one in which the node range was greater than the size of the network) and found to agree with values derived from Exhibit 8 in Chapter 4.

Because WinneuRFon is a research simulator, developed before a physical network was available, validation of it cannot be done by comparing its output with either the published results or the experimentally determined behavior of a physical network. WinneuRFon was designed with the relatively narrow goal of modeling the behavior of a cluster network employing the distributed MD, so several potential network nonidealities (listed and discussed in Appendix B) are neglected, and some simplifications of network traffic are assumed. Network traffic simplifying may include defining all messages to be identical in length, twice as long as the beacon length, and generated by independent random variables in each node with a uniform random distribution. For many potential applications, such as wireless thermostats, this message model is a good approximation of reality; for others, such as security systems, which may have more bursty, correlated message generation, other message models would be appropriate. Future versions of WinneuRFon will consider these nonidealities and allow a more flexible description of message generation.

Three sample parameter sets for WinneuRFon, for three potential applications, are given next. Exhibit 7 lists parameters for a wireless thermostat application; Exhibit 8 lists parameters for a wireless supermarket price tag application; and Exhibit 9 lists parameters for a consumer electronic application. Unless otherwise stated, all simulations were performed with the input parameters in Exhibit 7.

5.5 RESULTS

5.5.1 Throughput (Cumulative Percentage of Messages Arriving at the DD versus Time)

Exhibit 10 plots the relationship between the cumulative percentage of messages arriving at the DD and time since network formation. Due to the nonzero message latency of the network, just after network formation many messages have been generated but not delivered, which lowers the throughput calculation. The latent messages become a smaller and smaller fraction of total messages generated over time, allowing the cumulative throughput to asymptotically approach 100 percent. As presented in Exhibit 7, the offered throughput was 0.5 messages/node/hour.

5.5.2 Throughput (Cumulative Percentage of Messages Arriving at the DD versus Node Level)

Exhibit 11 plots the relationship between the cumulative percentage of messages arriving at the DD and the node level at which the messages were

Exhibit 7.   Network Simulation Parameters for Wireless Thermostat Application

Parameter	Value
Physical area	40 × 40 meters
DD location	Center (20, 20)
Number of network nodes (including DD)	64
Range	10 meters
Beacon period	2 seconds
Beacon length	0.001 second
Average messages generated per node per hour	0.5
MD period range	1500–2500 seconds

Exhibit 8.   Network Simulation Parameters for Supermarket Price Tag Application

Parameter	Value
Physical area	50 × 50 meters
DD location	Corner (5, 5)
Number of network nodes (including DD)	200
Range	10 meters
Beacon period	30 seconds
Beacon length	0.001 second
Average messages generated per node per hour	0.25
MD period range	20,000–30,000 seconds

Exhibit 9.   Network Simulation Parameters for Consumer Electronic Application

Parameter	Value
Physical area	5 × 5 meters
DD location	Edge (0, 3)
Number of network nodes (including DD)	5
Range	10 meters
Beacon period	1 second
Beacon length	0.001 second
Average messages generated per node per hour	5
MD period range	100–300 seconds

generated. The node level is defined as the distance of the generating node from the DD, plus one; thus, the DD is level 1, its children are in level 2, etc. In this simulation, the physical network area was covered in four layers. As expected, the message arrival percentage falls slightly as one moves away from the DD, indicating that messages are still in transit.

5.5.3 Average Message Transmission Time

Exhibit 12 plots the relationship between average message transmission time and time since network formation. The network reaches a finite steady state value (of approximately 1100 seconds), indicating that it can support the offered message generation rate of 0.5 messages/node/hour.

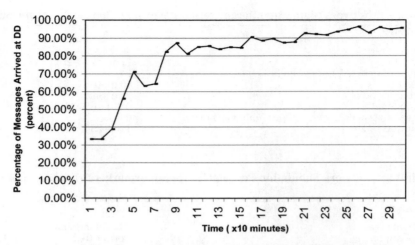

Exhibit 10. Cumulative Percentage of Generated Messages Arriving at the DD

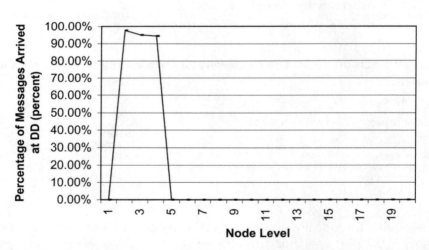

Exhibit 11. Cumulative Percentage of Generated Messages Arriving at the DD versus Node Level at Which They Were Generated

5.5.4 Average Message Latency versus Node Level

Exhibit 13 plots the relationship between the average message transmission time and node level. Although the average time to send a message from level 4 to the DD may seem long (1522 seconds), it is at the lower end of the range for a single MD period (1500–2500 seconds). Because an MD must be active for a communication link to be active, messages arriving

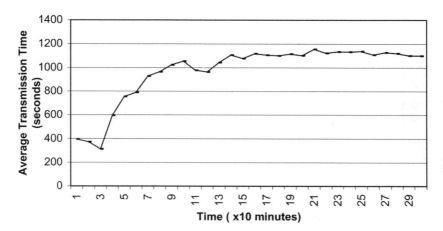

Exhibit 12. Average Message Transmission Time

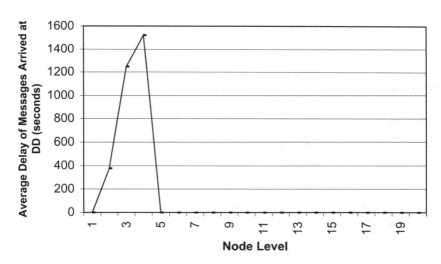

Exhibit 13. Average Message Latency versus Node Level

from level 4 had, on average, three MD interactions in 1522 seconds — indicating the value of the distributed MD protocol in a dense network.

5.5.5 Packet Collisions versus Time

Exhibit 14 plots the relationship between average number of packet collisions per node per hour and time since network formation. This is of interest due to the asynchronous nature of the nodes' beacons and the

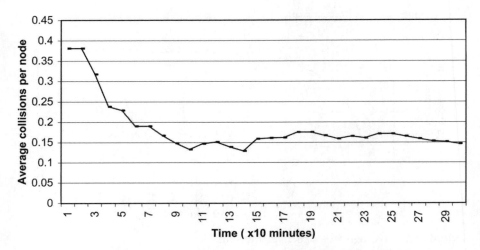

Exhibit 14. Average Number of Packet Collisions per Node per Hour versus Time

ALOHA signaling to the MD. The steady state value of 0.15 collisions/node/hour indicates that the network performance will not be limited by packet collisions.

5.5.6 Duty Cycle

Exhibit 15 plots the relationship between average node duty cycle (including both transmitting and receiving) and time since network formation. The input beacon duty cycle, from Exhibit 7, is 0.1 percent; with a beacon period of 2 seconds and an MD period range of 1500–2500 seconds, the duty cycle due to MD activities is also approximately 0.1 percent, for a predicted total duty cycle of 0.2 percent. The simulated steady state value is 0.195 percent. When used in Equation (2) of Section 3.3.2.2 with the given values of I_{on} = 19.5 mA and I_{stby} = 30 μA, the simulated value results in a node AAA battery life of 460 days, or more than 15 months.

5.5.7 Duty Cycle versus Level

Exhibit 16 plots the relationship between average node duty cycle and node level. The near-constant duty cycle between nodes in differing levels of this network hierarchy is contrary to that found in other hierarchical networks (cf. Heinzelman et al.[60]), and is explained as follows.

With the low data throughput of the network, the node duty cycle is dominated by beacon and MD activities. Unless they are in MD mode, the nodes transmit beacons with a constant duty cycle. When messages are

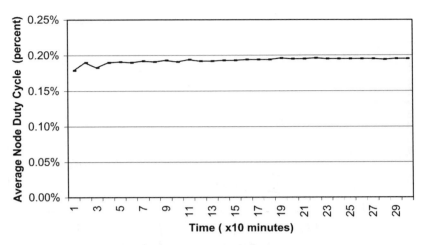

Exhibit 15. Average Node Duty Cycle versus Time

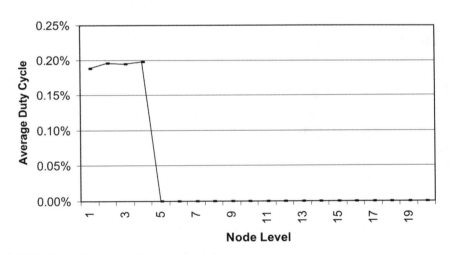

Exhibit 16. Node Duty Cycle versus Level

sent, the node replaces beacon transmissions with message transmissions, so the average duty cycle remains relatively unchanged. The occasion of a large amount of traffic at a node does not occur because (1) the nodes themselves are assumed to generate only infrequent messages, and (2) traffic between nodes can only occur due to an MD period of a nearby device. This limits the amount of traffic the network can transport because MD activity is infrequent, but also equalizes power consumption for all

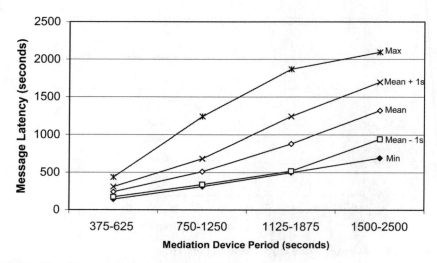

Exhibit 17.   Message Latency versus MD Period

nodes. Lower-level nodes, which carry more traffic in the cluster tree, are, therefore, not significantly penalized in power consumption.

5.5.8 Message Latency versus MD Period

Exhibit 17 plots the relationship between message latency and MD period, for 30 randomly generated topologies at each MD period value. The minimum, maximum, mean, mean minus one standard deviation, and mean plus one standard deviation curves are shown. Message latency is a strong function of the MD period because message relays can only occur due to MD activity.

5.5.9 Maximum Network Throughput versus MD Period

Exhibit 18 plots the relationship between maximum network throughput and MD period. As expected, network throughput increases as nodes become connected via MD action more often.

5.5.10 Maximum Network Throughput versus Node Density

Exhibit 19 plots the relationship between maximum network throughput and node density. This simulation was performed by keeping the number of network nodes constant at 64 and reducing the physical area of the network. The range of each node was kept constant, at 10 meters, which resulted in a flatter cluster tree (fewer average hops to the DD). A significant factor affecting message throughput is the increased average number

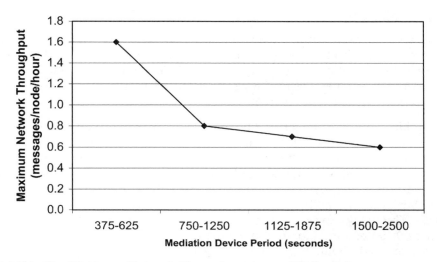

Exhibit 18. Maximum Network Throughput versus MD Period

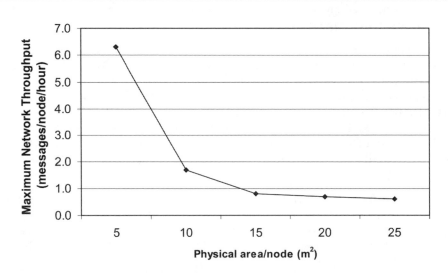

Exhibit 19. Maximum Network Throughput versus Node Density

of nodes operating in MD mode within range of communicating nodes. In the degenerate case of one node/meter^2 (all 64 nodes in a 64 m^2 area), a maximum network throughput of 85 messages/node/hour was achieved. In this case, all nodes communicated directly with the DD (i.e., the network operated as a star network), with the MD performing dynamic synchronization.

5.6 CONCLUSION

Operation of, and routing in, a wireless sensor network represent significant challenges to the network protocol designer due to the need for low duty cycle operation of network nodes. This low duty cycle requirement limits the allowable overhead for synchronization, negotiation, coordination, and other network layer activities.

In this chapter, a cluster tree network capable of multi-hop routing was developed and simulated. This network employed the distributed MD protocol, and was designed to address the problem inherent in networks of clusters, routing to non-neighboring clusters. A discrete-event network simulator was employed to evaluate the throughput, message latency, and number of packet collisions of the network, and the network node duty cycle. The network was found to be stable even with node duty cycles as low as 0.19 percent, and able to handle an offered data throughput of 0.5 messages/node/hour. The node duty cycle was shown not to vary significantly with position in the cluster tree hierarchy; this is due to the fixed (but low) duty cycle of the node beacons, which are simply replaced by data messages when sent, keeping overall duty cycle nearly constant with variations in message traffic.

Message latency was found to be a strong function of the MD period, rising from an average of 241 seconds with a 500-second MD period, to an average of 1321 seconds with a 2000-second MD period. This is as expected because communication between nodes can only take place after MD activity of a nearby node. The advantage of the distributed MD protocol in dense networks is also apparent: Even though the average message made at least two hops, the average message latency is approximately half of an MD period, indicating that multiple neighbors serve as MDs in the path of the average message.

The maximum network throughput was found to increase with node density, due both to the resulting flatter tree network and the more frequent MD availability to communicating nodes. Packet collisions were rare, validating the medium access function of the distributed MD protocol.

References

1. Hubert Zimmermann, OSI reference model — the ISO model of architecture for Open Systems Interconnection, *IEEE Trans. Commun.,* v. COM-28, n. 4, April 1980, pp. 425–432.
2. Andrew S. Tanenbaum, *Computer Networks.* 3rd ed. Upper Saddle River, NJ: Prentice Hall. 1996. p. 31.
3. Christine E. Jones et al., A survey of energy efficient network protocols for wireless networks, *Wireless Networks*, v. 7, n. 4, September 2001, pp. 343–358.
4. John Jubin and Janet D. Tornow, The DARPA packet radio network protocols, *Proc. IEEE*, v. 75, n. 1, January 1987, pp. 21–32.

5. Dennis J. Baker and Anthony Ephremides, The architectural organization of a mobile radio network via a distributed algorithm, *IEEE Trans. Commun.,* v. COM-29, n. 11, November 1981, pp. 1694–1701.

6. Dennis J. Baker, Anthony Ephremides, and Julia A. Flynn, The design and simulation of a mobile radio network with distributed control, *IEEE J. Selected Areas in Communications,* v. SAC-2, n. 1, January 1984, pp. 226–237.

7. Baker and Ephremides, ibid.

8. Mario Gerla and Jack Tzu-Chieh Tsai, Multicluster, mobile multimedia radio network, *Wireless Networks,* 1995, v. 1, pp. 255–265.

9. R. E. Bellman, *Dynamic Programming.* Princeton, NJ: Princeton University Press. 1957.

10. L. R. Ford Jr. and D. R. Fulkerson, *Flows in Networks.* Princeton, NJ: Princeton University Press. 1962.

11. Jie Wu, *Distributed System Design.* Boca Raton, FL: CRC Press. 1999. pp. 180–181.

12. Lance Eric Hester, A self-organizing wireless network protocol, Ph.D. dissertation, 2001, Northwestern University, Evanston, Illinois.

13. Leonard Kleinrock and Farouk Kamoun, Hierarchical routing for large networks, *Computer Networks*, v. 1, 1977, pp. 155–174.

14. A. Bhatnagar and T. G. Robertazzi, Layer net: a new self-organizing network protocol, Conference Record, IEEE Military Communications Conference, 1990, v. 2, pp. 845–849.

15. W. R. Heinzelman, A. Chandrakasan, and H. Balakrishnan, Energy-efficient communication protocol for wireless microsensor networks, Proc. 33rd Annual Hawaii International Conference on System Sciences, 2000, pp. 3005–3014.

16. Jay Lin Gao, Energy efficient routing for wireless sensor networks, Ph.D. dissertation, 2000, University of California at Los Angeles.

17. Jamil Ibriq, A reduced overhead routing protocol for ad hoc wireless networks, Master's thesis, 2000, Florida Atlantic University, Boca Raton, Florida.

18. Josh Broch et al., A performance comparison of multi-hop wireless ad hoc network routing protocols, *Proc. MOBICOM, 1998,* pp. 85–97.

19. E. M. Royer and Chai-Keong Toh, A review of current routing protocols for ad hoc mobile wireless networks, *IEEE Pers. Commun.,* v. 6, n. 2, April 1999, pp. 46–55.

20. Sung-Ju Lee et al., A performance comparison study of ad hoc wireless multicast protocols, *Proc. IEEE INFOCOM,* 2000, v. 2, pp. 565–574.

21. M. Scott Corson, Joseph P. Macker, and Gregory H. Cirincione, Internet-based ad hoc networking, *IEEE Internet Computing,* v. 3, n. 4, July–August 1999, pp. 63–70.

22. Charles E. Perkins and Pravin Bhagwat, Highly dynamic Destination-Sequenced Distance-Vector (DSDV) routing for mobile computers, *ACM SIGCOMM Computer Communication Review, Proc. Conf. on Communications Architectures, Protocols and Applications,* v. 24, n. 4, October 1994, pp. 234–244.

23. Vincent D. Park and M. Scott Corson, A highly adaptive distributed routing algorithm for mobile wireless networks, *Proc. IEEE INFOCOM,* 1997, v. 3, pp. 1405–1413.

24. Tsu-Wei Chen and Mario Gerla, Global state routing: A new routing scheme for ad-hoc wireless networks, Conference Record, *IEEE Int. Conf. Commun.,* 1998, v. 1, pp. 171–175.

25. Charles E. Perkins and Elizabeth M. Royer, Ad hoc on-demand distance vector routing, *Proc. Second IEEE Workshop on Mobile Computing Systems and Applications,* 1999, pp. 90–100.

26. A. Iwata et al., Scalable routing strategies for ad hoc wireless networks, *IEEE J. Selected Areas in Commun.,* v. 17, n. 8, August 1999, pp. 1369–1379.

27. Guangyu Pei, Scalable routing strategies for large ad hoc wireless networks, Ph.D. dissertation, 2000, University of California at Los Angeles.

28. Mario Gerla, Xiaoyan Hong, and Guangyu Pei, Landmark routing for large ad hoc wireless networks, *IEEE Global Telecommunications Conf.,* 2000, v. 3, pp. 1702–1706.

29. Robert Dunbar Poor, Embedded networks: Pervasive, low-power, wireless connectivity, Ph.D. dissertation, 2001, Massachusetts Institute of Technology, Cambridge, Massachusetts.
30. Ibid.
31. Robert Dunbar Poor, Hyphos: a self-organizing wireless network, Master's thesis, 1997, Massachusetts Institute of Technology, Cambridge, Massachusetts.
32. Suresh Singh, Mike Woo, and C. S. Raghavendra, Power-aware routing in mobile ad hoc networks, *Proc. MOBICOM*, 1998, pp. 181–190.
33. Jung-hee Ryu and Dong-Ho Cho, A new routing scheme concerning energy conservation in wireless home ad-hoc networks, *IEEE Trans. Consumer Electronics*, v. 47, n. 1, February 2001, pp. 1–5.
34. Jie Wu and Hailan Li, On calculating connected dominating set for efficient routing in ad hoc wireless networks, *Proc. 3rd Int. Workshop. Discrete Algorithms and Methods for Mobile Computing and Communication*, 1999, pp. 7–14.
35. Jie Wu and Hailan Li, A dominating-set-based routing scheme in as hoc wireless networks, *Telecommunication Syst. J.*, special issue on wireless networks, v. 3, 2001, pp. 63–84.
36. Jae-Hwan Chang and Leandros Tassiulas, Energy conserving routing in wireless ad hoc networks, *Proc. INFOCOM*, 2000, pp. 22–31.
37. Henry Petroski, *To Engineer is Human: The Role of Failure in Successful Design*. New York: St. Martin's Press. 1985. pp. 35–39.
38. C. F. Chiasserini and R. R. Rao, Routing protocols to maximize battery efficiency, *Proc. MILCOM*, 2000, v. 1, pp. 496–500.
39. Ibid.
40. B. Das and V. Bharghavan, Routing in ad hoc networks using minimum connected dominating sets, *IEEE Int. Conf. on Communications*, 1997, v. 1, pp. 376–380.
41. B. Das, R. Sivakumar, and V. Bharghavan, Routing in ad hoc networks using a spine, *Proc. Sixth Int. Conf. on Computer Communications and Networks*, 1997, pp. 34–39.
42. Jie Wu et al., On calculating power-aware connected dominating sets for efficient routing in ad hoc wireless networks, *J. Commun. Networks*, v. 4, n. 1, March 2002.
43. Ruud Schoonderwoerd et al., Ant-based load balancing in telecommunications networks, *Adaptive Behavior*, v. 5, n. 2, 1996, pp. 169–207.
44. Guoying Li, Subing Zhang, and Zemin Liu, Distributed dynamic routing using ant algorithm for telecommunication networks, *Proc. Int. Conf. on Communication Technology*, 2000, v. 2, pp. 1607–1612.
45. T. Michalareas and L. Sacks, Link-state and ant-like algorithm behaviour for single-constrained routing, *IEEE Workshop on High Performance Switching and Routing*, 2001, pp. 302–305.
46. Chalermek Intanagonwiwat, Ramesh Govindan, and Deborah Estrin, Directed diffusion: a scalable and robust communication paradigm for sensor networks, *Proc. Sixth Annual Int. Conf. on Mobile Computing and Networking*, 2000, pp. 56–67.
47. Ibid.
48. Wendi Rabiner Heinzelman, Joanna Kulik, and Hari Balakrishnan, Adaptive protocols for information dissemination in wireless sensor networks, *Proc. Fifth Annu. ACM/IEEE Int. Conf. on Mobile Computing and Networking*, 1999, pp. 174–185.
49. Ibid.
50. Y.-C. Cheng and T. G. Robertazzi, Communication and computation tradeoffs for a network of intelligent sensors, *Proc. Computer Networking Symp.*, 1988, pp. 152–161.
51. Y.-C. Cheng and T. G. Robertazzi, Distributed computation with communication delay (distributed intelligent sensor networks), *IEEE Trans. Aerospace and Electronic Syst.*, v. 24, n. 6, November 1988, pp. 700–712.

52. A description of the network at an earlier stage of design is available at http://ieee802.org.http://ieee802.org/15/pub/2001/May01/01189r0P802-15_TG4-Cluster-Tree-Network.pdf.
53. Hester, ibid.
54. The author is indebted to Masahiro Maeda, Motorola Japan Research Laboratory, Tokyo, Japan, for Exhibit 1.
55. Alan D. Amis et al., Max-min D-cluster formation in wireless ad hoc networks, *Proc. IEEE Infocom,* 2000, pp. 32–41.
56. http://www.opnet.com./nsnam/ns/.
57. http://www.isi.edu.
58. A paper describing some of the difficulties was presented at OPNETWORK 2001. See Jian Huang et al., Simulation of a low duty cycle protocol, presented at OPNETWORK 2001, Washington, D.C., August 2001. An OPNET simulation was later designed; see Yan Huang et al., OPNET simulation of a multi-hop self-organizing wireless sensor network, presented at OPNETWORK 2002, Washington, D.C., August 2002. Both are available at http://www.opnet.com/products/modeler/biblio.html. OPNET simulations of a similar network are available in L. Hester et al., neuRFon™ netform: A self-organizing wireless sensor network, *Proc. 11th Intl. Conf. Computer Communication and Networks*, 2002, pp. 364–369.
59. John Heidemann, Kevin Mills, and Sri Kumar, Expanding confidence in network simulations, *IEEE Network,* v. 15, n. 5, September/October 2001, pp. 58–63.
60. Heinzelman, Chandrakasan, and Balakrishnan, ibid.

Chapter 6
Practical Implementation Issues

6.1 INTRODUCTION

The design of a wireless sensor network is more than the design of a communication protocol. The design of the actual network nodes is critical to the success of the network (i.e., to how well the network meets the needs of the application). Several features of the design of wireless sensor network nodes are unique (or nearly so) to the design of small, battery-powered wireless devices, and are not often encountered in the design of other types of electronic equipment. For example, in the design of a cellular telephone or other land mobile communications system, the primary cost to be borne in mind during development is the cost of the system infrastructure — the base stations, towers, switching equipment, etc. — both as an initial capital outlay and then as monthly rental, depreciation, and maintenance expense. To make the system economically viable, this equipment must service the largest possible number of users to defray these fixed costs among them to the point that the service is affordable. The cost of the individual mobile handset is, to a very large degree, irrelevant when compared with the infrastructure cost; not only is its cost relatively low when compared with the infrastructure cost, its maintenance cost is also negligible — and often borne by the user, in any event, in the form of electricity to recharge the battery.

Because the system design has a goal to maximize the number of users supported by the infrastructure equipment, sophisticated techniques (of modulation, coding, etc.) are employed to do this, even if the cost of the handset is raised significantly — because the total cost to the user, when the system infrastructure costs are factored in, will be lower.

The situation with wireless sensor networks is quite different. Because no infrastructure is associated with the ad hoc networks they employ, the economics of wireless sensor networks only consider the cost of the individual network node. This greatly changes the design goal from that of the

115

land mobile system designer. Instead of designing for the maximum possible system capacity, the designer must design for the lowest possible unit cost (consistent with the other network goals, of course). This places the design emphasis squarely on the product cost of the network node because its cost of operation (electricity, etc.) is trivial and focuses the designer's attention on some network node features that typically receive little attention in other services.

This chapter discusses a few of these more unique features, including some subtleties involved in deciding how much of the network node should be integrated (and into which chips), the importance of sensor interfaces, and the interrelationship between time base accuracy and the power consumption of the network node.

6.2 THE PARTITIONING DECISION

The conversion of a system block diagram into an end user product is a complex one, with trade-offs evaluated and decisions made at several levels in the design process. The decisions are especially critical in the design of a wireless sensor network node, due to the enormous attention paid to product cost. A too-conservative integration strategy can price a manufacturer out of the market, while a too-aggressive strategy may result in insurmountable technical difficulties, or (perhaps worse) a functioning product, but one useful only in a market too small to produce a useful return on the engineering investment. Although some flexibility may be retained by programmability, it nevertheless remains true that integration is a form of product specialization, and, just as in biology, there is an optimum amount of specialization for a given velocity of environmental change.

A block diagram of a generic wireless sensor network node is presented in Exhibit 1. The node consists of an antenna, a radio frequency (RF) transceiver, a communication protocol handler, an application processor (often application-specific), random access memory (RAM), read-only memory (ROM), and one or more transducers (i.e., sensors and/or actuators). In addition, the network node requires some form of power conditioning, usually some type of human interface (e.g., a display driver) and perhaps energy-scavenging circuitry (unless a more conventional power source is used).

The question facing the integrated circuit manufacturer that desires to enter the wireless sensor network node market is how to partition this block diagram into one or more integrated circuits. A number of factors should be considered:

- *Application computational requirements (RAM, ROM, and processor speed).* The amount of computational resources needed by the intended application(s) can vary by many orders of magnitude. Some appli-

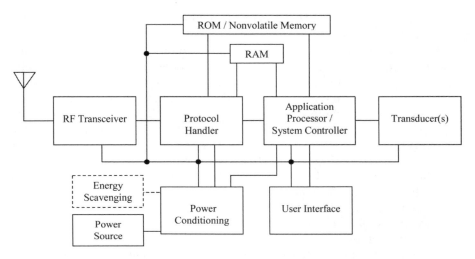

Exhibit 1. A Generic Wireless Sensor Network Node

cations may require almost none at all; inputs to and from the
transducer are simply passed directly to the protocol handler as data
with just a slight amount of formatting required. Other applications
may require extensive computational resources — for instance, if the
operation of the transducer is complex or if significant post-processing of the data (e.g., a fast Fourier transform) is required. Additional
resources are required if the wireless sensor network is required to
perform data fusion — the analysis and condensation of data received
from other network nodes — in addition to processing its own data. Finally, although the processing required of each transducer may be
simple, a single network node may support a large number of transducers. The device manufacturer must then choose between a relatively inexpensive design of limited capability, incorporating only a
small amount of RAM and ROM, or a more capable but more expensive
design incorporating more RAM and ROM and a faster processor.

- *Ability to use host resources.* Often a wireless sensor network node is
  incorporated in other, larger systems, such as a personal digital assistant (PDA), cellular telephone, or even a fixed asset such as a building
  environmental controller. In these applications, resources (especially
  computational resources) may be available from the host. Having
  such resources available means that they do not have to be supplied
  by the wireless sensor node itself; the designer may take advantage of
  this fact, remove duplicate resources from the network node design,
  and reduce its cost accordingly. The penalty, of course, is that the total

available market for this product has now shrunk to that with supporting hosts.

- *Desired market flexibility.* The integration of specific energy scavenging and power conditioning circuits reduces cost and improves efficiency, but limits the design to a particular power source. Separating the power source and conditioning circuits from the rest of the network node design enables a single design to be reused with multiple external sources and conditioning circuits, albeit at higher marginal product cost and lower overall efficiency. Integration of a particular transceiver requires a new chip if use in a new band, or a new protocol in an existing band, is envisioned. For example, the manufacturer of a fully integrated 900-MHz product line that wishes to expand its product line to include a 2450-MHz product must duplicate its entire line. If the transceiver is not integrated with the rest of the network node, the existing chips in the product line may remain, and only an interface-compatible 2450-MHz transceiver need be designed — a much easier task. Having market flexibility is especially important in emerging, immature markets, such as those for wireless sensor networks, in which the largest volume niches may not be proven and standard designs not yet adopted.

- *Product physical size goals.* Of course, a direct trade-off exists between the level of system integration and final product physical size; however, certain applications do not require the absolute minimum physical size of the network node circuits. For example, total system integration will often result in a design that is smaller than its user interface. It is relatively pointless to reduce the network node circuit area to 1 cm^2 if, for instance, a keypad and display are required for a user to enter a security code — especially if increasing integration is increasingly expensive due to the need for an expensive, specialized integrated circuit (IC) process. Similar arguments can be made concerning the relative sizes of batteries and antennas.

- *Required time to market.* A critical point in the electronics industry is time to market. A fully integrated wireless sensor network node is a complex, mixed-signal (i.e., containing both analog and digital circuits) design that takes time to design, test, and debug. However, its components — processors, transceivers, etc. — are either available off-the-shelf or are much simpler to design as stand-alone chips. Therefore, one desiring to enter the market and receive revenue quickly may choose a less integrated design to start, perhaps while working on a more integrated design to follow as the market matures.

- *Network heterogeneity.* Many wireless sensor networks support multiple types of network nodes, having different capabilities. For example, the Institute of Electrical and Electronics Engineers (IEEE) 802.15.4 standard[1] specifies two types of devices: a Full Function Device (FFD) and a Reduced Function Device (RFD). The FFD may be the master (called the "PAN Coordinator" in the IEEE 802.15.4 standard) of a star network and, therefore, must have enough memory to store address information of all devices in the network. In a star network, an RFD may be only a slave, and thus needs memory only sufficient to store the address of the network master. Most manufacturers would want to sell both types; to enable both FFD and RFD functionality, they may elect to design one chip with programming to support either of these functions. Because the RFD design then has the memory required for the FFD, however, its cost of production is not as low as an RFD-only design. To enable a low-cost RFD design, the manufacturer may elect to design a second fully integrated RFD chip; however, an alternative can be to design only one chip for both FFD and RFD, but not including the processing function. Instead, the chip can have an interface to an external microcontroller. This approach, although not having many of the benefits of full integration (small size, low cost, etc.) does have the advantage of flexibility; the external microcontroller can be matched to the processing needs at hand — whether for RFD, FFD, or even to support specific applications.

- *Digital logic.* To minimize costs, digital logic should be placed in the IC process with the highest possible complementary metal oxide semiconductor (CMOS) gate density. This process usually supports only CMOS logic, i.e., without additional features such as bipolar devices (BiCMOS), high-quality integrated analog passive components, or memory cells.

- *RF and analog performance.* IC processes optimized for RF and analog performance usually include bipolar transistors, high-quality inductors and capacitors, and large value resistors. Recent lithographic improvements in CMOS processes have made metal oxide semiconductor (MOS) devices useful at RF; however, their noise and device-matching behavior (two process features important for most RF and analog design) are still inferior to that of bipolar devices.

- *Need for high-voltage operation.* State-of-the-art CMOS processes have limited voltage capability; even a dual-oxide process often cannot support devices switching more than 3.6 V or so. Many circuits, particularly energy scavenging, power conversion, and user interface circuits, may need to support higher voltages.

- *Need for high-current operation.* Some transducers may need significant amounts of current to operate. In addition, some actuators may need to switch high currents. A specialized IC process and specialized packaging may be necessary to support these requirements.

- *Packaging and chip-to-chip interconnect.* Packaging and chip-to-chip interconnect is expensive. The fewer packages that must be used, the lower the material cost and the higher the overall yield of the factory process. Chip-to-chip interconnect requires at least two interconnection pads (one on each chip); such pads, whether conventional wire bond pads or pads for any of the direct chip attach (DCA) schemes, take up prodigious amounts of die area — especially when their electrostatic discharge (ESD) protection circuitry is included.

• *High-speed signals.* High-speed signals off-chip can consume a lot of power. Due to the greater capacitance of off-chip interconnect, when compared with integrated interconnect, designs using high-speed, off-chip signals can suffer a power consumption penalty.

With this many factors to consider, no one optimum design is recommended. Different markets and manufacturers weigh these factors differently, and so the definition of optimum varies by market and the goals of the manufacturer. One may evaluate several partitionings of the block diagram in light of these factors, however.

The degenerate partitioning, of course, is to place the entire system in a single chip — the system-on-a-chip (SoC) (Exhibit 2a). This design, because it has the fewest number of packages, likely will have the smallest implementation and the lowest component cost, and is usually optimal for stand-alone applications (e.g., a light switch). An additional advantage is the high level of communication that may take place between the different blocks; without the power drain and limited pinout concerns of interchip communication, information can be easily transported across the system. This can be useful, for example, for the protocol handler to inform the power conditioning circuit that the RF transceiver will soon be in a low-power-consumption state, enabling the power conditioning circuit to revert to a lower output power, but higher efficiency, mode. As discussed previously, however, the SoC design is usually optimal only for a specific market or market segment; the segment must be large enough to provide a return on the significant investment required to design it.

A second partitioning separates the RF transceiver into a separate chip (Exhibit 2b). This design may employ IC processes more suitable for the transceiver (e.g., BiCMOS) and the rest of the node, which is largely digital logic (e.g., CMOS), than a single process that would be a compromise for both. This design also can result in a reduced time to market, because the individual chips may be easier to design and may, in fact, be off-the-shelf components. In addition, placing the RF transceiver in a separate chip makes it relatively easy (assuming the interface has been standardized) to change transceivers for operation in different bands, for different power levels, or other physical layer changes without the need to redesign the integrated circuit. Finally, the RF transceiver may be held constant and an alternative protocol handler/application processor chip may be substituted for different applications. This enables a single RF transceiver design to be used for multiple applications. An example of this type of partitioning (albeit with a

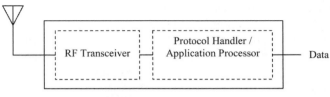

(a) One-chip (SoC) implementation

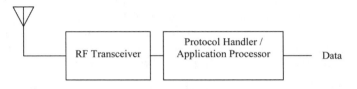

(b) Two-chip implementation

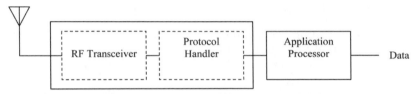

(c) Implementation with separated protocol handler and application processor

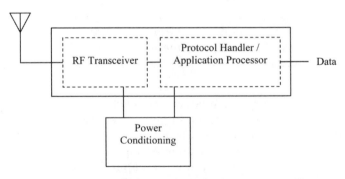

(d) "One-chip" implementation with external power conditioning IC

Exhibit 2.   Some Possible Network Node Partitions

CMOS RF transceiver) is the transceiver/microcomputer combination described in Appendix C.

A third alternative is to move the chip partition between the protocol handler and the application processor (Exhibit 2c). A multichip design based on this partition decouples the communication protocol and the application; a single application can operate on multiple protocols merely by replacing the communication chip. Conversely, a manufacturer can design a single communication chip, but optimize its solution for multiple applications by offering several application processors, each with varying amounts of memory and processing speed. A disadvantage of this approach is that the user must pay for two sets of memory: one for the communication chip and one for the application processor (unless an external memory arbitrator is employed, an expensive alternative); a single-chip approach using shared memory will usually be able to employ less total memory. Memory is often a very significant fraction of component cost in wireless sensor network nodes, therefore, the total amount of memory a user must purchase is often an important factor to consider.

The previous partitionings did not explicitly place the power conditioning circuits. An alternative partitioning is to separate the power conditioning circuits from the rest of the node chips, as shown in Exhibit 2d for the erstwhile SoC design. This approach has several advantages:

1. The power supply specifications of the rest of the chipset can be greatly simplified; they will receive an external supply with a standardized voltage provided by the power IC. This simplifies their design.
2. A single chipset can be supplied by the mains, multiple types of batteries, or various energy scavenging techniques merely by using the appropriate power-conditioning (and maybe scavenging) chip. This can open many markets without redesign of a large SoC.
3. The power-conditioning chip may be placed in an IC process best suited for its performance requirements, which may include relatively high voltages and the use of bipolar transistors and high quality passive components.

A disadvantage of this approach, however, is the communication bottleneck of interchip communication. At a minimum, the power-conditioning chip needs to understand the type of loads it must supply (which will vary with application processor types, for example), and the protocol handler and application processor need to know the capabilities of the power source to which they are connected (so that they will limit peak current requirements, for example). Further, to maximize the efficiency of the total system, the power-conditioning chip needs to react to the state of the network node; similarly, the network node needs to react to the state of the power-conditioning chip. For example, the power-conditioning chip can operate most efficiently if it is given warning of impending large current requirements (as may occur when the RF transceiver becomes active). In addition, the protocol handler can take advantage of a loss-of-battery indication from the power-conditioning circuit to cancel any activity proposed by the RF transceiver, and perhaps to store its present context in nonvolatile memory prior to exhaustion of the secondary storage element. This interchip information can result in the need for significant bandwidth at the power-conditioning chip interface, leading to power consumption problems or the need for an excessive number of chip interconnects.

The power conditioning and user interface circuits often have similar IC process requirements, including the need for relatively high voltages. This leads to another possible partition: the combining of the power conditioning and user interface circuits into a single support chip for the rest of the network node. This is a marriage born of practicality; other than the need for a similar IC process, the two functions have little in common. Certainly, from a marketing standpoint, it is rarely desirable to connect the user interface to the type of power supplying a network node; however, these factors often do not change. For example, it is entirely conceivable that a network node powered by a AAA cell, with a small liquid crystal display (LCD) for user interface, could be entirely satisfactory for a large portion of the consumer market. Viewed in this light, the development of a wireless sensor network node support IC can make economic sense.

A further partitioning decision concerns the location of RAM and ROM (nonvolatile memory). Many wireless sensor network protocols and the algorithms employed in processing sensor data are not yet standardized; updates may be issued while the sensor network is in the field. Further, it is often to maintain information, such as neighbor lists or routing tables, during brief unpowered periods (e.g., while changing batteries). For these reasons, simple ROM is not always desirable and some type of nonvolatile memory, such as flash memory, is preferred in wireless sensor network nodes. Although it is always desirable for power, cost, and size reasons to integrate as much as possible, the integration of flash and other types of electrically erasable memory represent a problem. Often, the IC processes with the densest general logic do not support embedded flash, due to its multiple gate and high voltage requirements. The processes that do support embedded flash are often significantly more expensive per logic gate than those that do not. It is therefore a quandary for the system designer; often, due to the commodity nature of flash memory, the economic solution is a compromise design in which the RAM is embedded but the flash is not. Magnetoresistive random access memory (MRAM), an emerging nonvolatile memory technology, may offer an alternative to embedded flash and RAM in many applications.

A complicating factor in the partitioning discussion is the use of a host processor. Many of the complications arise from the distribution of software between the wireless sensor network node and the host itself. Similar to hardware partitioning, the questions raised in software partitioning can be difficult.

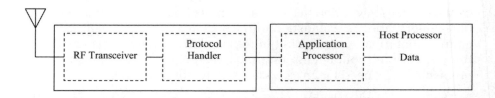

(a) Application processor integrated into host processor

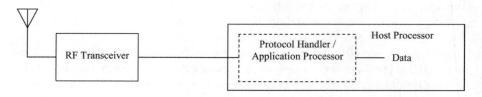

(b) Application processor and protocol handler integrated into host processor

Exhibit 3. Partitioning with a Host Processor

For example, the cost of the dedicated application processor may be avoided by performing this function in the host processor, as presented in Exhibit 3a. The application processor (and its associated memory) may then be removed from the network node chipset. This can reduce the component cost of the resulting system to the cost of an RF transceiver and protocol handler, especially if the host has such a large amount of memory for its own needs that no additional memory is needed, and the transceiver and protocol handler can be powered by the host. This is often the case, for example, when sensor networks are integrated into personal computers (PCs), as part of a communication link to computer peripherals. In this application, sensor information may be processed by the PC and then sent to the protocol handler for transmission. Similarly, the protocol handler may notify the host of incoming data by declaring an interrupt or may temporarily store the data pending a poll by the PC application processor. If both transmitted and received data may be buffered in this manner, this approach can be successful.

When one considers the addition of the protocol handler to the host, however, the situation changes (Exhibit 3b). Adding the protocol handling functions to the host may first be appealing, especially from the product cost viewpoint; because this approach shares memory, less must be purchased and the total cost of memory will be reduced. The real-time nature of communication processing can become a serious problem, however. Protocol handling cannot be delegated to most PC operating systems because they usually cannot guarantee protocol timing to the accuracy needed.

In non-PC implementations, software integration is also made more difficult by the addition of protocol handling, especially if the host is running other real-time (or near-real-time) processes. This is especially true of embedded systems, such as those found in PDAs, cellular telephones, and other battery-powered portable devices, which already have strict limits on their computational resources. Interrupt handling and context switching, for example, must be carefully evaluated to ensure that sufficient resources — in the form of RAM, nonvolatile memory, and instruction speed — are always available for all real-time processes, even in the worst-case scenario when the host processor is most heavily loaded. Because even the identification of the worst-case scenario may be difficult to determine, guaranteeing that resources are always available for multiple real-time processes can be difficult.

At the opposite extreme, a lightly loaded but quite powerful host processor is no panacea either. A powerful host processor that is frequently awakened to perform protocol-handling functions may have little trouble meeting the real-time requirements of the communication protocol. Unless special care is taken in its design, however, it is likely not to be a power-efficient implementation. A system that is lightly loading a powerful host processor is usually doing so to improve its average power consumption. This situation can develop, for example, when the processor is being used to service a second communication protocol, such as a cellular telephone protocol. Because the two protocols have independent timing requirements, one will interrupt the sleep periods designed into the other. This can result in the processor remaining active almost in perpetuity, and the average power consumption of the system rises dramatically.

Faced with these trade-offs, when a host processor is available, a compromise is often in order. An attractive approach may be to employ a dedicated protocol handler to off-load the host processor, but use the resources of the host processor for application data processing. This limits the timing requirements placed on the host as well as the total amount of processing required (because wireless sensor networks are assumed to have relatively low data throughput). This approach is also appealing in that it largely separates the data-generation function from the data-transmission function of the node. Legacy wired and wireless systems, for example, may already have application processing built into their code; employing a dedicated protocol handler enables a wireless sensor network node to be constructed with minimal attention paid to the existing application code. Not only does this represent less work that must be done by the software engineer, it also reduces the possibility that software bugs will be introduced into the existing code. Often, when faced with the choice between higher memory costs and higher software integration costs (with their associated quality concerns), the decision is to purchase memory.

The final partitioning decision concerns the location of the transducer(s). Their location, and the means by which their information is communicated to the rest of the wireless sensor network node, can be of special significance to the economic success of the design.

6.3 TRANSDUCER INTERFACES

For a manufacturer of wireless sensor network devices, the transducer interfaces can be a major economic problem. The manufacturer achieves profitability by achieving economies of scale (i.e., by manufacturing a large quantity of identical devices). In the electronics industry, this is done by semiconductor integration, which requires very large sales (on the order of millions of units) to defray the one-time costs of chip design, mask set generation, etc.; however, a nearly infinite variety of sensors and actuators are used. How may these be combined in a wireless sensor network product line so that the manufacturer achieves the sale of a large quantity of identical devices?

6.3.1 Integrated Sensors

One way of doing this is by identifying a selection of sensors that are capable of satisfying a substantial portion of the market, then integrating them along with the wireless sensor network transceiver. If, as is frequently the case, the associated "actuator" in the application is just an electrical switch, or information to be stored in RAM or passed on to a host computer, the node may be completely integrated. However, this approach has some difficulties:

- *The sensors may be difficult to integrate*. Ideally, one would like sensors that are integratable on the same die as the wireless sensor network transceiver, for a completely integrated solution; however, even if this is technically feasible, it may drive the cost of the fabrication process up to the point that it is not practical. An alternative solution is to place the sensor(s) in separate dice but in the same package as the transceiver — a method that attains much of the usefulness of total integration, if not all the cost benefits. In any event, many sensors still cannot be integrated at all.
- *The system is inflexible*. From the user's standpoint, a purchased network node is application-specific; it cannot be used for any other application. From the manufacturer's standpoint, if a different sensor is needed, even one of the same type but differing in specification (accuracy, response time, etc.), nothing can be done short of designing and manufacturing a new integrated circuit, a time-consuming and expen-

sive task. Because the actuators in this example are either simple switches or, in the degenerate case, information itself, it is relatively easy to reprogram the nodes to perform multiple functions based on the information received from the network (e.g., control the switches in a different manner). Even here, however, significant limitations exist. For example, the voltage and current capabilities of the switches in a completely integrated design will be limited to those compatible with a modern digital integrated circuit fabrication process. This means that, unless extraordinary measures are taken, the current will be limited to a few tenths of an ampere and the voltage to 3.6 V or so. This precludes the node from directly controlling household appliances, for example, although a logical output from the node can certainly control a semiconductor switch on a second die (made in a high-current, high-voltage process) that, in turn, can control a household appliance.

- *Legacy sensors, perhaps those associated with a previously installed wired sensor network, cannot be reused.* In some applications, for example, sensor networks in nuclear reactors, where access to placed sensors is difficult, this is a significant handicap.

Despite this (rather foreboding) list, some integratable sensors may find wide application. One example is a temperature sensor. Making a moderately precise and accurate temperature sensor of limited range in an integrated circuit is relatively easy; many possible designs are available, most of which revolve around the temperature dependence of the voltage drop across a forward-biased silicon diode, typically -2 mV/$^\circ$C.[2] Such a sensor may have wide application in areas as diverse as heating, ventilation, and air conditioning (HVAC) as well as industrial control, where the actuator can be a switch controlling a heater in a thermostat, for example, and in health monitoring, where the actuator may be a display, or simply data stored in a computer.

In most cases, however, it is not possible to integrate both sensor and actuator in all wireless sensor network nodes used in an application. This forces the node to have a transducer interface to the outside world. The question then becomes how to design the interface to achieve the maximum possible market penetration.

6.3.2 The External Interface

An old aphorism states, "If you try to please everyone, no one will like it," and nowhere is this more true than in the design of an external transducer interface. The manufacturer must put enough flexibility in the interface to meet the needs of a large segment of the market, without adding so much to the node that it becomes economically prohibitive to use in any one application.

To start, the manufacturer must decide if the transducer interface is to be analog or digital. The output of most sensor elements is inherently analog, while the wireless sensor network transmits only digital data; therefore, an analog-to-digital conversion must take place. Similarly, many actuators produce an analog output; digital-to-analog conversion is therefore necessary. Placing analog-to-digital and digital-to-analog converters (ADCs and DACs) on the network node may reduce the total cost of the system, especially if multiple sensors are attached to each node and the converters can be shared between them. Converters of more than moderate speed, accuracy, and precision are difficult to manufacture on a conventional digital IC process, however, and add power consumption and die area (and, therefore, cost). A manufacturer of wireless sensor network nodes may choose to reduce the cost of the product by incorporating neither ADCs nor DACs, instead using a digital transducer interface. This requires the user to employ transducers that have internal (or inherent) ADCs and DACs, which may significantly limit the market of this network node design.

If the decision is to employ a digital transducer interface, the next decision is to decide if the interface will be a completely logical one, or if special-purpose outputs (essentially very large pad drivers, used as switches, etc.) will be included. The addition of special-purpose outputs may make the network node capable of stand-alone operation in some high-volume but low-margin applications (e.g., a wireless light switch), by eliminating the need for a second, special-purpose switching transistor in the system design. A node with special-purpose outputs also may be capable of satisfying some low-volume but high-margin applications (e.g., military sensing). The special-purpose outputs can greatly increase the size of the integrated circuit, however, driving up costs for those applications that do not use them. The manufacturer must decide which market is to be addressed.

One compromise is to place a purely logical interface on the network node IC, but place a second, special-purpose die controlled by the logical interface, in the same package. In this way, the network node IC may be produced in high volume, while one of several relatively cheap (and easy to design and fabricate) special-purpose chips may be mated to it as the application requires. This approach offers market flexibility at the cost of making multiple special-purpose chips, and the cost associated with placing multiple chips in the same package. A variation on this theme is employed by thin-film ceramic module manufacturers, which may use the same network node IC, but customize each design by modifying the surrounding components as needed for the application.

If the manufacturer decides that the transducer interface is to be a completely logical one, decisions must still be made. The first decision is whether the interface should be in series or parallel. This decision hinges on the relative value in the implementation of speed versus size.

A parallel interface (an external bus) has the advantage of speed; with eight data pins, a byte (eight bits) can be transmitted in each clock cycle. Alternatively, at the same data rate, an eight-bit parallel interface may be clocked at one-eighth the clock rate of a serial interface; this may be an advantage if the interface is a source of RF interference. (However, the larger number of circuit board traces used by a parallel interface may negate this advantage.) If the network node contains a microcontroller, the microcontroller's general-purpose input/output (GPIO) pins can be efficiently used in a parallel interface. In addition, sensors designed for an eight-bit microcontroller bus can be quite common; however, these benefits come at some cost. The parallel interface requires ten pins on the network node (eight data pins, one clock, and one control or enable pin), which also must be routed to the sensor(s). In very small implementations, routing these lines may be difficult, or require a more expensive multilayer circuit board. In very low-end applications, especially those in which a custom integrated circuit (as opposed to an off-the-shelf microcontroller) is used, the ten pins needed for the sensor interface may double the pin count of the chip. Although the extra pins alone will drive up the product cost (due to the cost of wire bonding and packaging), the die area of the chip may now be pad-limited (i.e., determined by the number of wire bonding pads needed around its periphery, instead of by the area of the integrated circuits themselves). This will establish a lower limit on the size of the chip, regardless of the fabrication process employed.

A serial interface has the advantage of size. With a single data pin (plus one each for the clock and the control or enable pin), a serial interface is small, with a minimal effect on both chip pad count and circuit board area. This makes it ideal for very small, low-cost products, although it also has its drawbacks. Because the serial interface trades speed for size, it is slower; a careful analysis should be done to ensure that sufficient data can be transferred across it under worst-case conditions. These can occur, for example, in an emergency condition, when the wireless sensor network is requesting frequent sensor updates or giving frequent commands to an actuator. Several types of serial interfaces are also available on the market, not all of which are compatible. This can be a serious problem when field-replaceable transducers are replaced by those from a different vendor. Often, the interfaces may appear to be compatible, but may differ in some detail, such as implied addressing (i.e., their behavior when sequential data packets are sent to the same address). In addition, some serial interfaces are proprietary. To avoid these issues, many transducer manufacturers do not manufacture products with serial interfaces; this limits the market of a serial-interface wireless sensor network node (unless an external series-to-parallel converter is employed). Another alternative is the use of a mixed-mode interface, such as pulse-width modulation. Neither fully analog nor fully digital, pulse-width modulation encodes the transferred data in the duty cycle of a square wave.

Once the series/parallel interface decision is made, the remaining issue is one of selecting speeds and standards. Speeds, when not set by the standard, are usually limited by power, electromagnetic compatibility (EMC), and the application requirements.

Leakage current aside, the power consumed by a switching circuit element is

$$P = CV^2f$$

where

P = the power consumed by the switching element (in W)
C = the capacitance of the element (in F)
V = the voltage range over which the element is switching (in V)
f = the frequency of the switching (in Hz)

Because the voltage term is squared, a voltage reduction has the greatest effect on power consumption of an interface. However, to be compatible with industry (i.e., available transducers), the selection of available voltages is limited to a few choices: 2.7 V, 1.8 V, or, in a few cases, 1.0 V. The capacitance is typically unknown to the chip designer. Because the transducer to be used is typically not known (unless the chip is being designed for a specific transducer), its load capacitance is also not known; to make matters worse, the capacitance of the interface lines between the chip and the transducer, which can be significant (several tens of picofarads), is unknown as well.

This leaves the frequency of operation as the only parameter over which the designer has significant control. To minimize both power consumption and the potential for the production of electromagnetic interference (see Chapter 9), the designer typically will attempt to minimize the required maximum frequency of operation of the interface, consistent with meeting application requirements. The interface capacitance is unknown, therefore, the designer can only state a maximum total capacitance that the chip will support (based on the size of the driver circuit on the chip) and the maximum frequency of operation under that condition.

6.4 TIME BASE ACCURACY AND AVERAGE POWER CONSUMPTION

A subtle relationship exists between the accuracy of the time base used in a synchronous communication system, such as a wireless sensor network employing beacons, and the minimum attainable power consumption of that network. This relationship develops because of the time uncertainty that develops between the transmitting and receiving nodes that, in turn, requires the receiving node to be active a larger fraction of time to ensure that it is active when the message is sent from the transmitting node. For wireless sensor networks incorporating a relatively short beacon period, for example, this time uncertainty may be negligible when compared with the receiver warm-up time and the transmission time of the beacon itself. For networks that must employ a longer beacon period to reduce their average power consumption, however, the effects of finite time base accuracy produces a lower limit on the duty cycle of the receiving device and, therefore, its average power consumption.

The analysis that follows begins by assuming a simple, fixed, difference in period between the time base of the transmitter and receiver; any variation in period (i.e., jitter) is neglected, as is any change in the time base over time (i.e., drift). Later, the effect of these complicating factors is added to the model. Further, it is assumed that receiving nodes are capable of synchronization to transmitting nodes, so that the beacon timers of receiving nodes are reset upon successful reception of a transmitted beacon, and any timing error between the nodes does not accumulate beyond a single beacon period.

Consider a transmitting node with a stated time base accuracy of ε_t ppm. That is to say, if the desired beacon period is T_1, the node will transmit beacons with a constant period somewhere in the range $(T_1 \times (1 - \varepsilon_t), T_1 \times (1 + \varepsilon_t))$. If a receiving node has just received a beacon from this transmitter, it sets a timer to wake itself up in time to receive the next beacon. The issue is to determine what value the receiving node should program in its timer, not only because of the transmitter time base error, but also because of the time base of the receiving node itself is imperfect, with a stated time base accuracy of ε_r ppm. If the receiving node is expecting the next beacon to occur at time T_1, the situation is as shown in Exhibit 4.

The beacon time uncertainty, as observed by the receiving node, is the sum of both transmitting and receiving node time base uncertainties, placing the beacon in the range $(T_1 \times (1 - (\varepsilon_t + \varepsilon_r)), T_1 \times (1 + (\varepsilon_t + \varepsilon_r)))$. Note that this does not include the time of the actual beacon transmission, T_c, nor any warm-up time associated with the receiver, T_{wr}. When these factors are considered, the active time required of the receiving node to ensure reception of the transmitted beacon is $T_{wr} + 2T_1 \times (\varepsilon_t + \varepsilon_r) + T_c$, as illustrated in Exhibit 5.

As an example, suppose a system were designed to have a beacon transmission length of $T_c = 100$ μs and a period of $T_1 = 5$ s, using hardware that requires a warm-up period of $T_{wt} = 150$ μs prior to each transmission. The transmitted duty cycle is then $(T_c + T_w)/T_1 = 250$ μs/5 s = 50 ppm.

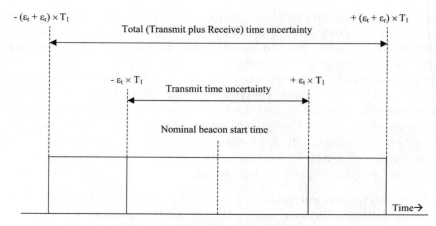

Exhibit 4.   Beacon Start Time Uncertainty Stack-Up

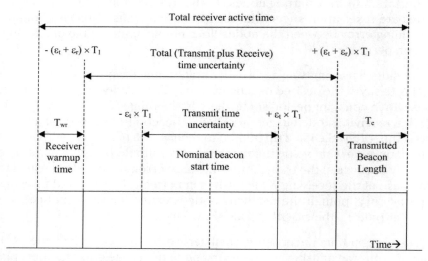

Exhibit 5.   Receiver Active Time

However, suppose further that the transmitting and receiving nodes each have a stated time base accuracy of ± 100 ppm, and consider the duty cycle of the receiving node. The time uncertainty of the receiving node is $2T_1 \times (\varepsilon_t + \varepsilon_r) = 2 \times 5s \times 200$ ppm $= 2$ ms. This is far larger than the receiver warm-up time and the transmitted beacon length, combined, and is, therefore, the primary factor limiting the duty cycle, and, therefore, the power consumption, of the receiving node.

The foregoing analysis is somewhat pessimistic because it overestimates the average time the receiving node must be active. Several techniques can be employed to reduce this time.

The approach that may come to mind first is to employ a more accurate reference at the receiver (and the transmitter, if possible). This will certainly reduce the time uncertainty at the receiving node; however, the more accurate reference may be expensive, so alternative techniques that do not raise hardware costs are preferred.

The first technique recognizes that the receiving node does not have to be active for the entire time the beacon may be transmitted. It may be put back to sleep once the beacon is received. If this is done, and if the distribution of both transmitting and receiving node time base errors is symmetrical about the desired value (i.e., no systematic offsets), the average active time of the receiver is $T_{wr} + T_1 \times (\varepsilon_t + \varepsilon_r) + T_c$; the beacon is received on average halfway through the total time uncertainty range at the receiving node. Because the time uncertainty at the receiving node has the largest effect on the duty cycle (and, therefore, the power consumption) of the receiving node, this is a significant improvement. In the example at hand, moving to this algorithm reduces the average receiver active time from 2.25 ms to 1.25 ms, a reduction of 44 percent.

A further reduction may be realized by a more sophisticated algorithm at the receiver that attempts to model the effects of inaccuracy of both transmitter and receiver time bases. This can be done by recognizing that a high degree of correlation often exists between the apparent periods of sequential beacons, especially if the jitter of the time base references is low, as is usually the case. Continuing the example, suppose that the time base of the receiving node is 10 ppm slow. If the time base of the transmitting node has no error, the beacon will appear to arrive 10 ppm early to the receiving node. If the time base of the transmitting node is 50 ppm fast, however, the beacon will appear to arrive $10 + 50 = 60$ ppm early; sequential beacon periods will appear to change only as the time references drift — if the time bases of both transmitting and receiving nodes are stable, all received beacons will appear to be 60 ppm early. The receiving node may recognize this correlation and compensate for it, by modifying the value programmed into its wakeup timer, so that it awakes just prior to the (apparently early) beacon. (This is equivalent to automatic frequency control, that is, to regulating its own time base to that of the transmitter.) In this case, the average active time of the receiving node can be reduced to $T_{wr} + T_{buffer} + T_c$, where T_{buffer} is a small time buffer that accounts for any possible drift between the transmitter and receiver time bases during the preceding beacon period. Such drift may occur, for example, by a temperature change of one or both of the node time base references, by a mechanical shock to one or both of the references, or other means. Higher-order modeling of the drift may be employed, if desired, to minimize T_{buffer}.

Note that this algorithm requires a learning period; at first, the receiver does not know the relative time base error of the transmitter, and so T_{buffer} must be very large so that a very wide time window is open to ensure reception of the beacon. Once a series of beacons is detected, however, T_{buffer} may be greatly reduced and the window significantly narrowed.

Continuing the present example further, if T_{buffer} is (arbitrarily and rather conservatively) set to 50 μs, accounting for a 10 ppm drift between time bases in one beacon period, the average active time of the receiver is T_{wr} + T_{buffer} + T_c = 150 μs + 50 μs + 100 μs = 300 μs. The duty cycle of the receiving node has been reduced from 2.25 ms/5 s = 500 ppm, using the initial unsophisticated algorithm, to 300 μs/5 s = 60 ppm, an improvement of 88 percent. Because the active power consumption can be three orders of magnitude greater than the sleep power consumption, this represents a significant average power consumption savings.

From this analysis, it is clear that to attain very low duty cycles and, therefore, very low average power consumption in receiving nodes, the design of the time bases used by both transmitter and receiver, as well as the design of the timing algorithm used by the receiver, are of critical importance, especially in systems requiring minimum implementation cost.

An interesting strategy to minimize both duty cycle and implementation cost is to specify a maximum any time reference value in the network of the drift allowed during one beacon period. Note that this is slightly different than a simple time base accuracy specification; a time base accuracy specification considers the time base error from a desired value, while a drift specification considers the change in time base error between consecutive beacon periods. This is done in the Bluetooth™ specification, for example. The strategy here is to enable the use of untuned or uncompensated reference elements to minimize the product cost, while still enabling receiving nodes to track transmitting nodes (by limiting how much they may drift within a beacon period). Interestingly, there are bounds over which this type of specification is useful: for very short beacon periods, any possible drift is negligible, while for very long beacon periods, the drift specification may be more strict than the time base accuracy specification. For example, one may specify a maximum reference drift of 10 μs and a reference accuracy of 100 ppm. For a beacon period greater than 10 s, the drift specification is the more stringent. This is especially significant in wireless sensor networks, which may have lengthy beacon periods, but the nodes of which may be very small, with low thermal inertia. With such devices, a change of temperature may be reflected in the reference within one beacon period.

Because of these considerations, a time base drift specification should be carefully evaluated by the system designer to ensure its usefulness and to avoid unintended consequences.

6.5 CONCLUSION

The design of wireless sensor network nodes has a larger effect on the economic potential of the resulting network than does the design of nodes for networks containing infrastructure, such as a cellular telephone network. This is because, although the cost of nodes in the telephone network may be justified by the value of the network provided by the infrastructure, the value of a wireless sensor network is determined solely by the nodes themselves. Node implementation is, therefore, of primary importance in the design of a successful wireless sensor network.

The partitioning decisions that must be made at the start of a wireless sensor network design effort are complex, and involve market forces, technological capabilities, and organization strategy and business relationships; there is likely a market for which every possible partition is optimal. The task facing the designer, then, is to ensure that the partition chosen is suitable for the market at hand. Often, the goal is to achieve utility in as many markets as possible, to maximize return on the engineering design time invested; this multiple-market requirement only adds to the difficulty of the partitioning decision. In addition to the components of the communication transceiver itself, the designer must consider the presence of any host processor, which may affect hardware/software trade-offs, the type of power source(s) available, and the type of interface(s) needed to communicate with sensors and actuators.

One important but often overlooked factor in the design of wireless sensor network nodes is the effect finite stability of the node time base has on the minimum attainable duty cycle, and, therefore, the minimum attainable average power consumption, of the network node. The resulting trade-off between the cost of the node time base and the attainable life of the battery (or other power source) must be made properly to achieve the desired market success.

References

1. Institute of Electrical and Electronics Engineers, Inc., IEEE Std. 802.15.4-2003, IEEE Standard for Information Technology — Telecommunications and Information Exchange between Systems — Local and Metropolitan Area Networks — Specific Requirements — Part 15.4: Wireless Medium Access Control (MAC) and Physical Layer (PHY) Specifications for Low Rate Wireless Personal Area Networks (WPANs). New York: IEEE Press. 2003.
2. Paul R. Gray and Robert G. Meyer, *Analysis and Design of Analog Integrated Circuits*, 3rd ed. New York: John Wiley & Sons. 1993. p. 16.

Chapter 7
Power Management

7.1 INTRODUCTION

Because low implementation cost and low power consumption are the raisons d'être of wireless sensor networks, it is of great importance to consider the design of the power supply system in wireless sensor network nodes.[1] A poorly designed system can rob the node of the low power consumption potential given by the communication protocol, move the cost of network nodes out of the desired market, or both.

The power management problem associated with wireless sensor network nodes is conceptually a simple one, based on supply and consumption: given a supply of energy, the node must consume it at the lowest possible rate (i.e., power). The problem becomes complicated, because the form of energy provided by the primary source (i.e., electrical energy at a certain voltage and available current) is rarely optimal for the load; further, although the power available from the source and the consumption of the load vary significantly over time, the load must be supplied properly at all times for correct operation.

Because of the many available power sources and wide variations in device loads, no single power supply system is optimal for wireless sensor network nodes. Instead, the system design needs to match the power source to the load circuits. When electrochemical cells (i.e., batteries) or other sources with a finite supply of energy are employed, the system must be designed to operate in a manner that maximizes the power available from the source, thereby maximizing the life of the node. When energy-scavenging techniques (or batteries with nontrivial internal resistance) are employed, the power source typically has a finite power limit; if power greater than this value is required at any time, some form of power-conditioning, usually requiring secondary storage of energy, is needed. This is often the case for wireless sensor network nodes, due to their low average power but relatively high peak power requirements. Of course, the cost and inefficiency of the secondary storage system must be included in the design of the network node.

This chapter discusses a few of the power management features unique to wireless sensor networks. Available power sources, including mains power, batteries, and a wide range of energy scavenging systems, are described. Ways to minimize the power consumption of loads are covered, including the use of low supply voltages. An outline of voltage converters and regulators that may be used in power-conditioning systems is presented. The chapter concludes with a discussion of power management strategy.

7.2 POWER SOURCES

The design of any power supply system begins with the choice of power source. Often a particular source is assumed a priori; however, in the case of wireless sensor networks there is often a choice to be made. Considerations in the selection of a power source include its availability (both at time of manufacture and for field replacement, as appropriate), initial and operating costs, time between servicing, internal resistance, available voltages, size, and ecological concerns. A surprisingly large selection of compact electric power sources may be considered for low-power applications;[2] even when limited to human-powered sources, the list is lengthy.[3,4]

Obviously, the availability of a power source is of primary importance because an ideal but unobtainable source is of little use. It is also important, however, to consider both the availability and cost of the proposed power source at the location of the network, not at the location of the designer. This is especially true in the case of batteries, which are trivially inexpensive in most areas of the world, but worth a month's salary of the average worker — if available at all — elsewhere. The time between servicing is important, especially for large networks; a good metric to apply is to divide the mean time between servicing each node by the number of devices in the network (all of which will need service) to obtain the mean time between servicing for the network as a whole. This metric focuses the designer on the user's viewpoint.

The internal resistance of the source is important, because the current drain of wireless sensor network nodes is typically quite bursty in nature, with relatively high peak currents (e.g., when the transceiver is active) separated by very low constant currents. In addition, sensors and actuators are often inactive for long periods, followed by high-current periods of activity. Sources with high internal resistance may be able to meet the average current drain needs of the network node and its sensors and actuators, but may not be able to source the much higher peak currents required. The power source must be able to supply the high currents necessary, without a significant drop in terminal voltage — a drop that may cause a microcomputer reset, loss of data, frequency synthesizer unlock, or other deleterious behavior. Sources with inherently high internal resistance may need conditioning, often in the form of a large capacitor or even an intermediate, low internal resistance battery, to support these peak currents, even if they can supply the average current with ease. Sources with internal resistance that rises with time (e.g., many battery chemistries) may have their lifetime limited not by their energy capacity, but by their inability to support the required peak currents.

The voltage available from a given source is, of course, important because it must be matched or otherwise converted to the voltage needed by the network node circuits. This is significant because any conversion circuits introduce inefficiency and power loss, additional cost, and may create an electromagnetic compatibility (EMC) problem. In addition, in many cases, the available source voltage can directly affect product size — for example, if a 1.8-V circuit were to be supplied by batteries, two cylindrical carbon zinc cells (each with an end-of-life voltage of 0.9V) would be required, although a single cylindrical lithium cell could supply the same voltage. Because wireless sensor network node circuits and sensors can be very small, the power source can represent a significant fraction of the total product size.

Due to the potentially large number of nodes in a wireless sensor network, and the expected ubiquity of wireless sensor networks around the world, the ecological effect of their power source must be evaluated. When used in such high volume, disposable and nonrecyclable power sources (e.g., primary batteries) can represent an environmental blight, one that would force the consideration of alternatives.

7.2.1 Mains

If mains power is available, and is compatible with the desired application, it is nearly always the preferred choice. In most areas of the world, it is more reliable than other sources; it is practically inexhaustible and, considering the very low power consumption of wireless sensor networks, has a trivial operating cost. It has its drawbacks as a supply for wireless sensor network nodes, however, not the least of which is the inherent contradiction of connecting a wire to a wireless device. Mains power will require a significant amount of conversion to make it compatible with the needs of typical wireless sensor network node circuits — it must be converted from AC to DC, and in voltage from one or two hundred volts to a few volts. This conversion can be expensive; it will require specialized hardware, and even a wall plug-mounted power converter is a significant size and cost expense for a wireless sensor network node. The effective internal resistance seen by the load (the network node) will be the internal resistance of these conversion circuits. In addition, the many different standards used around the world, with their differences in voltage, frequency, and hardware, can add significant complexity to a design intended for worldwide use or distribution. Similarly, obtaining safety regulatory approval can be challenging.

7.2.2 Batteries

For many applications, mains power is not an option, and batteries are used. Many types of batteries are available nearly worldwide and avoid many of the hardware costs associated with the use of mains power, but they, too, require design considerations.

7.2.2.1 Lifetime. First on the list of design considerations for battery-operated equipment is a finite lifetime, which requires that batteries be replaced or recharged. Primary cells, which cannot be recharged, can represent a significant operating expense if used in a large network; the majority of this expense is often the labor associated with the replacement, instead of the cost of the batteries themselves. Secondary cells, which can be recharged, avoid the purchase of new batteries in normal operation; however, the labor cost associated with recharging them remains, and they have a lower energy density than primary cells, so for a cell of a given size, recharging of a secondary cell must be performed more often than replacement of a primary cell.

7.2.2.2 "Low Battery" Detection. Because the cells have a finite lifetime, to avoid node failure in operation, there must be some indication available when the battery is nearing its end of life, so it may be replaced. This is a subtler problem than may appear at first; each battery type has a different type of discharge behavior, so the method of determining its remaining energy must be matched with the battery chemistry used and the type of load the wireless sensor node presents to it. For example, the terminal voltage of most lithium battery chemistries drops monotonically as the battery discharges; this convenient feature allows the battery voltage to be used directly as an indicator of battery life remaining. The terminal voltage of a Zinc–Air (ZnO_2) battery is nearly constant over most of its service life, however, as is its internal resistance. Determining the remaining life of a Zinc–Air battery is a much more difficult problem. Because the determination of remaining battery capacity is specific to the type of battery used, it is also important that the use of battery types other than that for which the system is designed be prohibited in some manner. This is usually accomplished by mechanical means, although some battery sizes, such as AA cells, are available in several chemistries. In these cases, it is up to the designer to ensure proper operation with each battery's chemistry or, at a minimum, ensure that the use of an improper battery does not result in an unsafe condition.

The determination of remaining life is not only a function of the battery, however; it is also a function of the load. Strictly speaking, the load defines the battery end of life, by defining a point in time at which the battery no longer supplies something needed by the load. Usually, this "something" is terminal voltage, but the load determines at which current this terminal voltage is to be maintained and, as discussed in Chapter 3, the current drawn by wireless sensor network nodes is very bursty, with relatively high peak currents separated by relatively long periods of low current consumption. Although this current profile does extract the maximum lifetime from the battery for a given average current drain,[5] the varying load on the battery means that the terminal voltage of the battery is likely to drop below a threshold acceptable to the load during the high peak current periods (instead of the lower current periods) due to the internal resistance of the battery.

It is, therefore, important that the "low battery" detection circuit respond to these transient threshold crossings, instead of the average terminal voltage. One way to accomplish this is to activate the "low battery" detection circuit only during the latter portion of the high peak current periods, when the battery voltage is typically at its lowest value. For example, the "low battery" detection circuit can be operated as the first stage of a "warm-down" procedure, used to take the network node from active to sleep mode. One advantage of this approach is that it minimizes the power consumed by the detection circuit itself, which, although small, can be significant when compared to the average power consumed by the network node. Another advantage is that the battery voltage measurements will have less variation because they are taken at the same point in the node's cycle of operation; this should lead to a more reliable operation of the "low battery" alarm.

To further extend the battery life, the "crest factor" or peak-to-average ratio of the current profile can be reduced near battery end of life, so that the peak currents are reduced. This can be done, for example, by staggering the operation of high-current circuits (including their warm-up transients) so that they do not occur simultaneously. By lowering the peak current load on the battery, the instantaneous voltage drop across the battery's internal resistance can be lowered, raising its terminal voltage above the "low battery" threshold.

It is important that the "low battery" detection circuit has an accurate threshold. A wireless sensor network node may have a battery lifetime of a year. If over this year, its terminal voltage drops 0.5 Volts, and if (for simplicity) one assumes a linear drop in terminal voltage over time, the terminal voltage drops 1.37 mV/day, or just under 10 mV/week. A positive threshold error (i.e., making the actual threshold greater than the desired value) can directly reduce the perceived lifetime of the network node by activating the "low battery" alert before it is necessary. A negative threshold error (i.e., making the actual threshold less than the desired value) can directly extend the perceived lifetime of the network node — until the error is such that the node fails before the alert is given. One view of the threshold is, therefore, seen as a marketing trade-off between obtaining the maximum apparent battery life, and giving the user sufficient time to respond to the "low battery" alert. Another view is an engineering tolerance analysis among the absolute minimum supply voltage at which the network node circuits are guaranteed to function, the amount of time desired to allow the user to replace the battery (as represented by a drop in battery voltage), and the accuracy with which the "low battery" threshold can be manufactured and operated. This analysis should consider manufacturing tolerances, the effect of temperature variations, etc.

Hysteresis in the "low battery" detection circuit is also important. A voltage drop of less than 10 mV/week is a very slow change to reliably detect, especially with low-cost system-on-a-chip (SoC) technologies. To complicate things further, the battery voltage is likely to have a significant amount of noise associated with it, caused by the many transients present in digital circuits. This noise may result in a flickering of the "low battery" alarm when the battery voltage nears its threshold value. Because the drop in battery terminal voltage is so slow, this flickering can persist for a week or more, something clearly undesirable. A solution that latches the "low battery" alert at the first transient crossing of the threshold is equally undesirable; not only does this reduce the apparent battery life of the product, but it does so in an unpredictable manner, determined by noise. A solution to this problem is the use of hysteresis in the "low battery" detection circuit. This hysteresis can take at least two forms: continuous-time or sampled hysteresis.

Continuous-time hysteresis is accomplished by having the "low battery" detection threshold be an analog voltage that varies with the state of the "low battery" alarm. If the "low battery" alarm is off, indicating a charged battery, the threshold is lowered somewhat; if the "low battery" alarm is on, indicating a discharged (low) battery, the threshold is raised somewhat. For example, if the "low battery" threshold is desired to be 1.1 V when the "low battery" alarm is off, the actual threshold is set to 1.05 V; when the "low battery" alarm is on, the actual threshold is set to 1.15 V. In this way, flickering or jitter of the "low battery" alarm is avoided. As the battery discharges, the "low battery" alarm remains off until the "low battery" detection circuit senses the battery voltage (plus any noise) to be below 1.05 V. Once this happens, the "low battery" alarm turns on, and the threshold moves to 1.15 V; the measured battery voltage (plus noise) must exceed this value to turn the "low battery" alarm off again. Operation of the "low battery" detection circuit is, therefore, protected against any noise having a peak-to-peak voltage less than 0.10 V.

A differential transconductance amplifier with continuous-time hysteresis, suitable for use as a "low battery" detection circuit, is illustrated in Exhibit 1. Transistors Q_1 and Q_2 form a conventional differential amplifier. Transistors Q_3 through Q_6 form two cross-coupled current mirrors, forming

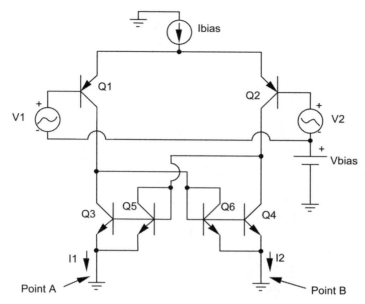

Exhibit 1.   A Differential Transconductance Amplifier with Continuous-Time Hysteresis

a differential current amplifier. The current transfer function of the amplifier is determined by the gain of the two current mirrors; the transfer response of the complete circuit is the product of the transfer functions of the Q_1-Q_2 differential amplifier and the Q_3-Q_6 differential current amplifier.

If both current mirrors have a gain greater than one (i.e., if the emitter area of Q_3 is greater than that of Q_5, and the emitter area of Q_4 is greater than that of Q_6), the Q_3-Q_6 differential current amplifier will have two stable states and a hysteresis characteristic. One of transistors, Q_3 and Q_4, will go into saturation and cause the other to turn off. All the bias current, I_{bias}, will subsequently flow either through output point A, or output point B. The circuit will remain in this state until the input voltage is adjusted to cause the Q_1-Q_2 differential amplifier output current ratio to exceed the gain of the saturated current mirror pair Q_3-Q_5 or Q_4-Q_6. At this time, the output current, I_{bias}, will switch to the opposite side of the circuit. For sufficiently large transistor beta, the current mirror gain will be substantially equal to the ratio of the corresponding transistor emitter areas. The transfer function of the Q_1-Q_2 differential amplifier will then result in the hysteresis trip voltages being a logarithmic function of the ratio of transistor emitter areas. For $\beta \gg 1$, the input voltage trip points, V_{tlo} and V_{thi}, are (relative to V_{bias}):

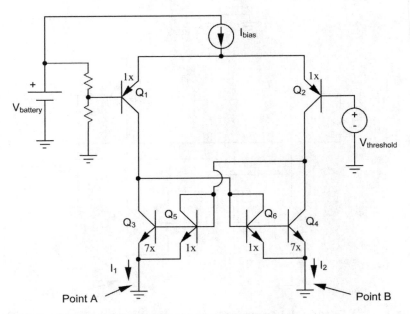

Exhibit 2.   Final "Low Battery" Detection Circuit

$$V_{tlo} = -(kT/q) \times \ln(A_1) \tag{1}$$

$$V_{thi} = (kT/q) \times \ln(A_2) \tag{2}$$

where

k = Boltzmann's constant,[6] $1.3806503 \times 10^{-23}$ J/K
T = absolute temperature, K
q = electron charge, $1.602\ 176\ 462 \times 10^{-19}$ C
A_1 = (Q_3 emitter area)/(Q_5 emitter area)
A_2 = (Q_4 emitter area)/(Q_6 emitter area)

For example, if ±50 mV of hysteresis is required around the "low battery" threshold, $V_{thi} = -V_{tlo} = 50$ mV. Then $A_1 = A_2 = \exp(qV_{thi}/kT) = \exp(0.05/0.0258) = \exp(1.94) \approx 7$. To achieve ±50 mV of hysteresis, Q_3 and Q_4 must have seven times the emitter area of Q_5 and Q_6, respectively. So that the "low battery" detection circuit can operate directly from a single C-Zn cell, the battery voltage is resistively divided down to a lower value for comparison. The final "low battery" detection circuit is illustrated in Exhibit 2.

Sampled hysteresis is accomplished by sampling the battery voltage at intervals, comparing the sampled values to a voltage reference, then storing the result of each comparison (for example, a "1" if the measured value were greater than the voltage reference, and a "0" if it were not) in a first in-first out (FIFO) buffer. The "low battery" threshold is then a count of "1" buffer entries. This threshold count can vary with the state of the "low battery" alarm: if the "low battery" alarm is off, the threshold is low; if the "low battery" alarm is on, the threshold is high. For example, if the FIFO buffer is 128 bits long and the "low battery" alarm is off, indicating a charged battery, the threshold count is set to 16, so that the buffer must contain less than 16 "1"s to activate the "low battery" alarm. If the "low battery" alarm is on, indicating a discharged battery, the threshold count is set to 112, so that the buffer must contain more than 112 "1"s to reset the "low battery" alarm. Note that in sampled hysteresis the "low battery" reference voltage is fixed, for example at 1.10 V; it is the threshold count of voltage samples above this value collected in a predetermined interval, instead of the value itself, that is varied.

Alternatively, sampled hysteresis can be implemented with an n-bit up-down counter, wherein a "1" is constantly input to the counter, the counter up/down control pin is connected to the output of the sampled value comparator, and the counter is strobed when each new sample is received. The value of the counter is a function of the last 2^n samples; the value can, therefore, be compared against a varying count threshold to produce the "low battery" alarm signal. Many other implementations are also possible.

Sampled hysteresis is not "true" hysteresis, in the sense that the voltage threshold is fixed, and does not vary with the state of the "low battery" alarm. It may be more properly described as an integration method that averages the samples over a long period, thereby removing the short-term effects of noise. The actual voltages at which the "low battery" alarm may be set and reset are functions of the noise waveforms present on the sampled battery voltage, and may not be easy to determine analytically. The system is easy to implement with the digital circuits available on SoC technologies, however, because it requires only one precise voltage reference and a simple comparator, and is accurate enough for most applications.

Circuits to monitor battery voltage have many other uses besides the generation of a "low battery" alert. In most wireless sensor network nodes, many circuits and devices, such as sensors, actuators, user interfaces (light-emitting diodes [LEDs], buzzers, etc.), and the transmitter power amplifier and receiver low-noise amplifier, draw significant current. When the battery is fresh, these circuits and devices may be operated with impunity because the terminal voltage is high. As the battery discharges, however, a point can be reached at which it is desirable not to operate certain loads in parallel, due to the excessive peak current drain (and associated drop in terminal voltage) from the battery. One may desire, for example, to offset the operation of the sensor and transmitter, so that they do not operate simultaneously. A circuit to detect the battery voltage at which this occurs is, therefore, useful to extend the life of the network node. There can be several such thresholds in a wireless sensor network node acting to control the operation of a collection of high-current loads as the battery discharges. The control of the power supply system of a battery-powered network node can, therefore, evolve into a sophisticated state machine.

7.2.2.3. "Low Battery" Alarm. Once the "low battery" alarm signal is generated, it must be directed to a user for remedial action. The familiar "low battery" icon seen on camera, cell phone, and pager displays can be used in applications, such as wireless personal computer (PC) peripherals, where a user is likely to see it. Notifying the user is a more significant problem in other applications, however, where network nodes may not have displays and may be located in places difficult to monitor routinely. Often, the desired alternative is to send a "low battery" alert message through the wireless sensor network to a node running a network status utility.

A network status utility is a portal with which users can monitor the status of the network. Depending on the application, the network status utility may be a:

- Background process running on a network node
- Process running on a dedicated piece of hardware (e.g., a laptop or personal digital assistant [PDA]) that periodically joins the network
- Process running on a gateway node (if one exists), where the user may access the utility via the Internet

In many applications, the network status utility is critical to the success of the network design. Although the network itself may be self-organizing and self-maintaining, users of large wireless sensor networks may need to be reassured that the network is functioning properly; further, in many applications, such as in emergency situations, the status of the network (in the form of functioning and nonfunctioning nodes) is useful in itself. Networks employing battery-powered nodes have a more pedestrian need for the network status utility because it is often a convenient way to identify to the user network nodes in need of battery replacement.

If the "low battery" alarm is to be sent via the wireless sensor network itself, the network node must be designed so that the transceiver function is the last node function left in operation. That is, other functions of the node, including the operation of sensors, actuators, and user interfaces, must be sacrificed under low battery conditions to ensure that the transceiver function remains in operation as long as possible and the "low battery" alarm is sent.

Several methods are available to the designer of a battery-powered wireless sensor network node so that the node responds to "low battery" conditions:

146

- The network node can send a single "change of status" message to a central location (e.g., a gateway node, or a node supporting a network status utility), only when the "low battery" alarm is enabled. This method minimizes network traffic overhead, but provides minimal information.
- The network node can include its battery status in a field included in the "network status update" message periodically sent to the network root (if one exists). This solution, although it slightly lengthens an existing network maintenance message, provides significant information to the user.
- If the network does not support "network status update" messages, the node may share its battery status information among neighboring nodes. The node may respond to queries concerning its status as long as it is able to do so. When its battery is exhausted, neighboring nodes may be queried to determine the reason for its silence.

7.2.2.4 Choice of Cell Chemistry. If the decision is to use batteries, the next design decision is the type of battery to employ, because there is a very wide variety from which to choose. Again, the task is to match the power source to the application; the designer should consider the intended lifetime of the network node, the load from the transceiver and transducer (which may be much greater than that of the transceiver), the physical size available, whether recharging is feasible, and other practical factors before making a battery decision.

7.2.2.4.1 Primary Cells. Of the primary (nonrechargeable) cells, cylindrical Alkaline Manganese Dioxide-Zinc (MnO_2-Zn) cells, such as AA and AAA cells, are inexpensive, have very wide availability, reasonably low (~200 mΩ) internal resistance (when fresh), a terminal voltage compatible with low-voltage CMOS circuit design, and reasonable gravimetric (measured in Wh/kg) and volumetric (measured in Wh/l) energy densities. They are not rechargeable; however, they have the disadvantage of being the same physical size as the older type, Carbon-Zinc (C-Zn) cells. This is a disadvantage because Carbon-Zinc cells have much lower energy density, much higher internal resistance (internal resistances greater than five ohms have been measured on new C-Zn cells), and are, in some areas of the world, still more widely available at lower cost than MnO_2-Zn cells. Users may be tempted to replace MnO_2-Zn cells with C-Zn cells when battery replacement time arrives (or may do so inadvertently), resulting in significantly shorter battery life or, worse, node failure due to the higher internal resistance of the C-Zn cells. Unsophisticated users may not appreciate the difference in batteries, blaming the fault on the equipment itself. To avoid this, the designer must design the network node to support the higher internal resistance of the C-Zn cells, which may be a significant design challenge if the transducer peak currents are high.

Zinc-Air (Zn-O_2) batteries have very high energy densities, a relatively constant terminal voltage (1.3 V) over their lifetime, and are available in a number of button cell formats of moderate capacity. Compared to MnO_2-Zn cells, however, their distribution is limited and their costs are high. Their internal resistance is often between five and twenty ohms, which limits their use to applications with low peak currents. (Unlike many other cell chemistries, however, their internal resistance remains constant for almost the entire life of the cell.) The most significant limitation of Zn-O_2 cells, however, is the limited temperature and humidity range over which they may operate: Performance degrades significantly below 4°C and above 40°C, and in both especially dry and especially humid environments. For maximum life, the amount of oxygen available to the cell may need to be limited, by placing it in an airtight enclosure pierced by pinholes of the appropriate number and dimension (determined by the load on the cell).

Several lithium chemistries must be considered for primary cells, including Lithium-Manganese Dioxide and Lithium-Thionyl Chloride.

Lithium-Manganese Dioxide (Li-MnO_2) cells are among the most inexpensive of the lithium cell chemistries and are available in a variety of cylindrical cell types. Although not hermetically sealed, they offer reasonably long shelf life (up to 10 years), operation over a wide temperature range, and good current-sourcing capabilities (low internal resistance). The terminal voltage of a Li-MnO_2 cell drops from the nominal 3.6 V, as it discharges and its internal resistance increases.

Lithium Thionyl Chloride ($Li/SOCl_2$) cells can have a very long shelf life (ten to twenty years), can operate over a wide temperature range, including the full military range of −55° to +125°C, and are available in hermetically sealed cylindrical packages. They have very high gravimetric and volumetric energy densities and, like most lithium chemistries, have a high open-circuit voltage when fresh, in this case, approximately 3.6 V; however, $Li/SOCl_2$ cells have limited current-sourcing capabilities. They are often used in applications having very low average power consumption and are, therefore, very useful in wireless sensor applications, such as remote outdoor utility meter reading that require 10 to 20 years of unattended operation, can amortize their higher cost, and are compatible with their more limited distribution.

7.2.2.4.2 Secondary Cells. In some applications, it is preferable to recharge batteries, instead of replace them; for these applications, secondary (rechargeable) cells are used. Although it may appear that rechargeable cells should always be used instead of primary cells (that must be replaced), secondary cells have some disadvantages when compared to primary cells. Principal among these is limited energy density; secondary cells typically have significantly lower gravimetric and volumetric energy densities than primary cells. This means that a larger secondary cell (as measured by weight or volume) is needed to attain a given capacity. In addition, secondary cells tend to have higher rates of self-discharge than primary cells; this limits their shelf life, and limits their service life under low-load conditions.

These limitations notwithstanding, secondary cells can be used to advantage on occasion. Secondary cells can offer lower internal resistance than primary cells; this can be a significant benefit in high-current pulsed applications. They can also be used in combined systems, in which a primary cell is used to trickle charge a secondary cell, which is used to supply the load. The load receives the benefit of the low internal resistance of the secondary cell, while the user avoids the purchase of a battery charger. The penalty for such a combined system is that the leakage current of the secondary cell discharges the primary cell, so the user does not get service life as long as if the secondary cell were not used.

Cylindrical Lithium-ion cells are the most common secondary cell type in use in portable communication devices today. They have very high gravimetric and volumetric energy densities and low internal resistance, making them suitable for high peak current applications.

Nickel-metal hydride (Ni-MH) cells offer good energy and power density, compared to other secondary cells, at the cost of somewhat higher leakage current that can make it unsuitable for many wireless sensor network applications. Ni-MH cells have low internal impedance, making them suitable for high peak current applications, such as land mobile radios and cellular telephones. In these applications, they replaced nickel-cadmium (Ni-Cd) cells, due to environmental concerns associated with the use of cadmium (and were, in turn, largely replaced by Lithium-ion cells).

Lead-acid (Pb-PbO) batteries have very low internal resistance, making them useful in high peak current applications. Especially useful are sealed cells in which the hydrogen released during charge is contained within the cell, and the sulfuric acid electrolyte has been placed in a gel form ("gel cells"). Lead-acid batteries tend to be relatively heavy, with relatively poor gravimetric energy density.

7.2.3 Energy Scavenging

In some cases, it is possible for the wireless sensor network node to obtain energy sufficient for its operation directly from its environment. Such methods have been termed "energy scavenging" or "energy harvesting,"[7] and include obtaining energy by photovoltaic, mechanical vibration, and other means.

Network nodes that employ energy scavenging "live off the land," so to speak, instead of using a self-contained energy source, as battery-powered nodes do. If properly designed, energy scavenging nodes can be very small because they do not need to carry their energy with them; however, similar to anyone living off the land, the success of the endeavor is tied to achieving regular production, which, for the network node, means understanding when energy will be available from the source. Energy available from the environment may vary greatly, and may be interrupted during certain periods of time. Often, these interruptions are deterministic and predictable, such as shift change periods on a piece of vibrating equipment; in other cases, the interruptions are stochastic. For the latter, a statistical approach is often used; the probability of a power outage of a given length is modeled. Because power from the environment sufficient for operation cannot be guaranteed at all times, the designer must choose a level of reliability based on the model of the scavengeable source, and design the node to meet that requirement.

For example, it is commonly appreciated that solar-powered systems must consider the effects of limited hours of daylight — some provision must be made for supporting nighttime operation, if such is desired. The designer must also consider the effects of operation at high latitudes, however, where in winter the hours of sunlight per day are greatly reduced (to zero in the polar regions), the variation of sun angle over the day and year, and extended periods of dense cloud cover, snow cover, etc. This means that the design of the solar-powered system must consider its location, and the weather at that location. The effects of location are deterministic, those of the weather, stochastic. A level of reliability, based on this location-based model of available sunlight, must be selected before the design can begin. An alternative to this design-by-location approach is to design the system for the worst-case scenario (long polar winters!), which may result in a system that is overdesigned for most other markets. A compromise is to design multiple systems for specific markets, which trades increased design, manufacturing, and distribution costs for improved product efficiency.

A common feature of nearly all sources of scavenged energy is that the average power available is low, typically 1 mW or less. Although the average power required by the wireless sensor network node may be less than the average power available from the scavengeable source, the peak power required by the node is nearly always greater than the minimum instantaneous power available from the source.

To meet this peak power requirement, and to support the varying output of most scavenged power sources, most scavenged power systems require some type of power-conditioning circuit.[8,9] A power-conditioning circuit sits between the power source and the load, and attempts to match the voltage and current available at its output with requirements of the load, while attempting to match the voltage and current requirements at its own input to those available from the source. Nearly all power-conditioning circuits require a secondary energy storage element, usually either a secondary (rechargeable) battery or some type of capacitor. The scavenged source essentially trickle charges the secondary storage element, which, in turn, supplies power to the network node. Secondary batteries have much greater volumetric energy densities than do capacitors, enabling them to supply higher peak powers for longer periods; however, they also have higher rates of self-discharge, which increases average power consumption. Depending on the technologies chosen, voltage conversion may take place either before the secondary storage element, after the secondary storage element, both before and after, or not at all.

The energy lost in the power-conditioning circuit, including that lost due to self-discharge of the secondary element, and inefficiencies in the secondary storage element charging, output regulation, and control circuits, can be a major factor in the total energy budget of the network node. This is so because the power-conditioning circuit must be constantly active; even in periods when the scavenged source is not providing energy, energy must be spent in circuits to detect its eventual return (to resume charging of the secondary element), and in the self-discharge of the element. The self-discharge of the secondary storage element in particular is often very significant, especially so because it cannot be overcome by clever circuit design.

A factor to consider in some detail during the design of systems employing energy scavenging techniques is their start-up and shutdown behavior. At start-up, the secondary storage element may be completely discharged, so the power-conditioning circuit will be unable to supply the load for a period, until it is at least partially charged. The difficulty is that, because the power-conditioning circuit has no or insufficient output during this time, the control circuitry that detects the start-up condition, monitors the state of charge of the secondary storage element, and finally connects the load to the output of the power-conditioning circuit, must operate from the scavenged power source directly. The scavenged power source can have wildly varying properties (thus the reason for the power-conditioning circuit in the first place), so the design of these control circuits can be especially challenging. Further, it is usually desired for the network node to continue to function for a period of time even if the scavenged power source is removed completely. The (now unpowered) start-up circuits must not adversely affect node function under this condition, yet the condition should be detected, so that a graceful shutdown of the system (e.g., by a gradual process of load shedding) can be performed while energy is still available in the secondary storage element. If the behavior of the scavenged source is such that direct operation of circuits from it is not possible, a small battery may be employed to supply the control circuits. Although it is a somewhat inelegant (and relatively expensive) solution, having a tertiary supply can greatly ease the system design of the power-conditioning circuit. In any event, the use of state machine design techniques is encouraged so that the behavior of the circuit is defined under all possible conditions and transitions, and one is not left in the embarrassing position of requiring a circuit to enable its own power supply.

7.2.3.1 Photovoltaic Cells. The most common energy scavenging technique is, of course, the use of photovoltaic cells to obtain power from ambient light, usually sunlight. For locations in which the availability of light to network nodes can be guaranteed to a sufficient degree, and for which mains and primary battery supply is impractical, this can be an excellent energy source; such locations include roadway signs; sailboats, buoys, and other marine locations; and remote areas.

The electrical power that can be extracted from a photovoltaic cell is proportional to the area of the cell and the intensity of the incident light. The terminal voltage of the cell resembles that of a semiconductor diode and is relatively insensitive to changes in light intensity, while the output current is directly proportional to light intensity. The output current of photovoltaic cells decreases with increasing temperature.

Although many types of photovoltaic cells are available, the most popular are silicon-based cells. These offer a reasonable price/performance ratio, are widely available, and are most sensitive to light with a wavelength near 800 nm. Three general types of silicon cells are used: monocrystalline, polycrystalline, and amorphous. Monocrystalline cells offer the highest efficiency (near 15 percent) and highest cost, amorphous cells the lowest efficiency (near 6 percent) and lowest cost, and polycrystalline cells are somewhere between (about 14 percent efficient). The combination of low cost and low power requirements of wireless sensor network nodes usually implies that amorphous cells offer the best value. Because the incident solar light power density at the earth's surface is typically 100 mW/cm^2, even a 4-cm^2 cell of 6 percent efficiency can produce almost 25 mW, or 50 mA of direct current, when fully illuminated. Indoor illumination typically has a much lower power density of 100 µW/cm^2 or less, however; the same cell would produce less than 25 µW under these conditions. The designer should use caution to ensure that the design suits the energy available at the application location.

The open-circuit voltage V_{oc} of Si cells is approximately 0.5 V. To be compatible with industrial standard supply voltages for integrated circuits, therefore, the cells must be placed in series to boost the output voltage. Because the output voltage is relatively constant with variations of light intensity, however, it is possible in some applications to use photovoltaic cells without further regulation. In most cases, however, power-conditioning must be used, not for voltage regulation per se, but to supply current in times when the photovoltaic cell is in darkness and unable to generate sufficient current to supply the load. The design of the power-conditioning circuit should recognize the fact that a photovoltaic cell can be modeled as a large forward-biased diode when not illuminated,[10] and ensure that a discharge path through the photovoltaic cell cannot exist under low illumination conditions to avoid draining the secondary storage element.

An example of wireless sensor networks employing photovoltaic cells is the Smart Dust program at the University of California, Berkeley.[11,12] The dust motes in this system communicate by optical, instead of radio frequency (RF) means.[13] Because communication is by line of sight, the additional requirement that the devices be powered by light shining on them (which may be other than solar radiation) is not a significant design issue.

7.2.3.2 Mechanical Vibration. Energy from mechanical vibration of a mass on a spring may also be scavenged effectively,[14] by use of one of three mechanisms:[15] piezoelectric (in which piezoelectric material converts strain on the spring into an electric charge), electromagnetic (in which a magnet attached to the mass induces a current in a coil as it moves), or electrostatic (in which a charge on the mass induces a voltage on a capacitor as it moves).

A rather substantial literature on the piezoelectric conversion of vibration to energy exists, based largely on the scavenging of energy from the human gait (e.g., Starner,[16] John Kymissis et al.,[17] and Shenck and Paradiso[18]). In these implementations, a piezoelectric material such as polyvinylidene fluoride (PVDF) or lead zirconium titanate (PZT) is placed in the heel or stave of the shoe. Some of the energy received by the shoe when it strikes the ground is converted by the piezoelectric element into an electric charge, which can then be stored via a power-conditioning circuit for use by a load. Although these systems are effective (and have received wide use as the power source for the novelty blinking lights in the heels of some high-end sneaker designs), the power-conditioning problem associated with piezoelectric transducers is nontrivial (see, e.g., Ross et al.[19]). In the two designs for the shoe application described in Shenck and Paradiso,[20] the energy is supplied in relatively short impulses of relatively high voltage (60 and 200 V) and relatively low power (15 and 75 mW), implying current pulses on the order of 250 and 400 μA. Nevertheless, average powers of 1.3 and 8.4 mW were generated, and a shoe-powered RF tag system was successfully demonstrated.

Strictly speaking, the piezoelectric shoe represents the extraction of energy from mechanical motion, instead of vibration, because no advantage is taken of the repetitive nature of forces on the shoe in the design. In this application, such advantage probably is not possible; not only does the frequency of the human gait vary by individual (and is rarely constant for an extended period of time, even for a single individual), the creation of a shoe-size piezoelectric structure resonant with the human gait (approximately 2 Hz) would be quite an achievement in its own right. (It is possible, of course, to create a secondary piezoelectric structure, resonant at a higher frequency, that is excited by the lower frequency impulses from the ground-shoe impact.) In applications such as rotating machinery, however, where vibrations of relatively high (~1–100 kHz) frequency may be present, direct resonant piezoelectric energy scavenging may be very practical.

Another application in which the piezoelectric generator has found commercial use is in a wireless light switch.[21] In this design, some of the energy dissipated in the toggling of a light switch is used to power a transmit-only wireless network node, communicating with a receive-only wireless node powered by the mains attached to the light.

Preliminary development of a miniaturized piezoelectric generator had been reported by Glynne-Jones et al.[22] and by Glynne-Jones, Beeby, and White.[23] A triangular, thin-film piezoelectric element was constructed of lead zirconium titanate (PZT) on a steel substrate. A resonance at approximately 80 Hz was observed, with a maximum load voltage of 1254 mV at a vibration amplitude of 1 mm. A maximum power output of slightly over 2 μW was produced in a load resistance of 333 kΩ.

Recently, an improved power-conditioning circuit for piezoelectric energy scavenging systems has been proposed that offers greatly improved efficiency over existing designs under sinusoidal vibration.[24,25] The authors of the circuit show that the peak available output power of a bridge-rectified piezoelectric element occurs when the rectifier DC output voltage is one-half the open-circuit voltage of the piezoelectric element. Because the voltage of the piezoelectric element varies with the level of mechanical vibration, however, the optimum output voltage will vary. To maintain the optimum value, a step-down DC– DC converter, operating in discontinuous current conduction mode (DCM) is introduced; it is demonstrated that the required duty cycle of the converter varies in an inverse manner with the open-circuit voltage of the piezoelectric element, asymptotically approaching a fixed value for large voltage.

The performance of this system is very impressive. The new adaptive power-conditioning circuit, employing the step-down converter, harvested more than four times the power of the same circuit when the converter was not used.[26] More than 70 mW was harvested from the new system, more than sufficient to power a wireless sensor network node, even in continuous receive mode.

Due to the complexity and associated power consumption of a discrete implementation of an adaptive control circuit needed to maintain the optimum duty cycle, a simplified converter employing a fixed duty cycle (the asymptotic value for large piezoelectric voltage) was also investigated.[27] For low values of piezoelectric element voltage, the converter was simply bypassed. Even this greatly simplified power-conditioning system produced more than three times the power of the same circuit when the converter was not used. This work is a significant step forward in the use of piezoelectric energy scavenging systems.

An alternative to the use of piezoelectrics in the scavenging of energy from vibration is the use of an electromagnetic transducer. An example of this type of scavenging is the work of Amirtharajah and Chandrakasan,[28,29] in which a miniature moving coil loudspeaker was used as the transducer. (A related design is simulated in M. El-hami et al.[30]) After passing through a 1:10 step-up transformer, half wave rectifier, and rectifier storage capacitor, the generator output drove a switching DC/DC converter and 1 V pulse width modulation (PWM) regulator (the power-conditioning circuit) that supplied power to a FIR filter. The output of the regulator was used to supply the converter as well as the load; the converter was bootstrapped to a backup battery, and switched to the regulator output via a control signal extracted from the regulator logic once the output voltage reached the desired value. Although it was calculated to be able to generate on the order of 400 µW, the load of the example circuit, including that of the power-conditioning circuit, was only 18 µW. (Because of this light loading, however, a single excitation of the generator produced 23 ms of finite impulse response (FIR) filter operation at 500 kHz, dissipating 114 nJ in the load.) 400 µW is a small amount of average power, to be sure, but it is sufficient to operate a low duty cycle wireless sensor network node, as discussed in Chapter 3.

A second example of electromechanical conversion of mechanical energy is described by Gerald F. Ross et al.[31,32] Here, a wireless sensor network was used to monitor door and window openings and closings. The network used the mechanical energy associated with the door or window opening or closing to drive a microminiature permanent magnet generator to power a transmit-only network node. A covert implementation is described that fits into a doorknob and is powered by a single turn of it. Interestingly, piezoelectric sources were investigated for this application but the power-conditioning requirements for them were found to be prohibitive.

A third example of electromagnetic conversion is the Kinetic wristwatch generator from the Seiko Epson Corporation.[33,34] In this design, used to power an electronic wristwatch, a moving mass is used to move a permanent magnet near a large coil, via a gear train. Because the energy of motion is derived from human activity, which may be erratic, a sophisticated power-conditioning circuit is used; the current developed in the coil is half-wave rectified, then used to charge a first capacitor. If the voltage across the first capacitor falls below the level needed to drive the watch, a voltage multiplier circuit is employed to charge an auxiliary capacitor. It is the voltage across the auxiliary capacitor that is used to power the watch circuitry.

A disadvantage of the use of a macroscopic moving coil in vibration energy scavenging is the difficulty of integrating the transducer — the Kinetic generator, for example, although a marvel of miniaturization, still requires many individually placed mechanical parts. One way around this problem is to employ a microelectromechanical system (MEMS) transducer, which may be integrated with the rest of the system electronics. An analysis of a MEMS moving-magnet generator[35,36] results in the following observations:

- The fundamental equation is $P_g = m\zeta\omega_n^3 Z_0^2$, where P_g is the generated power (W), m is the mass of the moving mass (kg), ζ is the damping coefficient, ω_n is the resonant angular frequency (radians/second), and Z_0 is the distance moved by the mass (m).
- The amount of power generated P_g is proportional to the cube of the vibration frequency ω_n. Therefore, the most appropriate use of such a generator is in applications with a high frequency of vibration.
- The moving magnet mass m should be as large as possible within the available volume of the device.
- The maximum displacement Z_0 of the moving magnet mass should be as large as possible in the space available.
- The spring should be designed so that the resonant frequency of the device matches the vibration frequency of the application.
- At resonance, Z_0 is inversely proportional to the damping coefficient ζ, so the generated power is inversely proportional to ζ. Because the electrical load impedance also is inversely proportional to the damping factor ζ, the electrical load impedance on the moving mass should be as large as practical to maximize generated power.

A MEMS-based electromagnetic microgenerator was described in Shearwood and Yates[37] and Williams et al.[38] The generator developed a maximum of 0.3 µW at a frequency of 4.4 kHz. A power output of 6 µW was predicted theoretically; the low power output, plus output power hysteresis as a function of vibration frequency, were ascribed to stretching of the membrane spring (upon which the magnetic mass rested) at large deflections.

The authors stated that this "spring stiffening," a nonlinear effect, was correctable by patterning the membrane.

The third type of mechanism to extract energy from mechanical vibration is by the use of an electrostatic transducer. This type of transducer takes advantage of the capacitor equation $Q = CV$, where Q is the charge on the capacitor (Coulombs), C is the capacitor capacitance (F), and V is the voltage across the capacitor (V). Operation begins by injecting a charge onto a mechanically variable capacitor when the capacitor value is highest (i.e., when the capacitor plates are closest together). External mechanical vibration is then allowed to perform work on the two plates, driving them apart, decreasing the capacitance while increasing both the voltage across the capacitor and the energy stored in it (electrostatic forces, of course, operate to force the plates together). At the point of maximum plate separation (minimum capacitance), the charge is removed from the capacitor at a net increase in energy.

The design of a MEMS-based electrostatic micro-generator was described in Meninger et al.[39,40] and Amirtharajah et al.[41] Although the MEMS device itself was not fabricated, the rest of the system operation was described. The expected power output was 8.6 µW, limited by the maximum operating voltage of the integrated circuit (IC) process.

Despite these advances, the power output of both electromagnetic and electrostatic MEMS-based transducers seems to be an order of magnitude too low at this stage of their development to be suitable for the needs of conventional wireless sensor networks. The use of piezoelectric transducers, and macroscopic electromagnetic transducers, seems much more promising in the short term. Even if MEMS-based energy scavenging does not become practical for wireless sensor network nodes, however, MEMS should still find wide application in them, as sensors and as filters and resonators in their wireless systems.[42,43]

7.2.3.3 Other Scavengable Energy Sources. Other sources of energy have been proposed for use in low-power portable electronic equipment. Although none of these appears to be practical for wireless sensor networks at present, they are presented as technologies to watch for development.

The production of useful amounts of power from temporal variations of air pressure and temperature has a surprisingly long history. Pierre de Rivaz successfully built and demonstrated a clock in 1740 powered entirely by such variations, and indications are that others before him had considered the idea.[44] Although many others experimented with the idea in the following years, however, the state of the clockmaker's art apparently limited commercial success.

Jean Léon Reutter developed and patented the first commercially successful clock powered by temporal variations of air pressure and temperature in the late 1920s.[45–47] His clock first employed a column of mercury, but the design soon evolved into one employing a small amount of ethyl chloride inside a sealed bellows. As the relative pressure outside the bellows changes, due to either atmospheric pressure changes or a change in temperature of the ethyl chloride (which then evaporates or condenses accordingly), the bellows expands and contracts, a mechanical motion that is used to wind the mainspring (the mechanical equivalent of the secondary source in a power-conditioning circuit), which drives the clockwork. The clock requires a temperature variation of only 1°C over 2 days to remain in operation indefinitely. Reutter spent many years perfecting the atmospherically powered clock and variations of it (e.g., Reutter[48]); under the name "Atmos," the clock remains in production to this day. More than 500,000 have been built.[49]

Recently, a wristwatch powered by temporal variations of temperature was patented by Steven Phillips.[50] Due to the size of the bellows required to produce enough power to drive the clockwork, a wristwatch based on the Reutter design is not practical. Phillips' approach is to employ a temperature sensitive bimetallic coil. In this design, one end of the coil is fixed and the other is allowed to move with changes in temperature. As in the Reutter design, this movement is used to wind the mainspring.

It is interesting to evaluate the potential of temporal variations of air pressure and temperature as energy sources for wireless sensor network nodes. The Reutter design is said to be able to generate 36 gram-centimeters of energy with either a 1°C variation in room temperature or a 3 mm Hg variation in atmospheric pressure.[51] In this context, one "gram" is a mass of one gram under gravitational acceleration (i.e., a weight or force). The available energy in mks units is therefore

$$E_{avail} = 36 \text{ gram} \cdot \text{centimeters} = 36 \times 10^{-5} \text{ kg} \cdot \text{m} \times 9.8 \text{ m/s}^2 =$$
$$3.5 \times 10^{-3} \text{ kg} \cdot \text{m}^2/\text{s}^2 = 3.5 \times 10^{-3} \text{ J} \qquad (3)$$

where

9.8 m/s^2 is the gravitational acceleration.

An efficient wireless sensor network node may have an average power consumption of 50 μW = 50 μJ/s. Under these conditions, and powered by the Reutter source, this node would operate only

$$\frac{3.5 \cdot 10^{-3} \text{J}}{50 \cdot 10^{-6} \text{J}/\text{s}} = 70\text{s} \qquad (4)$$

after the 1° temperature change, before requiring an additional change of temperature or pressure. This is a lot of temperature variation, to be sure, but not unusual for some applications; it is less than 1°C per minute of operation.[52] Viewed another way, the node would require an average room temperature variation of

$$\frac{86400s / day}{70s / C} = 1234C / day \qquad (5)$$

to remain in operation. To the author's knowledge, no attempts have been made to produce electrical power from temporal variations in atmospheric pressure or temperature, by either the Reutter or Phillips technique. An additional consideration for their use in wireless sensor networks is the cost of the energy scavenging mechanism; both designs are conceptually simple, but quite sophisticated in their implementations due to the need to minimize frictional and other losses. Miniaturization is also a concern — in fact, the development of the Phillips design was apparently motivated by the limited miniaturization possible with the Reutter design (the output power of which is a function of the enclosed volume of the three-dimensional bellows).

Fleurial et al.[53,54] describe the use of a thermoelectric generator based on the Seebeck effect to extract energy based on temperature difference between a hot and cold surface to power small consumer devices such as wristwatches. The power generated, however, is marginal for the power needs of a wireless sensor network node at human body temperature differentials, although it may have much utility in industrial environments where much larger temperature differentials exist. Kishi et al.[55] describes a thermoelectric wristwatch, marketed in December 1998; a maximum of 22.5 μW was developed by the thermoelectric generator at an output voltage of 300 mV, which was multiplied to charge a 1.5-V Lithium-ion cell.

A 1 mW, 300 MHz wireless transmitter powered by the ambient heat of a human hand is described in Douseki et al.[56] The system employs a thermoelectric generator and a switched capacitor voltage converter composed of silicon metal-oxide semiconductor fluid-effect transistors (MOSFETs) integrated on a 0.8 μm silicon-on-insulator (SOI) process. The thermoelectric generator can supply 1.6 mW at an output of −0.7 V, which the converter converts to a +1.0 V output with an efficiency of 71 percent. To enable the use of both positive and negative temperature differentials, produced by sources either hotter or colder than the reference temperature of the thermoelectric generator, the converter can convert either positive or negative voltage sources to a +1.0 V output.

Stevens[57] reports a program to develop a small thermoelectric generator for sensor use, extracting scavengable energy from the temperature difference between the air and ground. The advantages seen for this thermoelectric generator include high reliability, long life span, low visibility, nighttime power production, and ruggedness. Thermoelectric generators have also been proposed for use in cooking stoves and other high-temperature applications, where large temperature differentials enable the production of useful amounts of power despite their poor efficiency.[58]

A variant of thermoelectric generation that may have wide applicability in industry is thermophotovoltaic generation.[59] In this generation method, thermal radiation from a hot source is used to drive a photovoltaic cell optimized for conversion efficiency at the frequency of the thermal radiation.[60] Despite the relatively low efficiency it shares with conventional thermoelectric generation, thermophotovoltaic generation can be useful for the generation of small amounts of power, suitable for wireless sensor networks, in industries where waste heat is common. The glass industry has been proposed as one candidate where thermophotovoltaic generation may be used effectively.[61]

A novel use of piezoelectric polymers as energy scavengers is the Energy Harvesting Eel (Eel), a device to convert the mechanical flow energy of water to electrical energy.[62] The Eel is a ribbon of piezoelectric polymer (e.g., PVDF), placed behind a bluff body in flowing water. The bluff body sheds vortices that cause the Eel to flap behind it like a flag, in a motion analogous to that of a swimming eel. The motion of the Eel produces mechanical strain on the piezoelectric polymer, producing electrical power. The maximum power scavenged occurs when the vortex shedding frequency of the bluff body matches the flapping frequency of the Eel, a condition that produces the tightest coupling from water flow to the Eel. This implies that the Eel design must be optimized for a particular flow to scavenge the maximum possible power. The authors believe that very respectable amounts of power can be scavenged by the Eel — as much as 1 W with a 1 m/s flow.

Obviously, this amount of power would meet the requirements of almost all wireless sensor network applications. The limitation is the need to place the generator near flowing water; although applications in oceanography and on bridges, sailboats, and the like are the first to consider, it is also worthwhile remembering that many industrial environments contain flowing liquids from which energy can be scavenged using this technique. For example, many industrial processes employ cooling water; energy to monitor the process may be derived from the flow of this water. This approach has a fail-safe safety benefit: should the cooling water flow stop for any reason, the network node power is cut off and the node will disappear from the network, signifying trouble. Smaller amounts of power can be extracted from airborne variants of the Eel, in which the bluff body is placed in a wind stream. These can generate sufficient power to supply wireless sensor network nodes in many transportation applications, such as trucks and railroad cars.

Ranging further afield, significant amounts of power may be obtained from biological sources. A recent paper by Reimers et al.[63] notes that significant amounts of power, on the order of 10 mW/m^2, may be harvested from a salt marsh by placing a first graphite or platinum electrode at a depth of 2–4 cm in marine sediment, and a second electrode in the seawater just above it. A potential of 0.3–0.4 V develops almost immediately upon placing the electrodes, a value that rises to approximately 0.7 V over 1 to 2 days. Bond et al.[64] identifies the microorganisms responsible for this feat as several species of the family *Geobacteraceae,* particularly *Desulfuromonas acetoxidans* and *Geobacter metallireducens.* This is particularly interesting in that *Geobacteraceae* are known to degrade organic material, and Bond et al. speculate that *Geobacteraceae* could be used to harvest energy from organic waste materials. From the point of view of the wireless sensor network designer, the most significant point about this discovery is the relatively large amount of power that may be generated essentially in perpetuity (due to the large amount of organic material on the sea floor); both Reimers and Bond comment on the possibility of powering oceanographic instruments by this means. The disadvantage, of course, is that one must be near the shore, or be willing to establish and maintain a marine aquarium of suitable size. The size needed to power a wireless sensor network node may be quite reasonable, however; at 10 mW/m^2, 100 μW may be obtained from an area of only 100 cm^2.

These microorganisms form a natural "microbial fuel cell," in that they produce electricity directly from chemical action in the electrolyte, without consumption of the electrodes. The concept of a microbial fuel cell is not a new one; it has been proposed to power autonomous robots.[65] Such a robot would power itself by the consumption of "real" food, such as dextrose, employing a microbial fuel cell to convert the dextrose into energy. The Ecobot project,[66] at the University of the West of England, Bristol, has produced a light-seeking robot powered solely by microbial fuel cells.[67]

7.3 LOADS

For optimum performance and efficiency, a power source and its load should be matched in both voltage and current requirements. For example, a 1.5-V primary power source should be used to supply a circuit requiring 1.5 V, and if the circuit requires 20 μA of continuous current, the primary power source should be able to produce that current. When either the voltage or current requirements are not met, a conversion stage must be introduced, the inefficiency of which causes an increase in average power consumption over what otherwise might be obtained. In many low-power applications, this efficiency loss may equal the power consumption of the load because the conversion circuits often include fixed overhead (i.e., power-consuming items such as voltage and current references, storage capacitor leakage currents, etc. that do not change with power supplied to the load). Much discussion often takes place on the need for power sources with specific capabilities; in this section, the tables are turned, and an examination is made of what opportunities exist to adapt the circuits and other loads in a wireless sensor network node to particular power sources.

When voltage is considered, "adapting the circuit to the power source" usually means identifying ways to lower the required supply voltage for the load, for, with the exception of lithium cells and piezoelectric generators, nearly all sources of energy available to the wireless sensor network node designer have voltage quanta substantially less than standard CMOS logic levels. This is especially true for scavenged energy sources; for example, silicon photovoltaic cells have an open-circuit terminal voltage on the order of 500 mV or so. Thermoelectric generators may have a Seebeck coefficient (the ratio of voltage generated per unit temperature differential across the thermoelectric element) as low as 200 µV/K, requiring hundreds of elements to be placed in series to develop a useable output voltage.[68] If circuits could be operated directly on a 500-mV supply generated from a photovoltaic cell, or on a 300-mV supply generated from a thermoelectric generator, without the need for voltage multiplication, total system efficiencies could be greatly increased. Existing energy scavenging methods could be implemented more easily, and the wireless sensor network market would grow.

7.3.1 Power Consumption of Analog Circuits

Before embarking on the low-voltage endeavor, it is prudent to consider the effects of low supply voltage on both analog and digital circuits, and to identify fundamental limits.

After Vittoz,[69] the minimum possible power consumed by a pole of analog filtering is

$$P_{min,a} = 8fkT(S/N) \qquad (6)$$

where

$P_{min,a}$ = the minimum possible power consumed by a pole of analog filtering, W
f = the signal frequency bandwidth, Hz
S = signal power, W
N = thermal noise power, W

If an allowance is made for their voltage gain, Equation 6 also applies to amplifier circuits. Note from Equation 6 that $P_{min,a}$ varies directly with signal frequency and the S/N ratio in the circuit. To minimize power consumption, one should, therefore, limit the peak dynamic range requirement of the analog circuit because its minimum power consumption increases directly with S/N ratio, an increasingly expensive trade as the required S/N ratio increases. In receivers, this means the aggressive use of automatic gain control (AGC) and other "floating point analog" systems to limit the instantaneous dynamic range required of analog circuits. The system designer should distinguish between the instantaneous S/N ratio required of analog circuits and the total dynamic range required of the system.[70] In some telecommunication systems, such as cellular telephone systems, the required system dynamic range can be limited by regulatory means (e.g., by limiting the allowable power present in a receiver's adjacent channel). This can be an effective technique to reduce the minimum possible power consumption, and minimum possible supply voltage, of receivers.[71] Unfortunately, in the unlicensed spectrum used by wireless sensor networks, the receiver designer is not afforded that luxury. No limits are placed on the size of signals that may be presented to the receiver's filter circuits. If a specification is not otherwise set (by a standard, for example) the designer must set one, by making a trade between power consumption and the loss of system reliability that results from strong signal effects (adjacent channel selectivity, blocking, etc.). An advantage in this regard held by wireless sensor networks is that they usually do not support isochronous communication and, therefore, permit the use of packet retransmissions (and occasionally deletions). A temporary system failure due to a strong interferer, therefore, is less of a concern than it might be in other types of networks, and this flexibility can be traded for reduced receiver power consumption.

A second effect of Equation 6 on receivers is the sideband noise of oscillators. If the sideband noise of a receiver local oscillator is too low, reciprocal mixing can occur in the receiver mixers, resulting in poor selectivity and spurious performance. (This is often a serious design concern in receivers employing synthesized local oscillators, which may have sideband noise several orders of magnitude above the thermal limit.) Equation 6 represents a lower bound on power consumption of an oscillator for a given sideband noise specification relatively far away in frequency from the desired output; sideband noise closer to the carrier is dominated by nonthermal effects, including the effects of flicker ($1/f$) noise in the active devices.

In transmitters, the effect of Equation 6 is to limit the attainable transmitted spectral purity. With a fixed thermal noise floor, the level to which the transmitted signal may rise above the floor is limited by the power consumption of the transmitter circuits. Or, viewed differently, the minimum attainable power consumption of the transmitter circuits is limited by the minimum required transmitted S/N ratio. Regulatory limits on transmitted spectral purity will usually set this requirement, although lower limits may be desirable when the collocation of transceivers from different services is contemplated because the transmitted noise from the first service may desensitize the receiver of the second. This can happen, for example, when a Wireless Personal Area Network (WPAN) transceiver is placed in a cellular telephone handset — the noise transmitted by the WPAN transmitter can raise the apparent noise floor of the cellular telephone receiver. For cases such as these, when the noise affects a service in a different frequency band, RF filtering at the transmitter output may be employed to maintain the low transmitter power consumption, at the cost of additional components. If desired receiving frequency of the affected service is too close to the desired transmitter frequency of the WPAN, however, filtering is impractical, and other means must be employed, including those that increase the power consumed by the transmitter.

Note that Equation 6 is a lower bound, and does not take any sources of noise, other than those due to thermal effects, into account. Nor does it consider inefficiencies in the active devices, or the power consumption of support circuitry, such as regulators or bias networks.

It is often stated that it is not possible to make useful systems employing low voltage analog and RF circuits. Simply stated, this is not true. Given the S/N limitations discussed, many examples of transceivers operating at a 1-V supply exist. For example, most radio pagers employ receivers operating from a single-cell supply, linearly regulated to 1 V.[72] These systems are fully functional frequency-shift keying (FSK) receivers, employing low-noise amplifiers, mixers, oscillators, intermediate-frequency (IF) amplifiers, filters, and detection circuits, and have been sold by the tens of millions for more than 15 years. The 930-MHz receivers employing a synthesized local oscillator have a total active power drain of 7 mW or less; 150-MHz receivers with a crystallized local oscillator can have active power drains below 3 mW. The RF specifications of these receivers, which must perform in the land mobile communication environment, far exceed those needed for wireless sensor networks. (Manku, Beck, and Shin[73] have proposed a modified cascode RF amplifier design that is suitable for supply voltages below 1 V.) The baseband analog functions are often designed using current-mode techniques;[74] although there is no fundamental reason why current-mode circuits should draw less power than voltage-mode circuits, in practice it is often easier to employ current-mode techniques when working with a 1-V supply.

A criticism of the previous examples can be that they are composed largely of bipolar transistors, and do not integrate well with modern very large scale integration (VLSI); however, the reduction of CMOS gate lengths due to improved lithographic techniques has increased the maximum frequency of operation of MOS devices far into the microwave region; submicron MOS is now widely used for integrated RF circuits.[75-78] Abidi[79] describes many CMOS RF circuits capable of operation from a 0.9-V supply. A transceiver suitable for a wireless sensor network node that operates from a 1.2-V supply in 0.5 μm CMOS is presented in Porret et al.[80] The list is not limited to RF circuits; for example, Serra-Graells and Huertas[81] describes a CMOS proportional-to-absolute temperature (PTAT) voltage and current reference operating from 0.9 V and dissipating less than 5 μW.

7.3.2 Power Consumption of Digital Logic

Turning to digital circuits, one notes that the dynamic range of digital systems is determined by the bit width of the signal path (i.e., the number of bits N_{bit} used to represent a signal); the power consumed by the system is thus only logarithmically related to the signal-to-noise ratio (SNR). Vittoz[82] observes this implies that, for low signal-to-noise ratios, analog signal processing is inherently lower power than is digital signal processing and that, for sufficiently high signal-to-noise ratios, the rapidly rising analog power consumption of Equation 6 overtakes the logarithmic curve of digital signal processing power consumption. The majority of digital circuits employed in wireless sensor network nodes often do not directly represent either transmitted or received signals, however; they are employed in command and control functions, and in baseband signal processing and protocol handling. For these activities, a measure of signal-to-noise ratio is not really relevant; instead, what is relevant is the total number of gates present, how much energy is used per transition, and how often transitions occur. The effect of leakage in nonideal gates must also be considered. The power consumed in a digital circuit therefore can be expressed as the sum of dynamic and static powers[83]

$$P_d = P_{dyn} + P_{stat} \tag{7}$$

where

P_d = total power consumed by the digital circuit, W
P_{dyn} = dynamic power, consumed due to gate transitions, W
P_{stat} = static power, consumed without gate transitions, W

Equation 7 is quite general, and applies to both hardware logic and software logic running on a microcontroller.

Assuming the rail-to-rail transitions of standard CMOS logic, the dynamic power P_{dyn} is determined by the energy consumed by a gate transition, the frequency of such gate transitions, and the number of gates undergoing transition cycles:

$$P_{dyn} = m\,f\,E_{tr} \tag{8}$$

where

m = the number of gates undergoing a transition cycle
f = the number of transition cycles per second, Hz
E_{tr} = $C_g V_{dd}^2$ = energy consumed per transition per gate, J
C_g = gate equivalent capacitance (including both interconnect and load), F
V_{dd} = gate supply voltage, V

In a wireless sensor network node, the number of gates undergoing a transition cycle at any one time can be very low, on average, due to the very low duty cycles supported by their communication protocols. (Of course, this does not consider any computation required on payload data.) The relative effect of P_{dyn} on P_d is reduced by this activity factor, such that P_{stat} is the dominant term in Equation 7 in many implementations. This relationship should be remembered during the determination of the optimum V_{th} and V_{dd} (described next).

The minimum value of E_{tr} is often defined as $8kT$, due to the need to provide immunity to thermal noise; however, the much higher value of $C_g V_{dd}^2$ is used here as a more realistic value for CMOS.

P_{stat}, the power consumption term due to the gate leakage current, is determined by the equivalent amount of that current per gate, the supply voltage, and the total number of gates powered:

$$P_{stat} = N_g I_{stat} V_{dd} \qquad (9)$$

where

N_g = number of powered gates not undergoing transitions
I_{stat} = equivalent amount of leakage current per gate, A

Substituting Equation 8 and Equation 9 into Equation 7,

$$P_d = m f C_g V_{dd}^2 + N_g I_{stat} V_{dd} \qquad (10)$$

An inspection of Equation 10 can provide considerable insight into power reduction for digital circuits. The first point to note is that the dynamic power consumption is a function of the supply voltage squared, and, therefore, reducing the supply voltage is the most important factor for low power digital operation.[84] In theory, V_{dd} may be reduced until E_{tr} approaches $8kT$, a fundamental limit imposed by thermal noise; in practice, V_{dd} may be reduced only until the gate delay rises to an unacceptable value. Reducing the supply voltage also reduces the static power consumption, although this relationship is linear instead of quadratic. The supply voltage should be reduced in all cases to that value that provides the minimum gate speed consistent with proper operation of the system. Because the speed requirements of the system vary with time, the supply voltage can be designed to vary as well, rising during computationally intensive periods, and then falling to a lower value during sleep or standby periods.[85] (V_{dd} may also be adjusted to compensate for IC process variations, by controlling the frequency of a ring oscillator via a phase locked loop, then using the loop control voltage as the supply voltage of the digital circuits.) In addition, the digital circuit block may be partitioned into sections, which may have their supplies gated or separately controlled. This last option is somewhat less useful, in general, due to the level shifting required to move between sections, but can have significant benefits when a small number of circuits operate significantly faster than the rest of the chip. An example in some systems is the signal processing logic, which typically operates at multiples of the symbol rate; the protocol handling logic typically operates much slower, at the packet or message rate. The protocol handling logic can then operate from a lower supply voltage than can the signal processing logic; because the interface between the two usually is relatively simple, only a few level-shifters are required.

Second, the other variables in Equation 10 may be classified as either process- or design-related: C_g and I_{stat} are directly related to the integrated circuit process itself, while the remaining terms m, f, and N_g are related to the particular design at hand.

As the minimum feature size of the lithographic process is reduced, C_g, and, therefore, dynamic power consumption, can be reduced accordingly. The situation with I_{stat} is more complex. The subthreshold current of an individual transistor is affected by the MOS threshold voltage in an inverse exponential manner:[86,87]

$$I_{ds} = \frac{W}{L} I_s e^{\left(\frac{V_{gs}-V_{th}}{av_T}\right)} \left(1 - e^{-V_{ds}/v_T}\right) \tag{11}$$

where

I_{ds} = drain-to-source current of MOS device in subthreshold region, A

W = effective width of the transistor (that contributes to the leakage current), m

L = length of the transistor, m

I_s = MOS device constant, approximately 0.3 µA/µ device width

V_{gs} = voltage of the transistor gate relative to the source, V

V_{th} = threshold voltage of the transistor, V

a = a constant slightly larger than 1

v_T = kT/q, the "thermal voltage," approximately 25.852 mV at 300 C

V_{ds} = drain-to-source voltage of MOS device in subthreshold region, V (assumed equal to V_{dd})

Even for V_{dd} = 0.5 V, $-V_{ds}/v_T$ = −19.3, so the effect of V_{ds} on I_{ds} is neglected in the following analysis. For a single transistor, the leakage current I_{stat} is defined as I_{ds} when V_{gs} = 0, or

$$I_{stat} = \frac{W}{L} I_s e^{\left(\frac{-V_{th}}{av_T}\right)}$$

(12)

To minimize I_{stat}, one wants to maximize the threshold voltage V_{th}; however, doing so comes in conflict with the need to reduce V_{dd}, which must remain greater than V_{th} in order for one CMOS gate to drive another. In fact, V_{dd} often must remain significantly greater than V_{th} to produce logic with speed commensurate with proper operation of the system. This then sets up a conflict between minimizing P_{stat} and minimizing P_{dyn}; said another way, for a required value of $V_{dd} - V_{th}$ (the so-called "gate overdrive"), there exists a unique value of V_{dd} that minimizes $P_d = P_{dyn} + P_{stat}$. In fact, the supply voltage for optimum energy efficiency of a CMOS gate has been demonstrated[88] to be approximately 0.5 V, although it is very sensitive to small changes in V_{th}. This last consideration is an important one; the optimum V_{th} may be only 200 mV or so, yet the threshold control of a typical IC process may vary by ± 100 mV — a very large effect because the drain current of a MOS device is an exponential function of V_{th}. It is, therefore, important to have good threshold control in a low-power CMOS process, or employ some type of feedback to adjust either the apparent V_{th} or the value of V_{dd} in response to process manufacturing variations.

Of the design-related factors, m, the number of active gates, and f, the frequency at which they transition, clearly should be minimized to reduce P_{dyn}. The number of active gates can be minimized by good logic design, the use of clock gating (i.e., disabling the clock signal to unused logic circuits), and by the use of asynchronous logic (to avoid the use of a clock signal altogether). Good logic design is sometimes overlooked, but it can be important: an inefficient implementation can increase not only the number of active gates, but also the frequency at which they must be clocked to produce the desired output, leading to a quadratic penalty in P_{dyn}.[89]

The gate transition frequency, f, is often a critical design choice that balances the system throughput requirement with the need to minimize power consumption. The nature of most wireless sensor network nodes, however, is to have short periods of relatively intense computation, interspersed by long periods of inactivity. Maintaining a single, relatively high, clock frequency suitable for the high-throughput computation periods, during the inactive periods is clearly not an optimum solution from a power consumption point of view; it is better to match f (i.e., the microcomputer clock frequency) to the computational load at hand, raising and lowering it as necessary. This can be done in any of several ways; multiple oscillators may be used, or a frequency synthesizer may be employed to generate high-frequency clock signals as desired from a low-frequency reference.[90]

The underappreciated factor, however, is often N_g, the number of gates *not* undergoing transitions. Wireless sensor network nodes spend a great deal of time asleep, or in very low power consumption modes — the protocols enable very low duty cycle operation for just this reason. Therefore, their standby power consumption can represent a large factor of their time-average power consumption, and it is important that it be minimized. Because P_{stat} is directly proportional to N_g, to minimize the standby power consumption the number of logic gates in the network node should be minimized. This is often done, of course, for cost reasons, but with logic densities surpassing 300,000 gates/mm^2 in a modern process, the designer may become undisciplined if the total size of the implementation is, say, only 50,000 gates. It is important to realize when employing a low-V_{th} process to minimize power consumption, that the power consumed by leakage current can be cut in half by cutting the number of gates in half — which may enable a lower V_{th}, which will reduce P_{dyn}. If the designer has no control over the IC process used, this may be the only control he has over P_{stat}.

The conclusion is that the designer should make every effort to lower the supply voltage as much as possible, and to use as few transistors as possible; however, the designer has other opportunities to reduce digital power consumption, especially if he is willing to generate his own standard cell logic library. Designers of low power logic systems must very often do this because the libraries published by most IC manufacturers usually are designed for speed, instead of low power. Significant improvements in power consumption can often be obtained if one is willing to devote the time to creating a specialized library, then characterizing it at the desired low supply voltage (or having the manufacturer do so). A specialized library often has other advantages as well; for example, the cells in most libraries are designed to be as small as possible (naturally enough); however, in doing so, the library authors, who must design for general use, make decisions that may not be in the best interest of the design at hand. One decision of this type is the handling of substrate contacts. In mixed-signal designs, it is often better if NMOS source contacts in standard cell logic gates are connected to a separate V_{ss} node, instead of to the global "ground" substrate contact. Having a return path for digital current that is separate from the substrate contact (to the extent of having separate package pins, when this is feasible) can reduce noise coupling to sensitive analog and RF circuits on the same chip, at the cost of a little die area. This cost can be more than repaid in reduced time-to-market, by avoiding a lengthy troubleshooting and redesign period.

One of the most promising and least exploited power reduction opportunities is the use of silicon-on-insulator (SOI) processes.[91] Instead of placing transistors on the surface of a bulk wafer of silicon, the SOI process places transistors on a thin layer of silicon over a relatively thick insulating layer of silicon oxide (over a bulk wafer for mechanical strength). If the silicon layer is thin enough that it becomes fully depleted at threshold, the device is termed a fully depleted SOI (FD-SOI) device; if it is thicker, so that some neutral silicon remains at threshold, the device is termed a partially depleted SOI (PD-SOI) device. The advantages of SOI are many, but the main ones relevant to low-power design are:

- The reduction of parasitic capacitance to the substrate (now separated from the transistor by the oxide layer), which lowers C_g
- The more ideal subthreshold slope, which improves operation in the MOS weak inversion region
- The ability to contact the body of each MOS device independently

This last advantage is employed in dynamic threshold MOS (DTMOS[92,93]), which is a CMOS-like PD-SOI logic family in which the gate and body contacts of each MOS device are connected together. This has two desirable effects: When $V_{gs} = 0$ V, the apparent value of V_{th} rises and when $V_{gs} = V_{dd}$, the apparent value of V_{th}, falls. Transistor leakage is, therefore, lowered when the device is supposed to be off, and current drive is increased when it is supposed to be on, making the transistor a better switching device. An apparent disadvantage to the body contact, other than the slightly larger transistor size that results, is that the body-to-source and body-to-drain pn junctions may become forward biased unless the supply voltage is kept below that necessary to do so. This implies that the supply voltage must be kept below 0.5 V or so, which coincidentally is the point of optimum MOS energy efficiency.

Whether or not SOI is used, a process with multiple thresholds can be used to advantage. Low- (or even zero-) threshold devices can be used for the fastest circuits, while higher-threshold devices can be used elsewhere in larger numbers to minimize leakage current, both at a constant supply voltage. FD-SOI is particularly suitable for multiple-threshold, low-voltage, low-power processes,[94] due to its nearly ideal subthreshold slope (reducing leakage), and its low parasitic capacitance, due to the inherent oxide isolation of devices (no junction capacitance) and lack of wells (small device area). As examples of the performance attainable in multithreshold FD-SOI CMOS, two papers were presented at the 2003 Institute of Electrical and Electronics Engineers (IEEE) International Solid-State Circuits Conference employing it: a 0.5-V, 400-MHz reduced instruction set computer (RISC) processor consuming 3.5 mW,[95] and a 0.3 V, 3.6 GHz, 0.3 mW frequency divider.[96]

The use of DTMOS, or either of the preceding multiple-threshold SOI processes, matches well with the voltages available from most energy scavenging techniques, including photovoltaic cells and thermoelectric generators. As already noted, this is very desirable from a system standpoint because it simplifies or eliminates any power-conditioning of the power sources, which improves overall efficiency and lowers average power consumption.

Even when a buck converter is required to lower the voltage from, say, an alkaline cell primary source, and the overhead from the converter must be considered, the use of a 0.5-V supply for digital logic circuits results in significant overall power savings. From Equation 8, P_{dyn} of a digital logic circuit operating from a 0.5-V supply is only one-quarter that of the same circuit operating from a 1.0-V supply and, from Equation 9, P_{stat} of the 0.5-V circuit is one-half that of the 1-V circuit. The design of a buck converter for such an application, again using multithreshold SOI, has already been reported;[97] the demonstrated circuit had a conversion efficiency of 91 percent with an output power of 10 mW. Because the reduction of digital logic power consumption is limited only by process and circuit and system design considerations (until the far-away $8kT$ thermal limit is approached), and not, as is the case of analog circuits, by attainable signal-to-noise ratio, there is no reason that the supply voltage of wireless sensor network node logic should not be below 1 V. The one exception to this statement is that of nonvolatile programmable memory, such as flash memory.

Nonvolatile memory represents a significant problem for the designer of a wireless sensor network node. Nodes require nonvolatile memory to store data, routing tables, etc., but the memory must also meet the node cost and power consumption requirements. Flash memory is usually suggested as the best fit for wireless sensor network nodes, but it is not without its problems. Due to the physical processes used in its operation (hot-electron injection for programming and Fowler–Nordheim tunneling for erasure, although some designs use Fowler–Nordheim tunneling for programming as well), flash memory requires that relatively high voltages (6–20 V) be available. These voltages are usually generated on-chip, in a small voltage multiplier in the flash memory array itself. Standard flash cells, operating from a 1.8-V supply, need only moderate amounts of multiplication, but operation from a 0.5–1-V supply requires a much larger multiplication chain, which can be difficult to design and manufacture reliably if integrated into a standard digital IC process. This is due not only to the need for a second, floating, gate in the memory cell, but also to the containment and control of the very high voltages involved, in a process that may be designed for 1.2–1.8-V operation.

The result of this situation is that integrated flash memory (so-called "embedded flash") is often not available in IC processes that are otherwise suitable for a true SoC design. The next-best solution for wireless sensor network designs places the nonvolatile memory in a chip separate from that of the rest of the digital logic, a chip manufactured in a specialized IC process more suitable for its requirements. This solution increases the physical size, EMC concerns, and power consumption (the latter two because external memory address and data lines must be driven) of the network node over a completely integrated design. It also creates a physical security issue if the external memory bus is not encrypted because node information (including security keys, node and network addresses, payload data, etc.) may be easily observed there. The multiple-chip design is often the preferred approach, however, because the embedded flash IC process is likely to be extremely costly — if it is available at all — and the resulting total system cost analysis usually will favor the multiple-chip design. It is likely that the nonvolatile memory IC will require its own supply voltage — at a value higher than that of the rest of the network node.

7.3.3 Power Consumption of Other Loads

The power consumption of the portion of a wireless sensor network node devoted to communication can be dwarfed by the power consumption of other circuits attached to its supply, including those of sensors, actuators, and user interface devices such as LEDs and audio transducers. Although the node designer may not have complete control of such external components because their power consumption may be large, it is incumbent upon the designer to exert what control is available.

When a wireless sensor network node may be used in a plurality of applications, it is usually not possible to optimize its design for any one of them. Instead, it is often best to design as much flexibility as possible into the node, and to place as much control as possible of this flexibility into the hands of the original equipment manufacturer (OEM) designing the node application or, as a last resort, into the hands of the actual user. The last option is to be used with caution because one of the goals of wireless sensor networks is to avoid any network administration, but it is sometimes necessary.

Because sensors and actuators can consume great amounts of power when active, their operation should be scrutinized closely for opportunities to minimize the node average power consumption. Many wireless sensor networks replace wired sensor networks already in place; their sensors, originally designed for wired operation, may be constantly powered, even though they may only need to provide data infrequently. In such cases, and where the accuracy of the sensor is not affected, power may be saved by simply turning the sensor off when an updated value is not needed, then turning it back on immediately prior to a wireless transmission. This is one example where flexible control comes into play; a wireless sensor network node that has an external control interface can synchronize the operation of the sensor with that of the node's communication network, enabling a significant overall power reduction. Other methods, in which the sensor is self-timed and strobes in a manner asynchronous to the network, may also be useful; in this case, the sensor may simply load a data register on the network node for later transmission.

User interfaces on wireless sensor networks should be used with caution. Because network nodes are usually designed for extended operation (months or years), it is likely that the great majority of time no one will be observing the user interface; any power expended in the unobserved interface during this time is therefore wasted. Because the average power consumption of a wireless sensor network node (excluding power consumed by sensors and actuators) can be less than 100 μW, the average power consumption of user interfaces should be less than 10 μW or so to avoid affecting this significantly. This 10-μW limit requires the duty cycle of a typical 2-V, 20-mA LED to be less than 2.5×10^{-4}, or a 30-ms flash of light every 2 minutes. The effect of the 20-mA pulse — and when it occurs relative to the current pulse associated with transceiver operation — must also be considered in the design of the power-conditioning circuit. Audible transducers (alarms) can be equally power-hungry; electromechanical transducers will operate from a 1-V supply, but may require 50 mA or more; piezoelectric transducers are usually more efficient, but require a special circuit to generate the high drive voltages they require — voltages that may be incompatible with most integrated circuit processes. Should transducers be used, they should be used in the most power-efficient manner; bridge drivers should be employed to produce the highest sound pressure level from the available supply voltage, and, if a volume control is needed, pulse width modulation techniques should be considered, instead of resistive dividers or linear drive voltage regulation. For power-constrained designs (i.e., those not supplied by the mains), it is usually best to stagger visual and audible signals so that the power source has some time to recover between them, and less secondary storage is required in the power-conditioning circuit.

As an alternative to LEDs for visual status indication, the use of cholesteric displays,[98] or other bistable (i.e., drawing no static power) displays, is suggested. Unfortunately, the voltage needed to write to these displays is typically 2.8–3.0 V (although falling as the technology matures); on the positive side, status information usually requires only infrequent updates, and can tolerate a hundred milliseconds or so of latency while a voltage multiplier is placed in operation prior to the update. With infrequent use, the power consumption associated with the updating of cholesteric displays can be quite compatible with the needs of wireless sensor network nodes.

Because some type of user interface is usually required, if for no other reason than to provide comfort to the user that all is well, an alternative to the constantly active user interface is to enable (or disable) the interface on demand. This can take the form of activating the user interface at initial power-on, then disabling it at some time after network association, or controlling it with a special network command.

7.4 VOLTAGE CONVERTERS AND REGULATORS

Despite the desire to do otherwise, more than one supply voltage must be generated in nearly all practical systems. Several reasons for this have already been mentioned: the different optimum supply voltages for analog and digital circuits and the higher voltages needed for nonvolatile memory, piezoelectric transducers, and other user interfaces. To meet the requirements of these circuits while minimizing the average power consumption of the total system, the wireless sensor network node designer should consider carefully the voltage generation strategy employed in the node. Not only should the type of voltage conversion and regulation circuits used be considered with care, but also the topology of the conversion and regulation system (i.e., what voltage gets converted or regulated to what voltage) needs to be considered with care. In nearly all cases, merely producing all needed voltages directly from the source voltage is not the lowest cost nor physically smallest implementation.

7.4.1 Types of Voltage Converters

Although a very large number of voltage converter types are used, and it is beyond the scope of this text to describe them in detail, a brief overview of the major classes is useful to outline the advantages and disadvantages of each. (Lenk[99] has a more detailed description of converter operation at the discrete component level.) In this discussion, a "converter" shall be any device that produces a DC output voltage different from a DC input voltage; a "regulator" shall be a converter in which a feedback circuit exists to control the output voltage to some desired value.

Two general classifications of voltage converters are available: linear and switching. The linear converter, which is nearly always used as a regulator, employs an active device (transistor) to provide what is essentially a variable resistance between the input voltage and the output voltage. The linear regulator must then be a down-converter (i.e., the output voltage must always be less than the input voltage). The current drawn by the output load passes through the transistor (the "pass transistor"), so the power dissipated in the pass transistor is the product of the load current and the difference between the input voltage and the output voltage. If this voltage difference is large, considerable power can be consumed by the linear regulator. At the other extreme, the input voltage must stay above the desired output voltage plus whatever minimum voltage is needed across the pass transistor for its operation (usually 20–50 mV or so). In low-power designs usually much effort is made to obtain a low value for this voltage because it represents not only a lower bound on power consumed by the regulator, but in the case of a linear regulator supplied by an electrochemical cell (battery), lowering this value extends the effective life of the cell, by lowering its end-of-life voltage. This can have a significant effect on useful battery life, depending on the cell chemistry used.

Because there is no switching of currents involved in its operation, the linear regulator can produce an output with much less noise and EMC problems than other types, although for very-low-noise applications, the shot noise associated with the pass transistor can be a problem. The lack of EMC problems often makes the linear regulator the circuit of choice for converters that must operate while sensitive RF and analog circuits are in operation. Because it does not require any reactive element (other than perhaps a compensation capacitor in the regulator feedback network), the linear regulator is quite suitable for integration.

The second major classification of voltage converters is the switching type. In this type, a reactive element, either an inductor or a capacitor, is switched in and out of the converter circuit. The energy stored in the reactive element is then used to produce a voltage that is either higher or lower than the input voltage; even voltage inverters, in which a negative voltage is produced from a positive voltage, can be made with this technique.

Of the switching converters, the inductive type is probably the most common. Many variations exist, but the boost (up-converter), buck (down-converter), and inverter (either a down- or an up-converter, and also known as a buck–boost converter), are most often seen. The circuit elements of each are common; the difference is in how the elements are arranged. The inductive converters need a switch, an inductor, a diode, and a filter capacitor.

The operation of the inductive boost converter is given in Exhibit 3. When the switch is closed, the charging current passes through the inductor L, creating a magnetic field around it. The diode is reversed biased, so there is no current through the diode to the output filter capacitor C (and, therefore, to the load itself).

When the switch is opened, the magnetic field around the inductor collapses, generating a discharging current passing through L in the same direction as the charging current that existed when the switch was closed. To generate this discharging current, the right side of L becomes more positive. When this voltage becomes high enough to forward bias the diode, the inductor discharging current passes through the diode, into the output

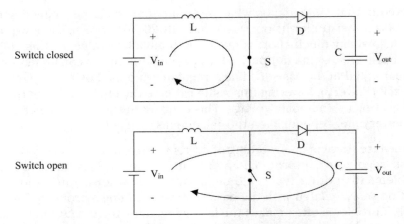

Exhibit 3. Operation of the Inductive Boost Converter

filter capacitor C and the load. This increases the output voltage above that of the input voltage.

This switching cycle repeats indefinitely. The output voltage may be regulated to a desired value by controlling the frequency and/or duty cycle of the switching waveform.

In the boost converter as presented, the power dissipated in the diode is equal to the load current times the diode forward voltage drop. To minimize this power loss, a Schottky diode, with a voltage drop of perhaps 0.2 V, is often used here instead of a conventional silicon diode, which may have a voltage drop of 0.7 V.

The output voltage of the boost converter is always greater than the input voltage due to the action of the diode. As the desired output voltage approaches the input voltage, the switch duty cycle (or frequency of operation, depending on how regulation is accomplished) approaches zero. Should the input voltage rise above the desired output voltage (plus the diode forward voltage drop), the diode will conduct, and continue to do so until the input voltage falls below the desired V_{out} again. Although this is usually not a significant design constraint for battery-powered converters, this scenario should be examined when energy-scavenging techniques are employed to supply the converter, and in other cases in which the input voltage is poorly regulated, to ensure that no unexpected behavior occurs.

The operation of the inductive buck converter is given in Exhibit 4. When the switch is closed, the charging current passes through the inductor L, creating a magnetic field around it. The diode is reversed biased, so

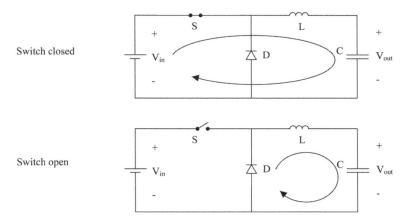

Exhibit 4.   Operation of the Inductive Buck Converter

there is no current through the diode to the output filter capacitor C (and, therefore, to the load itself).

When the switch is opened, the magnetic field around the inductor collapses, generating a discharging current passing through L in the same direction as the charging current that existed when the switch was closed. To generate this discharging current, the left side of L becomes more negative. When this voltage becomes low enough to forward bias the diode, the inductor discharging current passes through the diode, into the output filter capacitor C and the load.

The switching cycle is repeated. The output voltage may be regulated to a desired value by controlling the frequency and/or duty cycle of the switching waveform.

V_{out} must be less than V_{in} to create a charging current. If V_{out} ever approaches V_{in}, the charging current, and, therefore, the discharging current, approaches zero. This ensures that V_{out} remains below V_{in} under normal operation.

The operation of the inductive inverting converter (also called an inverter or a buck–boost converter) is given in Exhibit 5. When the switch is closed, the charging current passes through the inductor L, creating a magnetic field around it. The diode is reversed biased, so there is no current through the diode to the output filter capacitor C (and, therefore, to the load itself).

When the switch is opened, the magnetic field around the inductor collapses, generating a discharging current passing through L in the same direction as the charging current that existed when the switch was closed.

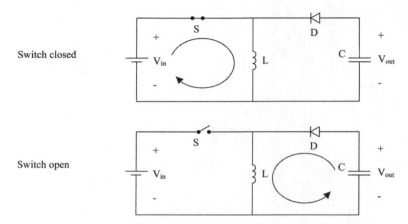

Exhibit 5. Operation of the Inductive Inverting (Buck–Boost) Converter

To generate this discharging current, the upper side of L becomes negative relative to the lower side. When this top side voltage becomes low enough to forward bias the diode, the inductor discharging current passes to the negative terminal of the output filter capacitor C (and the load), returning to the inductor through the diode. V_{out} is, therefore, negative with respect to V_{in}.

The switching cycle is repeated. The output voltage may be regulated to a desired value by controlling the frequency and/or duty cycle of the switching waveform. The absolute value of V_{out} may be greater than or less than that of V_{in}, but, in any case, V_{out} will be negative with respect to V_{in}. The inverting converter, therefore, offers greater design flexibility than other types; however, it has disadvantages. One disadvantage, of course, is the polarity inversion. Although this is often useful for the creation of a negative supply (needed, for example, to power some liquid crystal display [LCD] drivers and Gallium Arsenide (GaAs) field effect transistor (FET) circuits), in the general case, using it to generate a positive voltage complicates circuit board layout because there is no common voltage reference for all circuits. Shielding becomes more difficult, which can lead to EMC problems.

When $|V_{out}| < |V_{in}|$ (i.e., when the inverter is used to reduce voltage), a more subtle disadvantage of the inverter appears. In the buck converter design, current is supplied to the load during both cycles of operation (that is, during both charge and discharge). In the inverter, current is supplied to the load only during discharge. This means that, for a given load current, the peak current through the inductor in a buck converter is less than that of the inductor in an inverter. This has the following effects:

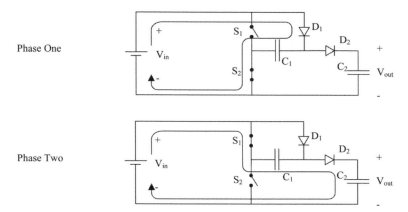

Exhibit 6. Operation of the Diode-Based Capacitive Converter

- The inductor of the buck converter can be smaller than that of the inverter, reducing size and cost.
- The switch switches lower currents in the buck converter, which reduces stress on it and may improve its reliability.
- Because lower peak currents are being switched through a physically smaller inductor, the ability to generate noise and to couple it to other circuits in the system, is reduced. The potential for EMC problems is reduced.

As a result, in most applications requiring a step down in voltage, the buck converter, instead of the inverter, is selected.

The advantage of the inductive converter is that it is very flexible and capable of converting relatively large amounts of power; however, because it switches relatively large reactive currents, operation at very low output currents is inherently inefficient. Its principal disadvantages, however, revolve around the inductor; it is difficult to integrate, expensive, physically large, and can be a source of EMC and reliability problems.

The second major type of switching converter is the capacitive converter. A capacitive doubler is illustrated in Exhibit 6. In this type of converter, a charge pump capacitor C_1 is first charged in parallel to the input voltage, less the forward voltage drop of diode D_1, by closing S_2 and opening S_1. C_1 is then switched to the output circuit in series with the input voltage, such that their voltages add; this voltage, less the forward voltage drop of diode D_2, charges the output filter capacitor C_2. It is, therefore, easiest to construct capacitive converters providing near-integral multiples of the input voltage. Nonintegral voltage multiplication values may be achieved by employing more complex switching structures.

Phase One

Phase Two

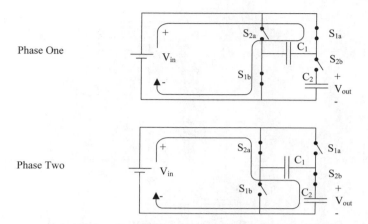

Exhibit 7.   Operation of the Switch-Based Capacitive Converter

An alternative structure for a capacitive converter is presented in Exhibit 7. In this structure, the diodes and their associated voltage drops have been replaced by two additional switches. In the first phase of operation, switches S_{1a} and S_{1b} are closed, and switches S_{2a} and S_{2b} are open. This charges charge pump capacitor C_1 to the input voltage. In the second phase of operation, the switch positions reverse: switches S_{1a} and S_{1b} are open, and switches S_{2a} and S_{2b} are closed. This places C_1 in series with the input voltage, and charges the output filter capacitor C_2 to twice V_{in}.

Regulation of capacitive converters may be done by adjustment of the switching frequency and/or duty cycle, or by controlling the series resistance of the switches.

The biggest advantage of the capacitive converter is that it can be integrated in CMOS. This is especially true for low output current applications where the capacitor values may be small. In an integrated implementation, the switches can be transmission gates. These features make the capacitive converter the converter of choice for applications such as memories and LCD display drivers. The most serious limitation to the capacitive converter is its limited current-sourcing capability; any output current must be transported via the switched capacitors and is, therefore, limited by the acceptable output voltage ripple and the frequency with which the switches are switched. Like the inductive converter, the capacitive converter has a tendency for EMC problems; however, the problems are often less severe due to the smaller physical size of capacitors. Because the capacitive converter does not employ inductors, its reliability can be higher than that of the inductive converter as well.

7.4.2 Voltage Conversion Strategy

For minimum power consumption, it is important for the system designer to understand the voltage and current needs of the load. Typically, some portion of the power overhead (i.e., power consumed by the converter from the source but not delivered to the load) of the converter is fixed, regardless of power delivered to the load, and some portion is variable, and is directly proportional (to a first approximation) to the power delivered to the load at any given time. The fixed portion of the overhead in most designs is commensurate with the maximum current capability of the converter. Operating a converter significantly below its maximum design value, therefore, represents a waste of power because the same output power could be delivered by a converter with a lower maximum power output capability and, therefore, lower fixed overhead. Importantly, as the power consumed by the load falls, the efficiency of the total system drops, because a greater fraction of total consumed power is consumed by the converter, instead of the load. Reducing the power consumption of the load becomes an effort with diminishing returns, for, although the absolute value of load power saved is directly reflected in a reduction in input power, the percentage improvement in input power becomes less and less as the load power becomes comparable with the converter overhead. In the limit, the minimum power the system can consume is that of the converter itself. This means, for example, that no matter how low-power the standby circuits (e.g., oscillators, timers) are in a battery-powered network node, the maximum lifetime of the node ultimately is determined by the overhead of its power converter. It is for this reason that it is desirable to avoid the use of power converters completely, by matching the current and voltage requirements of the load with those available from the source. When this is done, one gets the full benefit one expects from a reduction in load power consumption; a 40 percent power reduction of the load (achieved, say, by shrinking the size of the circuits in a smaller-geometry IC process), results in a 40 percent reduction in power consumed from the source and a 40 percent increase in battery life.

Should the use of a converter be necessary, however, a method to ameliorate its fixed overhead power consumption is to operate it in "burst mode" (i.e., cycling it on and off). When the converter is off, the load is supplied from the filter capacitor, the value of which usually is made as large as possible to extend the "off" periods. When the output voltage reaches a low threshold, the converter is restarted, increasing the voltage until an upper threshold is reached, when the converter is turned off again. Use of burst mode can be considered either a method of regulation (especially if no other form of regulation is present), or a strategy for power reduction (especially if a second type of regulation is present when the converter is active). It is especially useful for lightly loaded converters, which may otherwise operate very inefficiently due to their relatively large power overhead, and for loads (such as standard cell digital logic) that are insensitive to relatively large variations in their supply voltage. Burst mode is often employed during the sleep periods of wireless sensor network communication protocols, during which the converter load may consist of a single clock oscillator and counter, used to count down the time until the next active period.

The power consumption of wireless sensor network nodes typically varies significantly over time; it is very low while the node is asleep, rises several orders of magnitude for a brief period of transmission and/or reception, then returns to its low value as the node returns to sleep. For example, a node may have a constant 30-μA current drain while asleep, but have an additional 400-μs-long pulse of 20 mA every 2 seconds when the receiver is active. Supplying this load at all times from a single converter, capable of handling the maximum power sunk by the load at all times, would imply that for the vast majority of time, the converter was operating in a very inefficient mode; a converter capable of sourcing 20 mA may have a fixed overhead current of 20 μA. The average current drain of such a system would be

$$I_{avg\ 1} = 30\ \mu A + (200 \times 10^{-6}) \times 20\ mA + 20\ \mu A = 54\ \mu A \qquad (13)$$

A much more efficient system would be to employ two converters — a first converter (perhaps an inductive switching converter) providing power only when the node is active, followed by a second converter (perhaps an integratable capacitive switching converter) providing power during sleep mode. The first converter could be the same one used in the first design, and would be optimized to supply the 20-mA active current. The second converter would be designed for optimum efficiency supplying standby current drain values (e.g., an overhead of 3 μA while supplying 30 μA). The average current drain of this design is

$$I_{avg\ 2} = 30\ \mu A + 3\ \mu A + (200 \times 10^{-6}) \times 20\ mA + (200 \times 10^{-6}) \times 20\ \mu A = 37\ \mu A \quad (14)$$

The second design has reduced the average power consumption of the system by 31 percent.

The design of such multiple-converter systems results in additional system complexity; for example, the switching of the converters must not produce any momentary spikes or glitches in the supply, and the warm-up behavior of the second converter must be well-characterized so that it can be ready to supply the load of the active network node when required. A multiconverter design can be significantly more efficient than the single-converter approach, however, and can greatly extend the battery life of a network node. Burst mode may also be used in some applications to create a low-output mode from a high-output converter, with reasonably good efficiency, and thereby get the performance of two converters with only one set of components.

This quirk of converter behavior also must be considered in the design of application-flexible wireless sensor network nodes. When designing the power converter system for a network node for which the end application, and, therefore, the sensors and actuators that may be used, is not known, it is tempting to place a single power converter in the design, capable of handling the worst-case envisioned power consumption by the load. Unfortunately, this saddles the typical application with less-than-optimum overall power consumption. One solution to this problem is to offer multiple converters to supply the node transducers and ask the user (or OEM supplier) to select the proper one. This solution is expensive, due to the duplication of converters and their associated die and board area, chip pin-outs, and external components. A more economically attractive solution is to design a single, programmable converter, the maximum power capability of which may be adjusted via an external port. The use of such a converter is not a panacea, however; its design is nontrivial, and often only a limited range of power capabilities is possible without changing external components.

The problem of matching power converter efficiency with a time-varying load affects all supplied circuits. To optimize system power consumption, the network node designer must understand not just the maximum current required by all loads, but how each load's current requirements vary with time. The analysis of each load's current requirements can become complex, especially when multiple independent loads are driven from the same converter (e.g., an LCD driver and standby timing circuits), or when the outputs of multiple converters are associated (e.g., an LCD driver and its backlight). A thorough analysis that covers all potential states of operation should be performed to ensure that no case is overlooked and that the transitions between states are well understood.

In addition to these factors, the required temperature coefficient of the converter output must also be considered in a low-power design. For example, a wireless sensor network node may be required to operate over the industrial temperature range of −20°C to +85°C. A CMOS digital circuit in the network node may meet its speed requirements with a 1.0 V supply at a temperature of −20°C, but require a 1.2-V supply to achieve the same speed at +85°C. A converter with a temperature coefficient of zero ppm/C that supplies this circuit would require an output voltage of 1.2 V to ensure operation over the entire operating temperature range. This represents a waste of power over most of the temperature range, when a lower supply voltage would be sufficient. A better design uses a converter with a positive temperature coefficient that matches that of the load; such a converter would produce 1.2 V at +85°C, but a lesser voltage at lower temperatures, reducing the average power consumed by the system. Other loads, such as liquid crystal displays, can require a supply with a negative temperature coefficient. Matching the temperature coefficient of supplies with the needs of loads is important to minimize system power consumption, but this can be a complex task when multiple loads are present. In the extreme, one may see designs in which two or more regulators have the same nominal (room temperature) voltage, but different temperature coefficients.

When multiple source voltages are required, generating them by multiple switching converters can require a large number of components. It is often more practical to generate multiple source voltages by employing a single switching converter to generate a relatively high voltage, then using several linear regulators to create the required voltages from the switching converter output. This scheme is especially useful when the unregulated switching converter output voltage itself is used to supply a load, when the currents supplied by the lower voltage supplies are small in comparison to the current used by that load, and when the desired linear regulator output voltage is not too far from that of the switching converter output voltage. It must always be kept in mind that the current supplied by the linear regulators is very expensive in terms of the current drawn from the primary source (e.g., a battery) used to produce it, due to their derivation from the relatively high switching converter output voltage. This system can be much more economical to produce, however, and occupy much less physical area than an alternative design employing multiple switching converters.

Although in principle this concept could be extended, producing lower output voltages by placing linear voltage regulators as the loads of other linear regulators (with higher output voltages), each voltage converter has a power consumption of its own. The efficiency of the system is equal to the multiplication of the efficiencies of each of the converters, which makes such "daisy-chaining" of regulators inefficient when more than a very few are used. It is best when the minimum number of voltage conversions is made.

7.5 POWER MANAGEMENT STRATEGY

From the viewpoint of the load, all power management is based, in the final analysis, on three principles:

- Do not buy any unnecessary lights. That is to say, do not waste power on any unnecessary tasks. This is largely a function of the RF communication and sensor interface protocols to perform their tasks as efficiently as possible, by placing as little a load as possible on the hardware. An example of this is the trade-off between the energy costs of communication and computation. A typical wireless sensor network transmission and reception cycle may take 1 ms and require hardware that consumes 35 mW during that time, for a total energy consumption of 35 μJ — all to send perhaps 10 bytes of information, for an average energy consumption of 437.5 nJ/bit. (Due to fixed communication overhead, longer messages would be somewhat more efficient.) Computation on energy-efficient microcomputers requires on the order of 1–10 nJ/operation, however, so it is possible to do a significant amount of processing for the energy cost of one transmission. It is, therefore, energetically advantageous to perform a significant amount of raw data processing in the node itself, prior to transmission, if that processing either eliminates the need to transmit completely, or reduces the amount of data that is to be transmitted.

- Turn the lights off when you leave the room. In terms relevant to wireless sensor networks, do not waste power on any unused circuits. Turn off oscillators, pad drivers, in fact, all circuits when not in use. For CMOS logic circuits, this means to employ clock gating to stop the clocks to unused sections of logic; for low-voltage CMOS logic, where leakage current may be significant, it means employ multithreshold logic, so that the leakage may be reduced. For transceivers, this means to enable circuits in stages during warm-up, instead of all at once, and employ techniques to minimize warm-up time.
- Dim the lights when you are in the room. Use the minimum possible energy to complete the required task. For example, from Equation 8, the supply voltage and clock frequency of digital circuits may be dynamically varied over time to meet the application requirements — when the computational load is heavy, the supply voltage and clock frequency may be increased, then reduced when the computational load lightens. In RF transceivers, the transmit power may be reduced to the minimum necessary to complete the communication link; similarly, the RF amplifier in the receiver may be disabled and bypassed under strong signal conditions.

Viewing the system as a whole, however, the guiding principle is one of matching the (hopefully minimized) energy and power requirements of the load with the (hopefully maximized) energy and power available from the source. The better the source and load are matched, the less elaborate (and inefficient) the required power-conditioning circuits must be, and the more successful the overall design. One notes that the lower the power consumption requirements of the load, the more types of power sources, system designs, and applications will be compatible with it, and the greater its market flexibility. The load power consumption may then be used as a metric to estimate the economic viability of a given design.

7.6 CONCLUSION

Power management in wireless sensor network nodes is a critical, if not determining, feature of their design, yet the very wide variety of potential power sources, integrated circuit technologies, and system designs from which the designer may choose makes the power management problem exceedingly difficult to solve in general. A major challenge to network node designers is the right trade-off between optimization of the power management for a specific application and market, and the generalization of the design to serve as wide a market as possible (thereby generating the maximum return on the engineering design investment). This trade-off is eased by minimizing the power required by the load, something best achieved in these largely digital systems by low voltage (i.e., 1 V or less) designs.

References

1. Rex Min et al., Energy-centric enabling technologies for wireless sensor networks, *IEEE Wireless Communications,* v. 9, n. 4, August 2002, pp. 28–39.
2. Dwayne N. Fry et al., *Compact Portable Electric Power Sources.* Document number DE98054446 (ORNL/TM-13360). Springfield, VA: National Technical Information Service. 1997.
3. Millard Frank Rose, Ed., *Prospector IX: Human Powered Systems Technologies.* Document number ADA344414. Springfield, VA: National Technical Information Service. 1997.
4. Thad Starner, Human-powered wearable computing, *IBM Sys. J.,* v. 35, n. 3, 4, 1996, pp. 618–629.
5. C. F. Chiasserini and R. R. Rao, A model for battery pulsed discharge with recovery effect, *Proc. Wireless Commun. Networking Conf.,* 1999, v. 2, pp. 636–639.
6. A primary reference for this and other physical constants is: P. J. Mohr and B. N. Taylor, CODATA recommended values of the fundamental physical constants: 1998, *J. Phys. Chem. Ref. Data,* v. 28, n. 6, 1999, pp. 1713–1852.
7. Anantha Chandrakasan et al., Design considerations for distributed microsensor systems, *Proc. IEEE Custom Integrated Circuits Conf.*, May 1999, pp. 279–286.
8. Dwayne N. Fry, Alan L. Wintenberg, and Bill L. Bryan, Integrated power management for Microsystems, in M. Frank Rose, Ed., *Prospector IX: Human Powered Systems Technologies.* Document number ADA344414. Springfield, VA: National Technical Information Service. 1997.
9. Norio Hama et al., SOI circuit technology for batteryless mobile system with green energy sources, *Symp. on VLSI Circuits Digest of Technical Papers,* 2002, pp. 280–283.
10. David L. King, James K. Dudley, and William E. Boyson, PVSIM©: A simulation program for photovoltaic cells, modules, and arrays, *Conf. Record, 25th IEEE Photovoltaic Specialists Conf.,* 1996, pp. 1203–1206.
11. Brett Warneke et al., Smart dust: Communicating with a cubic-millimeter computer, *Computer,* v. 34, n. 1, January 2001, pp. 44–51.
12. Bryan Atwood, Brent Warneke, and Kristofer S. J. Pister, Preliminary circuits for Smart Dust, *Proc. Southwest Symp. Mixed-Signal Design,* 2000, pp. 87–92.
13. Brett A. Warneke, Bryan Atwood, and Kristofer S. J. Pister, Smart Dust mote forerunners, *Proc. 14th Annu. Int. Conf. Microelectromechanical Syst.,* 2001, pp. 357–360.
14. Chandrakasan et al., ibid.

15. C. B. Williams and R. B. Yates, Analysis of a micro-electric generator for microsystems, *Proc. 8th Int. Conf. on Solid-State Sensors and Actuators and Eurosensors IX,* 1995, v. 1, pp. 369–372.
16. Starner, ibid.
17. John Kymissis et al., Parasitic power harvesting in shoes, *Dig. of Papers, Second Int. Symp. on Wearable Computers.,* 1998, pp. 132–139.
18. Nathan S. Shenck and Joseph A. Paradiso, Energy scavenging with shoe-mounted piezoelectrics, *IEEE Micro,* v. 21, n. 3, May/June 2001, pp. 30–42.
19. Gerald F. Ross et al., *Batteryless Sensor for Intrusion Detection and Assessment of Threats.* Defense Nuclear Agency Technical Report DNA-TR-95-21. Springfield, VA: National Technical Information Service. 1 November 1995.
20. Shenck and Paradiso, ibid.
21. Markt & Technik, Fast wie ein Perpetuum Mobile (Almost like perpetual motion), n. 47, 15 November 2002, pp. 45–47. http://www.elektroniknet.de.
22. P. Glynne-Jones et al., A vibration-powered generator for wireless microsystems, *Proc. Int. Symp. Smart Structures and Microsystems,* 2000.
23. P. Glynne-Jones, S. P. Beeby, and N. M. White, Towards a piezoelectric vibration-powered microgenerator, *IEEE Proc. Science, Measurement, and Technology,* v. 148, March 2001, pp. 68–72.
24. Geffrey K. Ottman, Heath F. Hofmann, and George A. Lesieutre, Optimized piezoelectric energy harvesting circuit using step-down converter in discontinuous mode, *IEEE 33rd Annu. Power Electronics Specialists Conf.,* 2002, v. 4, pp. 1988–1994.
25. Geffrey K. Ottman et al., Adaptive piezoelectric energy harvesting circuit for wireless remote power supply, *IEEE Trans. Power Electronics,* v. 17, n. 5, September 2002, pp. 669–676.
26. Ibid.
27. Ottman, Hofmann, and Lesieutre, ibid.
28. Rajeevan Amirtharajah and Anantha P. Chandrakasan, Self-powered low power signal processing, *IEEE Symp. VLSI Circuits, Digest of Technical Papers,* 1997, pp. 25–26.
29. Rajeevan Amirtharajah and Anantha P. Chandrakasan, Self-powered signal processing using vibration-based power generation, *IEEE J. Solid-State Circuits,* v. 33, n. 5, May 1998, pp. 687–695.
30. M. El-hami et al., A new approach towards the design of a vibration-based microelectromechanical generator, *Proc. 14th European Conf. on Solid-State Transducers,* 2000, pp. 483–486.
31. Gerald F. Ross et al., *Batteryless Sensor Used in Security Applications.* U.S. patent 5,317,303. Washington, D.C.: U.S. Patent and Trademark Office. 31 May 1994.
32. Gerald F. Ross et al., *Batteryless Sensor for Intrusion Detection and Assessment of Threats.* Defense Nuclear Agency Technical Report DNA-TR-95-21. Springfield, VA: National Technical Information Service. 1 November 1995.
33. Masakatsu Saka and Kinya Matsuzawa, Seiko human powered quartz watch, in M. Frank Rose, Ed., *Prospector IX: Human Powered Systems Technologies.* Document number ADA344414. Springfield, VA: National Technical Information Service. 1997.
34. Yoshitaka Iijima, Microgenerator for the Wrist Watch, in Masayoshi Esashi and Shuji Tanaka, Eds., *Power MEMS: Workshop on Power Microelectromechanical Systems in Int. Symp. on Research and Education in the 21st Century (ISRE 2000),* 18 Aug 00, Sendai, Japan. Document number ADA384916. Springfield, VA: National Technical Information Service. 2000.
35. C. B. Williams and R. B. Yates, Analysis of a micro-electric generator for Microsystems, *Proc. 8th Int. Conf. On Solid-State Sensors and Actuators and Eurosensors IX,* 1995, v. 1, pp. 369–372.

36. C. B. Williams, R. C. Woods, and R. B. Yates, Feasibility study of a vibration powered micro-electric generator, *IEEE Colloquium on Compact Power Sources (Digest No.96/107)*, 1996, pp. 7/1–7/3.

37. C. Shearwood and R. B. Yates, Development of an electromagnetic microgenerator, *Electronics Lett.*, v. 33, n. 22, 23 October 1997, pp. 1883–1884.

38. C. B. Williams et al., Development of an electromagnetic micro-generator, *IEEE Proc. Circuits Devices Syst.*, v. 148, n. 6, December 2001, pp. 337–342.

39. S. Meninger et al., Vibration-to-electric energy conversion, *Proc. Int. Symp. on Low Power Electronics and Design*, 1999, pp. 48–53.

40. S. Meninger et al., Vibration-to-electric energy conversion, *IEEE Trans. Very Large Scale Integration (VLSI) Systems*, v. 9, n. 1, February 2001, pp. 64–76.

41. Rajeevan Amirtharajah et al., A micropower programmable DSP powered using a MEMS-based vibration-to-electric energy converter, *IEEE Solid-State Circuits Conf. Dig. of Technical Papers*, 2000, pp. 362–363, 469.

42. K. Najafi, Low-power micromachined Microsystems, *Proc. Int. Symp. Low Power Electronics and Design*, 2000, pp. 1–8.

43. Brett Warneke and Kristofer S. J. Pister, MEMS for distributed wireless sensor networks, *9th Int. Conf. on Electronics, Circuits, and Syst.*, 2002, v. 1, pp. 291–294.

44. Jean Lebet, *Living on Air: History of the Atmos Clock*. Le Sentier, Switzerland: Jaeger-LeCoultre. 1997. pp. 9–12.

45. Ibid., p. 24.

46. Jean Léon Reutter, Horloge à remontage automatique par les variations de température ou de pression atmosphérique *(Clock with automatic rewinding by the variations in temperature or atmospheric pressure)*. Swiss Patent CH130941. Berne: Swiss Federal Institute of Intellectual Property. 15 January 1929.

47. Jean Léon Reutter, Dispositif de remontage automatique pour mouvements de montres ou d'horloges et moteurs à resort ou à poids d'autres instruments ou appareils *(Automatic device for rewinding movements of watches or clocks, and engines with springs or weights in other instruments or apparatus)*. Swiss Patent CH144349. Berne: Swiss Federal Institute of Intellectual Property. 31 December 1930.

48. Jean Léon Reutter, *Moteur thermique utilisant les variations de température de l'air (Thermal engine using variations in temperature of the air)*. Swiss Patent CH175399. Berne: Swiss Federal Institute of Intellectual Property. 28 February 1935.

49. Lebet, p. 4.

50. Steven Phillips, *Temperature Responsive Self-Winding Timepieces*. U.S. Patent 6,457,856. Washington, D.C.: U.S. Patent and Trademark Office. 1 October 2002.

51. Lebet, p. 29.

52. The same 3.5 mJ, however, powers the Atmos clock for two days; this implies that the clockwork consumes energy at an average rate of 3.5 mJ/172800 s = 20 nW — a remarkable achievement.

53. Jean-Pierre Fleurial et al., *Electronic device featuring thermoelectric power generation*. U.S. Patent 6,288,321. Washington, D. C.: U.S. Patent and Trademark Office. 11 September 2001.

54. Jean-Pierre Fleurial et al., *Microfabricated Thermoelectric Power-Generation Devices*. U.S. Patent 6,388,185. Washington, D. C.: U.S. Patent and Trademark Office. 14 May 2002.

55. M. Kishi et al., Micro thermoelectric modules and their application to wristwatches as an energy source, *18th Int. Conf. Thermoelectrics*, 1999, pp. 301–301.

56. T. Douseki et al., A batteryless wireless system uses ambient heat with a reversible-power-source compatible CMOS/SOI DC–DC converter, *IEEE Int. Solid-State Circuits Conf. Dig of Technical Papers*, 2003, pp. 388–389, 501.

57. James W. Stevens, Heat transfer and thermoelectric design considerations for a ground-source thermoelectric generator, *18th Int. Conf. Thermoelectrics,* 1999, pp. 68–71.

58. Duncan Graham-Rowe, Third world TV runs on heat from the stove, *New Scientist,* v. 175, n. 2361, 21 September 2002, p. 21.

59. Millard Frank Rose, ed., *Prospector VIII: Thermophotovoltaics — An Update on DOD, Academic, and Commercial Research.* Document number ADA364420. Springfield, VA: National Technical Information Service. 1996.

60. Robert E. Nelson, TPV systems and state-of-the-art development, *Proc. Fifth NREL Conf. on Thermophotovoltaic Generation of Electricity,* 2002.

61. T. Bauer et al., The potential of thermovoltaic heat recovery in the glass industry, *Proc. Fifth NREL Conf. on Thermophotovoltaic Generation of Electricity,* 2002.

62. George W. Taylor et al., The energy harvesting eel: a small subsurface ocean/river power generator, *IEEE J. Oceanic Eng.,* v. 26, n. 4, October 2001, pp. 539–547.

63. Clare E. Reimers et al., Harvesting energy from the marine sediment — water interface, *Environ. Sci. Technol.,* v. 35, n. 1, 2001, pp. 192–195.

64. Daniel R. Bond et al., Electrode-reducing microorganisms that harvest energy from marine sediments, *Science,* v. 295, 18 January 2002, pp. 483–495.

65. Stuart Wilkinson, Gastrobots — benefits and challenges of microbial fuel cells in food powered robot applications, *J. Autonomous Robots,* v. 9, 2000, pp. 99–111.

66. http://www.ias.uwe.ac.uk.

67. Ioannis Ieropoulos, John Greenman, and Cris Melhuish, Imitating metabolism: energy autonomy in biologically inspired robots, *Proc. AISB, Second Int. Symp. on Imitation in Animals and Artifacts,* Aberystwyth, Wales, 2003, pp 191–194.

68. Kishi, ibid.

69. Eric A. Vittoz, Future of analog in the VLSI environment, *IEEE. Int. Symp. on Circuits and Syst.,* 1990, v. 2, pp. 1372–1375.

70. Eric A. Vittoz, Low-power design: ways to approach the limits, *41st IEEE Int. Solid-State Circuits Conf. Dig. of Technical Papers,* 1994, pp. 14–18.

71. Ed Callaway, 1-Volt RF circuit design for pagers, in Johan Huijsing, Rudy van de Plassche, and Willy Sansen, Eds., *Analog Circuit Design.* Boston: Kluwer Academic Publishers. 1999.

72. Ibid.

73. Tajinder Manku, Galen Beck, and Etty J. Shin, A low-voltage design technique for RF integrated circuits, *IEEE Trans. Circuits and Systems — II: Analog and Digital Signal Processing,* v. 45, n. 10, October 1998, pp. 1408–1413.

74. C. Toumazou, F. J. Lidgey, and D. G. Haigh, Eds., *Analogue IC design: the current-mode approach.* London: Peter Peregrinus. 1990.

75. Jan Crols and Michiel Steyaert, *CMOS Wireless Transceiver Design.* Dordrecht, The Netherlands: Kluwer Academic Publishers. 1997.

76. Samuel Sheng and Robert Brodersen, *Low-Power CMOS Wireless Communications: A Wideband CDMA System Design.* Boston: Kluwer Academic Publishers. 1998.

77. Thomas H. Lee, *The Design of CMOS Radio-Frequency Integrated Circuits.* Cambridge: Cambridge University Press. 1998.

78. Derek K. Shaeffer and Thomas H. Lee, *The Design and Implementation of Low-Power CMOS Radio Receivers.* Boston: Kluwer Academic Publishers. 1999.

79. Asad A. Abidi, Low-power radio-frequency IC's for portable communications, *Proc. IEEE,* v. 83, n. 4, April 1995, pp. 544–569.

80. Alain-Serge Porret et al., A low-power low-voltage transceiver architecture suitable for wireless distributed sensors network, *Proc. IEEE Intl. Symp. on Circuits and Systems,* 2000, v. 1, pp. 56–59.

81. Francisco Serra-Graells and José Luis Huertas, Sub-1-V CMOS proportional-to-absolute temperature references, *IEEE J. Solid-State Circuits,* v. 38, n. 1, January 2003, pp. 84–88.
82. Vittoz, Low-power design, ibid.
83. Ibid.
84. Ricardo Gonzalez, Benjamin M. Gordon, and Mark A. Horowitz, Supply and threshold voltage scaling for low power CMOS, *IEEE J. Solid-State Circuits,* v. 32, n. 8, August 1997, pp. 1210–1216.
85. Thomas D. Burd and Robert W. Brodersen, Design issues for dynamic voltage scaling, *Proc. Int. Symp. on Low Power Electronics and Design,* 2000, pp. 9–14.
86. Mark Horowitz, Thomas Indermaur, and Ricardo Gonzalez, Low-power digital design, *Digest of Technical Papers, IEEE Symp. on Low Power Electronics,* 1994, pp. 8–11.
87. Yannis Tsividis, *Operation and Modeling of the MOS Transistor.* New York: McGraw-Hill. 1987. pp. 136–140.
88. Gonzalez, Gordon, and Horowitz, ibid.
89. Horowitz, Indermaur, and Gonzalez, ibid.
90. Walter L. Davis, Barry W. Herold, and Wendell L. Little, *Synthesized Clock Microcomputer with Power Saving.* U.S. patent 4,893,271. Washington, D.C.: U.S. Patent and Trademark Office. 9 January 1990.
91. Jean-Pierre Colinge, *Silicon-on-Insulator Technology: Materials to VLSI.* 2nd ed. Boston: Kluwer Academic Publishers. 1997.
92. Fariborz Assaderaghi et al., A dynamic threshold voltage MOSFET (DTMOS) for ultra-low voltage operation, *Technical Digest, Intl. Electron Devices Meeting,* 1994, pp. 809–812.
93. Hendrawan Soeleman, Kaushik Roy, and Bipul Paul, Robust ultra-low power sub-threshold DTMOS logic, *Proc. Intl. Symp. on Low Power Electronics and Design,* 2000, pp. 25–30.
94. Takakuni Douseki, Junzo Yamada, and Hakaru Kyuragi, Ultra low-power CMOS/SOI LSI design for future mobile systems, *Symp. on VLSI Circuits Digest of Technical Papers,* 2002, pp. 6–9.
95. H. Kawaguchi et al., 0.5 V, 400 MHz, V_{DD}-hopping processor with zero V_T FD-SOI technology, *IEEE Int. Solid-State Circuits Conf. Dig. of Technical Papers,* 2003, pp. 106–107, 481.
96. T. Douseki, T. Shimamura, and N. Shibata, A 0.3 V 3.6 GHz 0.3 mW frequency divider with differential ED-CMOS/SOI circuit technology, *IEEE Intl. Solid-State Circuits Conf. Dig. of Technical Papers,* 2003, pp. 114–115, 482.
97. Tsuneaki Fuse et al., A 0.5-V power-supply scheme for low-power system LSIs using multi-V_{th} SOI CMOS technology, *IEEE J. Solid-State Circuits,* v. 38, n. 2, February 2003, pp. 303–311.
98. Xiao-Yang Huang et al., Reflective cholesteric displays: Development and applications, in Liquid crystal materials, devices, and applications VIII, Liang-Chy Chien, ed., *Proc. SPIE,* v. 4658, 2002, pp. 1–6.
99. John D. Lenk, *Simplified Design of Switching Power Supplies.* Boston: Butterworth-Heinemann. 1995.

Chapter 8
Antennas and the Definition of RF Performance

8.1 INTRODUCTION

The most underappreciated component of a wireless sensor network node is often its antenna. Most wireless sensor network nodes are designed to be physically small, with low power consumption and low product cost primary design metrics. These requirements make the design of the node's antenna system both important and difficult. This situation is often exacerbated by the need to adjust the antenna design to fit aesthetics of the network node's physical design, which often forces the antenna to be the last electrical component designed in the system.

This chapter discusses a few of the more unique features of network node antennas, including their behavior in the proximity of the network node itself and how they affect the design of the wireless sensor network. The chapter concludes with a discussion of the difficulties associated with conventional performance metrics when applied to wireless sensor network nodes, and the advantages of the use of message error rate, instead of bit error rate, in the definition of node performance.

8.2 ANTENNAS

A physically small antenna is not necessarily difficult to design. A resonant half-wave dipole, for example, is only 6.25 mm long — at an operating frequency of 24 GHz. The difficulty arises when an antenna must be physically small and be used at a relatively low frequency, so that a free-space wavelength is much larger than the physical space allotted for the antenna. Such an antenna is termed an electrically small antenna.

8.2.1 Antenna Characteristics

The directivity D of an antenna may be defined as the ratio of the antenna's maximum radiation density to the radiation density of the same radiated power P_{rad} radiated isotropically.[1] Expressed mathematically,

$$D = \frac{P_{d,\max} 4\pi r^2}{P_{rad}}$$ (1)

where

D = antenna directivity, numeric,
$P_{d,\max}$ = maximum radiation density, in any direction, W/m^2
r = radial distance from the antenna to the observation point, m
P_{rad} = total power radiated by the antenna, W

Because of the scattering of radio waves that occurs due to their inter-action with the environment, the direction in which maximum radiation density occurs is usually not of great interest in wireless sensor network design, especially if the network is to be used indoors. Exceptions to this are some outdoor applications that more closely approximate free-space propagation conditions (i.e., those with little or no scattering of the incoming wave), which can choose to take advantage of this fact by employing antenna directivity to increase range. Instead of directivity, often the parameter of most interest to the network node designer is antenna efficiency.

The efficiency of an antenna may be defined as

$$\eta = \frac{P_{rad}}{P_{accept}}$$ (2)

where

η = antenna efficiency (a numeric fraction)
P_{accept} = power accepted by the antenna from the source, W

The antenna efficiency η considers only power accepted by the antenna from the source. For maximum power transfer from transmitter to antenna, however, the impedance of the antenna must be correct.[2] The result of any impedance mismatch may be characterized by[3]

$$M = \frac{P_{accept}}{P_{avail}} = 1 - |s_{11}|^2$$ (3)

where

M = mismatch factor (a numeric fraction)
P_{avail} = power available from the source
s_{11} = s-parameter associated with the input reflection coefficient at the antenna port

Neglecting any polarization concerns, the gain G of an antenna then can be defined as the product of the antenna directivity, its efficiency, and any mismatch effects:

$$G = D\eta M = \frac{P_{d,\max} 4\pi r^2}{P_{avail}} = \frac{P_{d,\max} 4\pi r^2}{P_{rad}} \cdot \frac{P_{rad}}{P_{accept}} \cdot \frac{P_{accept}}{P_{avail}} \tag{4}$$

8.2.2 Efficiency and Antenna Placement

If not properly designed, electrically small antennas may be very ineffi-cient, meaning that much of the power sent to them by a transmitter may be dissipated as heat, instead of radiated into space. (By a reciprocal argu-ment, much of the received energy of an inefficient antenna does not reach the receiver, either.) Using a Thevenin equivalent circuit, antenna effi-ciency may be characterized as

$$\eta = \frac{R_{rad}}{R_{rad} + R_{loss}} \tag{5}$$

where

R_{rad} = Radiation resistance of the antenna, Ω
R_{loss} = Loss resistance of the antenna, Ω

The radiation resistance of an antenna is the ratio of the power radiated by the antenna to the square of the rms current at the antenna feed point,[4] that is,

$$R_{rad} = \frac{P_{rad}}{I^2} \tag{6}$$

where

P_{rad} = Power radiated by the antenna, W
I = rms current at the antenna feed point, A

For example, an electrically small dipole (defined as a dipole of length much smaller than a free-space wavelength) has a radiation resistance of[5]

$$R_{rad} = 80\pi^2 \left(\frac{L}{\lambda}\right)^2 \tag{7}$$

where

L = dipole length, m
λ = free-space wavelength, m

If $L = 0.0125$ m and $\lambda = 0.125$ m (i.e., operation at 2.4 GHz), $L/\lambda = 0.1$ and $R_{rad} = 7.9$ Ohms. Electrically small antennas may have values of radiation resistance of less than an Ohm, especially at lower frequencies of operation; the same $L = 0.0125$ m dipole antenna at 400 MHz has $R_{rad} = 0.22$ Ohm.

It is clear from Equation 3 that R_{loss} should be smaller than R_{rad} for good efficiency. As the frequency of operation increases (i.e., the wavelength of operation decreases), the electrical length of an antenna of fixed physical length increases, raising its radiation resistance. This makes it increasingly easy to reduce R_{loss} to the point that reasonable antenna efficiency is attained. In addition, the input impedance of an electrically small antenna of fixed physical length has a reactive component that increases as the frequency of operation is lowered; the reactance may produce large mismatch losses if some type of impedance matching network is not used. The increasing reactance makes the design of a low-loss impedance matching network increasingly difficult as the frequency of operation is lowered. These two factors encourage the designer of wireless sensor networks to employ relatively high frequencies of operation, especially for applications requiring a physically small implementation.

Poor antenna efficiency can limit the range of a wireless sensor network node; this loss of range can be recovered only be the use of additional transmit power or improved receiver sensitivity, both alternatives that can increase power consumption significantly. It is, therefore, important that the node designer understand the factors that affect R_{loss}.

Perhaps the most obvious source of loss reflected in R_{loss} is the finite conductivity of the antenna itself — as antenna current passes through materials of finite conductivity (nonzero resistivity), some input power is lost as heat and is not radiated. This loss can be minimized by using antenna materials of high conductivity and by minimizing the current density (e.g., by using wide conductors) where possible. In addition to this loss associated with the antenna conductors, however, there is a second loss reflected in R_{loss} — the loss associated with materials near the antenna. This loss can be greatly affected by how (and where) the antenna is placed in the network node.

Conductive materials, such as circuit board traces, placed near the antenna can be a source of resistive loss due to the large currents that may be induced in them. Conductive materials may also induce loss by coupling energy into other, more lossy, circuit components; however, dielectric (nonconductive) materials placed near the antenna can also be a source of loss. In this case, loss of energy is associated with high electric charge accumulation, instead of large current flow. Use of lossy circuit board materials (e.g., FR-4) may greatly enhance this loss mechanism.

From Maxwell's equations, the electric and magnetic fields associated with an infinitesimal current element, when viewed in spherical coordinates at a distance much greater than the length of the element itself, can be shown to be[6]

$$E_r = -j\frac{Idl}{2\pi}e^{-jkr}Z_0k^2\left[\frac{j}{(kr)^2}+\frac{1}{(kr)^3}\right]\cos\theta \tag{8}$$

$$E_\theta = -j\frac{Idl}{4\pi}e^{-jkr}Z_0k^2\left[-\frac{1}{kr}+\frac{j}{(kr)^2}+\frac{1}{(kr)^3}\right]\sin\theta \tag{9}$$

$$H_\phi = j\frac{Idl}{4\pi}e^{-jkr}k^2\left[\frac{1}{kr}-\frac{j}{(kr)^2}\right]\sin\theta \tag{10}$$

where

E_r = electric field along the r coordinate, V/m
E_θ = electric field along the θ coordinate, V/m
H_ϕ = magnetic field along the ϕ coordinate, A/m
I = current, A
dl = length of current element, m
k = wave number, = $2\pi/\lambda$, m^{-1}
Z_0 = wave impedance of free space,[7] approximately 376.73 Ohms
r = radial distance between the current filament and the point of observation, >> dl, m

The infinitesimal current element is assumed to be at the origin of the spherical coordinate system. A dual set of equations results from analysis of an infinitesimal current loop placed in the same location:

$$H_r = \frac{kIS}{2\pi}e^{-jkr}k^2\left[\frac{j}{(kr)^2}+\frac{1}{(kr)^3}\right]\cos\theta \tag{11}$$

$$H_\theta = \frac{kIS}{4\pi}e^{-jkr}k^2\left[-\frac{1}{kr}+\frac{j}{(kr)^2}+\frac{1}{(kr)^3}\right]\sin\theta \tag{12}$$

$$E_\phi = Z_0 \frac{kIS}{4\pi} e^{-jkr} k^2 \left[\frac{1}{kr} - \frac{j}{(kr)^2} \right] \sin\theta \qquad (13)$$

where

H_r = magnetic field along the r coordinate, A/m
H_θ = magnetic field along the θ coordinate, A/m
E_ϕ = electric field along the ϕ coordinate, V/m
S = surface area enclosed by the loop carrying current I, = dl/k, << r/k, m^2

Both sets of equations, Equation 8 through Equation 10 associated with a dipole antenna and Equation 11 through Equation 13 associated with a loop antenna, contain terms that vary as $1/kr$. These terms can generate a propagating electromagnetic wave in the far field (i.e., as r becomes very large, so that the $1/(kr)^2$ and $1/(kr)^3$ terms become negligible).

Closer to the antenna, however, the so-called "nonradiating" fields associated with the $1/(kr)^2$ and $1/(kr)^3$ terms are not negligible; in fact, as r becomes less than $\lambda/2\pi$, these terms quickly become much larger than the "radiating" $1/kr$ term. If lossy materials are placed in these intense fields near the antenna, these materials may dissipate energy that may otherwise be radiated. This is a significant problem for antenna design in physically small products, and is, in addition to conductor losses in the antenna itself, a major cause of poor antenna efficiency. Batteries, displays, sensors, circuit boards, and other materials having significant loss at RF must be kept out of this near-field region. Note that this also includes any circuit board traces; because they are usually quite thin, circuit board traces often have high RF resistance despite being composed of low-resistivity materials (e.g., copper).

This "keep out" region can be a significant problem for the wireless sensor network node designer. Following a strict rule to eliminate all material within a radius of $\lambda/2\pi$ quickly becomes impractical; even at 2.4 GHz, $\lambda/2\pi$ is approximately 2 cm. A keep-out region extending out from the antenna a radius of 2 cm in all directions is larger than many network node designs. Applying the rule for a 400 MHz antenna produces a keep out region with a radius of 12 cm! Fortunately, this analysis is conservative; mitigating factors are working in favor of the designer. First, the offending lossy material is often, although not always, physically small relative to the antenna. Being physically small implies that the material only interacts with a small fraction of the nonradiating fields and, therefore, produces proportionally less loss. Second, the loss is not uniform throughout the keep out region; the intensity of the nonradiating fields, and the power dissipation they induce in lossy materials, increases as the material approaches the antenna. Finally, the designer may have a materials choice, and can choose a material with low RF loss.

As a general product design procedure, therefore, the wireless sensor network node designer should start by placing the largest, most lossy materials (batteries are particularly egregious) as far away from the antenna as possible, given the physical design of the network node itself. Smaller and less lossy materials should follow, approaching the antenna as slowly as possible. A useful rule of thumb for induced loss is that no surface-mounted electronic components and printed circuit board traces should be placed within a radius of $\lambda/100$ from the antenna. This rule of thumb assumes a typical density of electronic components and trades a modicum of loss for reduced package size of the final product. As with all rules of thumb, the designer should modify this rule as appropriate to best reflect the needs of the product and experience with similar designs; when keep out regions of this size are not possible, the system designer must keep the increased losses — and their associated reduction in communications range — in mind.

The placement of the antenna in the network node has several subtle but important facets resulting from the fact that, in most cases, the outside dimensions of the network node are defined at the start of its development, and the antenna must be designed with these limitations in mind. From this given physical volume, the volume of the other components in the node (batteries, sensors, actuators, displays, circuits, etc.) is subtracted, leaving a substantially fixed volume for the antenna.

Assuming matched polarization, the effective aperture A_e of an antenna is defined as

$$A_e = \frac{\lambda^2}{4\pi} G \qquad (14)$$

where

A_e = effective aperture of the antenna, m^2

The effective aperture of an antenna is a useful concept because it relates the power available at the antenna terminals to the radiation density P_d of an incoming electromagnetic wave:

$$P_{term} = P_d A_e \qquad (15)$$

where

P_{term} = power available to the receiver at the antenna terminals, W
P_d = radiation density, W/m^2

Antennas may be placed into one of two categories: constant gain antennas, in which the gain of the antenna remains constant as the wavelength λ is varied, and constant aperture antennas, in which the effective aperture of the antenna remains constant. Common dipoles and loops are examples of constant gain antennas; the parabolic antenna is an example of a constant aperture antenna.

Consider the constant gain antenna. By Equation 14, the effective aperture of constant gain antennas varies as λ^2, that is, it varies inversely with the square of the frequency of operation. By Equation 15, however, this implies that the power available for delivery to the receiver P_{term} also varies inversely with the square of the frequency of operation.

For the constant aperture antenna, the situation is just the reverse; by Equation 14, the gain of constant aperture antennas varies as λ^{-2}, that is, it varies directly with the square of the frequency of operation. The power available for delivery to the receiver P_{term} then varies directly with the square of the frequency of operation.

It is often stated that lower frequencies have lower propagation loss, and are, therefore, more suitable for wireless sensor networks. This is not true; the radiation density P_d falls as r^2 regardless of the frequency of operation. Nonetheless, it is true that typical network nodes have longer range at lower frequencies than at higher frequencies. The effect is not due to propagation effects, however, but to the choice of antennas used in their design. The usual antenna for wireless sensor network nodes is a constant gain design, often a loop or dipole. As the frequency of operation increases, their effective area decreases, lowering P_{term} and producing the observed range reduction at higher frequencies. One may reverse the effect, producing greater range at higher frequencies, by the simple expedient of replacing the constant gain antenna with a constant aperture type. Alternatively, one may produce frequency-independent range by making either the transmit or receive antenna a constant gain type, and the other a constant aperture type.

When using constant aperture antennas, it is important to recall that one does not receive something for nothing; specifically, the increasing gain at high frequencies comes from the increasing directivity of the antenna, by Equation 4. Antenna directivity can be very useful for fixed links that require extreme range; however, excessive antenna directivity can become a problem in ad hoc networks. If multiple network nodes must be contacted, the antenna may have to be redirected to establish communication; if the antenna is pointed in one direction, communication links in an orthogonal direction may be broken. Even if there is only a single communication link to support, maintaining sufficient directional accuracy can become a burden for a network that is supposed to be self-maintaining.

The preceding range discussion omitted any discussion of efficiency. Real antennas, of course, are not perfectly efficient; in fact, when a fixed limit is placed on physical size, a lower frequency antenna is less efficient than a higher-frequency antenna, due to the higher circulating currents in the lower frequency antenna.

In practical designs, this effect is often exacerbated by the larger keep out regions required by lower frequency antennas. The radius of the keep out region is proportional to the wavelength employed; at lower frequencies, compromises must often be made due to the physical size of the network node. These compromises, in the form of lossy materials placed in regions of relatively large nonradiating fields, reduce the efficiency of antennas for lower frequency bands. Viewed another way, for a fixed physical volume inside the network node used for the antenna, lossy materials will be proportionally closer for a low-frequency antenna.

A third facet of antenna placement is that the other components of the network node are not necessarily benign. Often they can be sources of significant amounts of RF noise; this noise may limit the attainable receiver sensitivity. Paradoxically, under these conditions, enlarging the antenna may actually reduce, instead of increase, receiver sensitivity, by placing the antenna closer to a noise source. Methods for understanding this effect, and designing small products with this in mind, are covered in the next chapter.

8.2.3 Bandwidth

There is a relationship between the physical size, efficiency, and maximum instantaneous fractional bandwidth of electrically small antennas (i.e., those with maximum physical dimensions significantly less than their operating wavelength λ).[8] Specifically,

$$BW = \frac{2\left(\dfrac{2\pi d}{\lambda}\right)^3}{\eta} \qquad (16)$$

where

BW = maximum instantaneous fractional bandwidth (i.e., bandwidth as a fraction of center frequency)

$2d$ = maximum dimension of the electrically small antenna (i.e., the diameter of the smallest sphere that can completely contain the antenna).

As the efficiency of the electrically small antenna is increased, its maximum instantaneous bandwidth is reduced. This effect is important because it may lead to the requirement for antenna tuning.

Antenna tuning can be required for two reasons. The first is the need to operate efficiently on multiple frequencies (i.e., to be frequency agile). This "dynamic" need can arise in bands that are relatively wide, relative to their center frequency; an electrically small, fixed-tuned antenna with sufficient bandwidth to cover the entire band may have undesirably low efficiency. The solution to this problem is to tune the antenna in synchrony with the operating frequency; as the transceiver moves across the band, the antenna is retuned accordingly. This enables a relatively efficient antenna to be used over a large fractional bandwidth, overcoming the limitation of Equation 16. This is usually accomplished with varactor tuning with a control voltage, although at low frequencies, permeability tuning of an inductor with a control current is also possible.

The second reason for antenna tuning is the need to overcome manufacturing variations. If the fractional bandwidth of the antenna is low, it may be economically impractical to manufacture it with the precision necessary to place the antenna on the correct frequency of operation, and overcome normal production variations in component tolerances, dielectric constants of materials, component placement, etc. If the antenna can be tuned near the end of the production line, after the housing and other sources of variation are in place, overall production yields may significantly increase while reducing component cost (because high precision parts and manufacturing processes are not required). In this scenario, the fractional antenna bandwidth is sufficient for operation across the band without retuning, but its bandwidth is insufficient to compensate for the additional variation due to manufacturing variations. The antenna is tuned once during production, then its tuning is fixed. It is, of course, possible to use antenna tuning for both dynamic tuning and manufacturing yield reasons.

Antenna tuning is useful, but like all solutions to engineering problems, it is not without its drawbacks. Tunable elements, whether capacitors or inductors, are more expensive than their fixed counterparts. Dynamically tuned systems usually require additional components, including RF chokes and bypasses and, perhaps more importantly, a source of the control signal (usually a voltage) must be designed into the system. For receive-only operation the control signal often can be developed from the synthesizer control voltage; however, for transmit operation, a more sophisticated strategy is usually necessary, due to the possibility of transmitted energy traveling through the control circuitry and corrupting the synthesizer. The cost of this design effort, of course, must be amortized in the product cost. Tuning at production can often be done with a single tunable element (often either a manually or laser-trimmed capacitor), but tuning in production has its own costs, too, in the form of slower production and the need to develop a tuning algorithm. For automated production, which is a requirement for successful wireless sensor network markets, an automated trimming and testing station must be purchased.

Tunable antenna elements often have lower Q (i.e., greater loss) than their nontunable counterparts. If not designed with care, a tunable antenna may have sufficient loss with the addition of a tunable element that its bandwidth is increased to the point that tuning is not necessary! An inspection of Equation 16 reveals that the instantaneous bandwidth limit is proportional to the *cube* of the maximum antenna dimension expressed in wavelengths; when designing a product, one should always consider the alternative of a larger, fixed-tuned antenna, instead of adding tuning components to a smaller one.

8.2.4 Antenna Design Choices

The type of antenna employed in a wireless sensor network node depends greatly on its application. Usually an internal antenna is preferred, due to its inherent portability and protection from mechanical and environmental damage. Several types of internal antennas are available.

The first option is often the use of a circuit board trace as an antenna. This option has the advantage of low cost because no components must be purchased, handled, or placed on the board; however, because circuit board area is not free, its cost is not zero. It is also very thin, an advantage for low-profile network node designs. The most significant disadvantage to this type of antenna is its poor performance. Due to the thinness of the circuit board trace, its series resistance can be relatively high despite the low resistivity of copper; the low quality of the circuit board material often adds dielectric loss. Being in the circuit board itself also enables the circuit board trace antenna to more readily couple to lossy components and other circuit board traces, and to sources of noise. Finally, the tuning of circuit board antenna can be subject to significant variation caused by etching variations during the circuit board manufacturing process. This variation is often difficult to control without significant expense because the required etching accuracy is typically greater than that needed for simple connectivity of the circuit board.

A second alternative is the use of simple wire or metal strip antennas placed as components on the circuit board. These antennas can have remarkably improved performance over that of the circuit board trace antenna, due to their significantly lower loss and the fact that they are above the circuit board. (Being above the circuit board, however, they are susceptible to better coupling to some noise sources, such as the inductors in switching power supplies.) Wire antennas can be either dipoles or loops; nodes designed for on-body use, such as health monitoring devices, often employ loops as so-called H-field antennas (i.e., magnetic field probes) in a plane normal to the body. Wire antennas often require dielectric (e.g., plastic) supports to maintain their mechanical shape and, therefore, their frequency of resonance, to the required tolerance. They are often difficult to robotically place in an automated assembly line, and must be inserted and soldered by hand. Nevertheless, wire antennas are often a good compromise between the low cost and efficiency of the circuit board antenna and the relatively high cost and high efficiency of the external antenna.

A third choice is the use of specialized ceramic antenna components. These components are simple to place by automatic equipment, are physically smaller than wire antennas, and do not require tuning. They are usually more expensive than wire antennas, however, and are often available only for the most widely used frequency bands (e.g., the 915- and 2450-MHz Industrial, Scientific, and Medical [ISM] bands).

Because they usually are free of the size limitation placed on internal antennas and are removed from the noise sources present in the network node itself, external antennas can have very high performance. In applications where the utmost in range is desired and a directive antenna must be used, an external antenna is almost certainly required. Designers of nodes for use in white goods (refrigerators and other large household appliances) may prefer external antennas to avoid the shielding of their metallic housings. In applications in which several different antennas must be used, for example, to meet differing range requirements of multiple applications, external antennas are the obvious choice. Although external antennas offer the highest performance and design flexibility, they usually are the most expensive; not only must the antenna be purchased, but usually one or more high-quality RF connectors, as well. Losses associated with any feedline (usually miniature coaxial cable) between the network node and the antenna itself must be included in the network node design. A hidden cost not always considered is associated with the frequency selectivity of the antenna. Because it is under the control of the node designer, the selectivity of an internal antenna may be used as part of the transceiver system design (e.g., Zolfaghari and Razavi[9]). Because users may replace external antennas with those of unknown selectivity, when using an external antenna, the node designer may often be required to include additional RF selectivity in the node design, at additional product cost.

External antennas need not always be expensive, however; for example, Ross et al.[10] describes a covert wireless sensor network node (used in security applications) that uses the doorknob housing the network node as its antenna.

8.3 RF PERFORMANCE DEFINITION AND MEASUREMENT

8.3.1 Definition and Measurement

One of the more fundamental decisions to be made concerning the definition of receiver sensitivity and other RF performance parameters is the point in the system at which they are to be defined; specifically, on which side of the antenna this point should be. A "conducted" definition of RF performance, made at a point between the antenna and transceiver, may be made in terms of power (e.g., dBm); a "radiated" definition, made on the other side of the antenna (i.e., in free space) in terms of power alone is not meaningful. Placing the measurement point in space requires the definition of RF performance parameters in terms of electric field strength (e.g., dBμV/m) or, less conveniently, magnetic field strength (e.g., dBμA/m), at a specified distance. It also includes antenna performance in the measured results. The antenna performance can be significant, not only for "in band" RF parameters like receiver sensitivity and transmitter output, but also for out of band parameters like spurious responses and transmitted noise, because, in many of these low-cost designs, the selectivity of the antenna is used to meet system requirements.

A consideration in this decision is the relevance of the test itself to the parties involved. Chip manufacturers and their customers usually desire a conducted definition because the manufacturer usually has little control over the antenna used by their customers, and their customers usually want to decouple the performance of the antenna, which they are not buying from the chip vendor, from the performance of the chip itself, which they are. Original equipment manufacturer (OEM) network node manufacturers and their customers, however, may often prefer a radiated definition, because it includes the antenna performance and is, therefore, more representative of the node's performance in the field. Antenna performance is one way a network node manufacturer may distinguish its product from that of others using the same integrated circuits.

This decision is also influenced by the type of antenna used. If a removable external antenna is employed, an RF signal generator may be easily coupled into the receiver by removing the antenna and using the antenna connector, making the definition and measurement of RF parameters at this point straightforward and relatively simple. If an internal antenna is employed, however, some type of external connector must be added to perform measurements at this point, plus a switch to connect the transceiver to the connector for testing. This adds cost to what is usually a very cost-sensitive product, for a capability (performance testing) usually not performed by the end user. If the performance is specified to include the antenna, however, simple conducted RF measurements are not possible; instead, the network node must be placed on an antenna range, in an anechoic chamber, or, most practically, inside a transverse electromagnetic (TEM) cell, such as a Crawford cell[11] or gigahertz transverse electromagnetic (GTEM) cell,[12] to perform RF measurements. This can complicate product testing and development.

Often, both definitions are used, the conducted definition by the chip manufacturer and the radiated definition by the OEM network node manufacturer. This is possibly the most pragmatic approach, although it must always be kept in mind which definition is being used.

A second fundamental problem concerns the method by which the received data is extracted from the receiving node under test and, in this regard, the definition and measurement of receiver sensitivity of a wireless sensor network node is significantly different from that of a more conventional data receiver, such as one for Wireless Local Area Networks (WLAN) service. Unlike a WLAN device, a wireless sensor network node often has no "data output" port — its received data is used to perform functions, such as reading a sensor or operating an actuator. It, therefore, can be quite difficult to obtain the received data to compare it with the transmitted data and detect bit errors. Although several approaches can be used to get around this "data egress" problem, none is entirely satisfactory:

- A loop-back technique may be employed, wherein the received data is retransmitted back to the test system over the radio frequency (RF) transmitter of the network node. This is probably the most common technique, although it is not without its problems. The return RF link is subject to interference, especially in a factory in full production, where there may be many devices undergoing tests simultaneously. Expensive shielded boxes are often needed. Further, the data can usually be sent out only as fast as it can be sent in; this effectively doubles the time under test of the device. Finally, the network node must have temporary storage available to buffer the received data prior to retransmission. An advantage of this method is that the performance of the transmitter may be checked simultaneously.

- The data can be sent to a special output port of the network node. This requires some type of dedicated output, such as an infrared or acoustic link, or a physical connection, all of which add cost to the node. Further, if the output port is a physical connection to the node, the RF performance of the node (and, therefore, its measured sensitivity) may be affected, especially if the node employs an internal antenna. The output port may produce RF interference (electromagnetic compatibility (EMC) problems), which may affect measured sensitivity unless the received data is buffered during reception, and transmitted only out the port between received packets. In addition, if a physical connection is used, the method used to ensure proper connection and disconnection must be well designed — especially if the test system is automated.

- The external transducer interface, or some other existing network node output, may be used. Concerns here are the same physical connection and EMC problems of the dedicated output port approach, plus the fact that the factory must be designed for the test of nodes with specific node outputs. If a new node, perhaps one without an external transducer interface, is designed, a major testing program redesign is required before it can be placed into production.

- Finally, the comparison can be performed within the device itself, by prestoring the test data stream in memory. In this approach, the device under test compares the received data with that stored, and then signals the result back to the test station (via any of the communication methods outlined previously). The advantage is that much less information need be sent over the return link — just the number of bit errors detected, for example. This can greatly simplify the test procedure, and make otherwise impractical communication links, such as a node actuating a light, feasible. The disadvantages of this approach, however, are that the test data to be sent can be large (10^5 bits) and, therefore, expensive to store, due to the required memory. Further, the system is inflexible; a new collection of data must be stored if the test data is to be changed. In addition, no information is provided concerning where in the data the bit errors occurred, unless additional information is returned, at the cost of additional time and complexity. Without this information, troubleshooting and the correction of systematic measurement and testing errors can be difficult.

A second problem in the use of bit error rate for receiver sensitivity measurements is the small payload available in the data frames of most wireless sensor network communication protocols. In a conventional data receiver, large amounts of data are typically received (up to 1500 bytes per packet in TCP/IP service, for example). In the wireless sensor network receiver, however, due to the emphasis placed on the minimization of complexity and cost, the maximum length frame is often quite short. The Institute of Electrical and Electronics Engineers (IEEE) 802.15.4 Low-Rate WPAN standard, for example, has a maximum Media Access Control (MAC) frame (message) length, including the overhead associated with higher communication protocol layers, of only 127 octets (8-bit bytes). Of this maximum length frame, at most, 114 bytes are MAC payload data.

The conventional definition of receiver sensitivity for wireless data networks is the minimum input power level required to achieve a given average received bit error rate (BER), usually BER = 10^{-2}, 10^{-3}, or 10^{-5}. Assuming an additive white Gaussian noise (AWGN) channel, however, the production of bit errors is a stochastic process; to experimentally determine the average received BER a very large number of bits must be transmitted (and received). A useful rule of thumb is that one hundred bit errors must be received to produce a repeatable result; if the average received BER = 10^{-3}, 10^5 bits (12,500 bytes) must be transmitted to the receiver during the test, or 110 maximum-length IEEE 802.15.4 frames. If the average received BER = 10^{-5}, 10^7 bits (1.25×10^6 bytes) must be transmitted to the receiver during the test, or 10,965 maximum-length IEEE 802.15.4 frames.[13]

This large number of frames makes this definition difficult to put into practice. Sending 110 maximum-length IEEE 802.15.4 frames, for example, takes a minimum of 1.6896 seconds — quite a long time in a high-speed production line. Although this issue could perhaps be overcome by employing parallel test stations (at additional capital expense), other issues associated with the large number of required frames are more difficult to overcome.

For example, the definition overlooks frame and symbol synchronization. In most communication protocols, at the beginning of every frame is a special series of symbols sent to enable the receiver to identify the beginning and end of received symbols, and then to identify the start of the received frame. This process is necessary prior to the successful reception of any data; if either of these steps fails, the entire frame will be lost. (Symbol synchronization may, or may not, be updated throughout the frame, depending upon the implementation; if not, there is a greater likelihood of bit errors near the end of the frame due to symbol synchronization drift.) One may, therefore, consider two types of sensitivity — the signal level necessary for symbol and frame synchronization to occur at the start of a frame, and the signal level necessary for a certain average received BER of payload data once synchronization has been achieved. A well-designed communication protocol attempts to ensure that symbol and frame synchronization may be reliably performed at signal levels below that at which reliable data reception may occur, and a well-designed implementation of that protocol attempts to ensure that that is in fact the case. Nevertheless, when receiver sensitivity is defined by BER, one is interested only in reliable data reception; it is, therefore, important to consider the effect of possible synchronization problems because they can significantly affect the BER measurement.

If, as is often the case in wireless sensor networks, the average message is very short (perhaps one payload byte, or 10 to 15 bytes in total), the number of synchronization attempts per number of received bits will be much greater than if maximum length frames are used. Therefore, there will be more opportunities available for a synchronization failure. The use of maximum length frames in a BER-derived sensitivity definition can, therefore, underestimate the minimum signal level required for reception of shorter frames; significantly different experimental values for average received BER may be obtained simply by varying the length of the messages carrying the test data.

A similar situation can occur due to a failure of destination address correlation. Should an address correlation fail, an entire frame will be lost; however, transmitted addresses usually are not as well protected from error as symbol and frame synchronization information, although the effect of an error (loss of the frame) is the same.

A second issue associated with the use of BER as the basis for a receiver sensitivity definition is that most protocols (including IEEE 802.15.4) require some type of error detection in the frame, usually in the form of a cyclic redundancy check (CRC). Frames containing any errors at all will be dropped in the lower protocol layers under normal operation; to perform a received BER test, this feature must be inhibited. This requires the placement of the network node in some type of "test mode," which, although commonly done, requires care to ensure that the behavior of the node in that mode is representative of the behavior of the node in normal operation. Subtle EMC effects and software bugs, for example, can produce unexpected behavior; one must be especially wary of the effects resulting from the bypassing of the higher protocol layer functions. These functions may be performed by a separate processor than that performing the lower-layer processing; if the higher-layer functions are not performed during the sensitivity test, one may be unaware of interference generated by the higher-layer processor when the network node is placed into normal operating mode. The result can be an optimistic estimation of the receiver sensitivity.

One way to get around both the synchronization and CRC problems is to employ a specialized average received BER test mode, in which the synchronization header and destination address are sent only once, followed by a substantially endless stream of unformatted, random symbols, encoded and modulated as in the wireless sensor network protocol. In this scheme, the troublesome synchronization process is reduced to the minimum; because no message length limitations exist, the number of accumulated bit errors needed to ensure an accurate average received BER measurement may be obtained in the minimum amount of time. This solution requires a separate test of the device to ensure that it accepts and interprets formatted messages properly; however, this test may be performed implicitly elsewhere in the manufacturing process and may consist of very short and, therefore, very fast messages. Note that this solution to the synchronization problem does not overcome other receiver sensitivity problems, such as the data egress problem; it also is subject to EMC and software bug concerns because the message receiving routines the network node would use in normal operation are not being employed.

These difficulties may be overcome by returning to the fundamental purpose of the wireless sensor network node, and observing that its purpose is not, in fact, to transmit and receive data. Instead, its purpose is to transmit and receive messages, most of which are largely composed of header information and actually contain very little payload data; most of the problems just described defining receiver sensitivity result from trying to send enough payload data through the network node to get statistically significant results for an average received BER = 10^{-3} and below. Although it is clear that if the average received BER is known, the probability of receiving a frame in error may be determined, reversing the process and determining the average received BER from received frames is more difficult, for the reasons described previously. A similarity may be observed between the requirements of wireless sensor networks and those of radio paging, which are also concerned with the transmission of messages with relatively little payload data. It is instructional, therefore, to understand how the sensitivity of selective call receivers (i.e., paging receivers) is defined.

The fundamental unit of performance in paging is the "80 percent calling success rate," also known as the "80 percent message acceptance ratio" in the European Telecommunication Standards Institute (ETSI) ETS 300-133-5[14] and TBR7[15] standards and the "80 percent calling probability" in the International Electrotechnical Commission (IEC) 60489-6 standard.[16] It is the signal level required to produce an 80 percent probability that a transmitted message will be received correctly. This definition was developed to determine the signal level that produces a 50 percent probability that the receiving device would respond correctly to three calls in a row.[17] Specifically, if $p(call)$ is the probability of correct reception of a call, $(p(call))^3$ is the probability that three calls in a row will be received correctly. If $(p(call))^3 = 0.5$, $p(call) = (0.5)^{1/3} \approx 0.7937 \approx 0.80$. The calls are assumed asynchronous, that is, symbol and frame synchronization is to be obtained independently for each transmitted message. The transmitted message should be standardized, but one has the freedom to standardize it to a length common to the application at hand. One is, therefore, reassured that the network node under test is being tested in a manner similar to its normal operation.

Exhibit 1 is a curve of $p(call)$ versus input signal level for a typical selective call receiver. For this node, $p(call) = 0.80$ occurs at an input signal level of −90 dBm. Note, however, that because the curve is asymptotic in both directions, the device may occasionally respond to a message sent at signal levels below −94 dBm, and may miss a message sent at signal levels above −88 dBm. During this particular test run, however, the lowest signal level at which a message was received was −93.5 dBm, and no messages were missed above −88.5 dBm. Because this sensitivity curve describes the behavior of a random variable, $p(call)$, a relatively large number of mes-

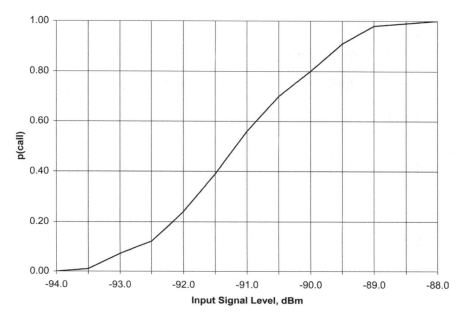

Exhibit 1.   A Selective Call Receiver Sensitivity Curve

sages must be sent at each input signal level to produce a reliable curve. One of the pleasing features of the sensitivity curve, however, is that it becomes steeper as the length of the test message is reduced; the use of shorter test messages leads to reduced measurement variation and an improved estimation of the signal level required for a given $p(call)$. Increasing the steepness of the sensitivity curve also increases the steepness of the risk curve associated with this device (Section 8.3.2).

Although the generation of complete sensitivity curves is seldom necessary in a production environment, it is often necessary to determine the required input signal level for a given $p(call)$. A method to determine the required input signal level for $p(call) = 0.80$ is the so-called "20-page" method[18] (actually a misnomer), variations of which are defined in both the ETSI ETS 300-133-5 and IEC 60489-6 standards.[19] The procedures in the ETSI ETS 300-133-5 and IEC 60489-6 standards, although similar, are not identical; the following procedure is taken from ETSI ETS 300-133-5.

The 20-page method to determine $p(call) = 0.80$ is as follows:

- A very low input signal level, such that the device correctly responds to less than 10 percent of the transmitted test messages, is first used.
- Test messages are repeatedly sent to the device under test, while observing in each case whether or not the device responds correctly.

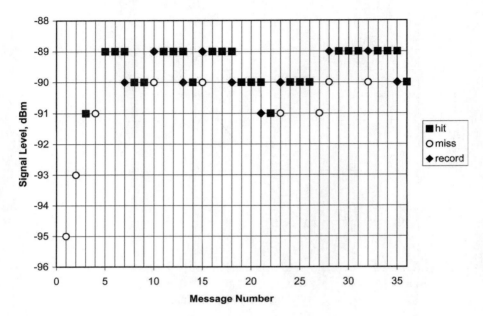

Exhibit 2. The 20-Page Algorithm

The input signal level is increased 2 dB for each occasion in which a correct response is not obtained. This process continues until three consecutive correct responses are obtained.

- Once three consecutive correct responses are obtained, the input signal level is reduced 1 dB, and this new value is recorded. Test messages are then repeatedly sent. For each message to which the device responds incorrectly, the input signal level is increased 1 dB, and this new value recorded. If three consecutive correct responses are obtained, the input signal is decreased 1 dB, and this new value recorded. The process continues until ten signal level values have been recorded.

- The measured value of $p(call)$ is the arithmetic mean of the ten recorded signal level values.

A typical 20-page test is plotted in Exhibit 2. The recorded values, in order, are −90, −89, −90, −89, −90, −91, −90, −89, −89, and −90 dBm, the arithmetic average of which is −89.7 dBm. The input signal level required for $p(call)$ = 0.80 for this device is, therefore, determined to be −89.7 dBm.

Note that, although 36 messages were required in this instance to determine $p(call)$, the messages were not maximum-length messages. The messages used are those typically used when the device is in service. The algorithm itself is quite simple, and easily automated; correct message reception may be detected by either an actuator operation or a transmitted reply.

The 20-page method can be modeled as a Markov chain. Any change in signal level depends only on the immediately preceding transmitted message.[20] Because one can arrive at it only from either missing three consecutive calls at signal level n, or by missing any one of three calls at signal level n + 2, the probability of being at signal level n + 1 dB is

$$P_{n+1} = P_n \cdot \left[1 - \left(P_n(call) \right)^3 \right] + P_{n+2} \cdot \left(P_{n+2}(call) \right)^3 \qquad (17)$$

where

P_n, P_{n+1}, and P_{n+2} are the probabilities of being at the input signal levels n, n + 1, and n + 2 dB, respectively.

$P_n(call)$ and $P_{n+2}(call)$ are the probabilities of successfully receiving a single message, out of one transmitted, at input signal levels n and n + 2 dB, respectively.

These values may be obtained from the device sensitivity curve (e.g., Exhibit 1).

Equation 17, when taken with a device sensitivity curve and the fact that the sum of all signal level probabilities $P_i = 1$, enables one to predict the repeatability of the 20-page method. The analysis presented in Annex E of IEC 60489-6 places the 95 percent confidence limits of the 20-page method at ± 2 dB of the true value.

The 20-page method may be used in the measurement of receiver parameters other than sensitivity. For example, a blocking test may be performed by combining an undesired (interfering) signal at a fixed level, with a desired signal, the level of which may be varied. A 20-page procedure is then performed on the desired signal, resulting in the desired signal level needed for $p(call) = 0.80$ in the presence of the undesired signal. The blocking performance is then calculated to be the difference between the level of this desired signal and that of the undesired signal. Alternatively, one may define the blocking test to include a desired signal at a fixed level (usually the value needed for $p(call) = 0.80$ without the presence of the undesired signal, plus three dB), and an undesired signal of variable level. In this method, one then performs the 20-page procedure with the undesired signal, with the distinction that the directions of signal level change are reversed (i.e., one starts with a very strong interfering signal, reducing its level until three consecutive messages are received, etc.). The blocking value is then calculated to be the difference between the undesired signal level determined by the 20-page method and the desired signal level.

Either one of these approaches is technically correct; the latter approach is used in the ETSI TBR 7 and IEC 60489-6 standards, although some Japanese specifications call for the former approach. The former approach has the advantage that the blocking performance in the presence of interferers of a specific level can be determined; because blocking behavior is typically a very nonlinear process, it is often difficult to predict performance under one undesired signal level from the performance obtained at another. Intermodulation, of course, is another very nonlinear process that also behaves in this manner. For these nonlinear processes, performing both types of tests, including a selection of very high level fixed undesired signals, can often provide a more complete view of the receiver's behavior than can either type of test alone.

At the other extreme, adjacent channel rejection, when dominated by filter selectivity (as opposed to reciprocal mixing or other nonlinear effects), is a relatively linear process. Once nonlinear effects are ruled out (usually by experience with a given design), the adjacent channel rejection behavior of a receiver is usually sufficiently described by the ETSI TBR 7 and IEC 60489-6 test methods alone.

8.3.2 Production Issues

One may distinguish between two types of sensitivity testing: a sensitivity measurement, in which one attempts to determine the signal level required to produce a desired value of $p(call)$, and a sensitivity compliance test, in which one simply attempts to determine the sensitivity of the device under test relative to a given signal level threshold (i.e., a specification). Because the output of a sensitivity measurement is an estimate of required signal level, it usually takes longer to complete than the compliance test.[21] Fortunately, in most production applications, the compliance test is all that is required.

In a compliance test, one must consider the risk curve, a curve describing the probability of passing the test, as a function of the true value of the measured parameter relative to the test threshold. One desires the risk curve to be a step function occurring at the threshold value; however, due to the dispersion present in all experimental data, some devices with parameters exceeding the threshold will fail, and some devices with parameters below the threshold will pass. The probability of passing a device the performance of which is below the threshold value is termed a type I error; the probability of failing a device the performance of which is above the threshold value is termed a type II error. Depending upon the penalty resulting from each of these types of errors, a factory test program may be designed with a bias against one or the other type of error; it may also be unbiased, so that $p(type\ I\ error) = p(type\ II\ error) = 0.5$ at the threshold specification. The goal in a factory test setting is to have the risk function approach the step function as closely as possible without unduly extending the time needed to conduct the test. One then must establish goals for what percentage of false positives, and what percentage of false negatives, are acceptable, and at what distance away from the threshold value.[22] An example of how these factors affect factory testing is perhaps instructional.

Similar to the BER-based definition of sensitivity, the message error rate (MER)-based definition is statistically based, and the binomial distribution applies. At a given signal level,

$$P_{test}(call) = {}_nC_k \left(p(call)\right)^k \left(1-p(call)\right)^{n-k} \tag{18}$$

where

$P_{test}(call)$ = probability of successfully receiving k messages, of n transmitted

${}_nC_k$ = $n!/[(n-k)!k!]$, the binomial coefficient associated with n things taken k at a time

$p(call)$ = probability of successfully receiving one message, of one transmitted, at the given signal level

For example, suppose the device of Exhibit 1 is being manufactured. It has $p(call) = 0.80$ at an input signal level of −90 dBm. Suppose further that the factory test consists of sending ten messages to the device under test, and passing any device that responds correctly to at least eight of the ten messages. The probability of receiving at least eight out of ten transmitted messages (that is, receiving either eight, nine, or ten of the ten transmitted messages) at that signal level is

$P_{test}(call)$ = p(receiving 10 of 10) + p(receiving 9 of 10) + p(receiving 8 of 10)

= ${}_{10}C_{10}(p(call))^{10}(1 - p(call))^{10-10} + {}_{10}C_9(p(call))^9(1 - p(call))^{10-9} + {}_{10}C_8(p(call))^8(1 - p(call))^{10-8}$

= $(10!/[(10-10)!10!])(p(call))^{10}(1 - p(call))^0 + (10!/[(10-9)!9!])(p(call))^9(1 - p(call))^1 + (10!/[(10-8)!8!])(p(call))^8(1 - p(call))^2$

= $(p(call))^{10} + 10(p(call))^9(1 - p(call)) + 45(p(call))^8(1 - p(call))^2$

= $(0.8)^{10} + 10(0.8)^9(1 - 0.8) + 45(0.8)^8(1 - 0.8)^2$

≈ $0.107 + 10(0.134)(0.2) + 45(0.168)(0.04)$

≈ $0.107 + 0.268 + 0.302$

≈ 0.68

Therefore, if the factory test were to send ten messages to the device at −90 dBm, and only pass those that received eight or more of them, $(1 - 0.68) \times 100 = 32$ percent of the devices would fail the test, even though, by

Exhibit 3.   Example Risk Curve Values

Signal level, dBm	$p(call)$	$p_{test}(call)$
−94.0	0.00	0.00
−93.5	0.01	4.4×10^{-15}
−93.0	0.07	2.3×10^{-8}
−92.5	0.12	1.5×10^{-6}
−92.0	0.24	3.1×10^{-4}
−91.5	0.39	0.01
−91.0	0.56	0.11
−90.5	0.70	0.38
−90.0	0.80	0.68
−89.5	0.91	0.95
−89.0	0.98	0.9991
−88.5	0.99	0.9999
−88.0	1.00	1.00

assumption, $p(call)$ = 0.80 at that signal level. This is, of course, a very large number of type II errors for a production facility.

To reduce the number of type II errors to an economically acceptable level, the production test can employ a somewhat higher signal level. If the input signal level is increased 1 dB (i.e., from −90 to −89 dBm), from Exhibit 1, $p(call)$ = 0.98. The calculation then becomes

$p_{test}(call)$ = p(receiving 10 of 10) + p(receiving 9 of 10) + p(receiving 8 of 10)

= $(p(call))^{10} + 10(p(call))^9(1 - p(call)) + 45(p(call))^8(1 - p(call))^2$

= $(0.98)^{10} + 10(0.98)^9(1 - 0.98) + 45(0.98)^8(1 - 0.98)^2$

≈ 0.81707 + 0.16675 + 0.01531

≈ 0.9991

The probability of making a type II error is now less than 0.1 percent. A similar analysis for the other values of $p(call)$ found in Exhibit 1 results in the values given in Exhibit 3. These values form the 8-of-10 test risk curve for this product plotted in Exhibit 4, along with the values of $p(call)$ for comparison.

Note that the 8-of-10 test risk curve rises more steeply than the sensitivity curve, moving from $p_{test}(call)$ = 0.05 to $p_{test}(call)$ = 0.95 in less than 2 dB of signal level change. The steepness of this curve will increase with the number of test messages sent to the device, approaching a step function as n increases; it is up to the manufacturer to determine how many messages in the factory test are economically justifiable.

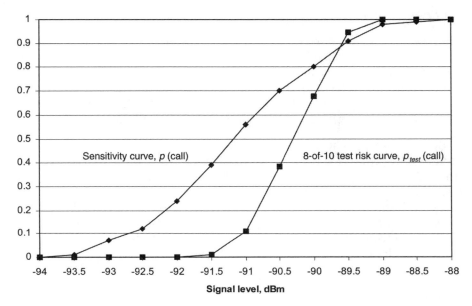

Exhibit 4. Risk and Sensitivity Curves

Note that the identical risk curve is produced if the values of the sensitivity curve are interpreted not as the probability of calling of a single device over varying signal levels, but the probability of calling of many devices at a fixed signal level (due to manufacturing variations between devices). This interpretation is often more useful to the manufacturing engineer, because it enables the risk curve to be viewed as the yield curve of the factory test.

The advantage of using a MER-based definition of receiver sensitivity for wireless sensor network nodes, instead of a BER-based definition, comes from the fact that the MER-based method tests the node in its typical operating mode. That is, a MER-based test can be performed on a node connected to its sensor or actuator, and the test will exercise substantially all of the signal and control paths the node will use in the field. By testing complete system functionality, the MER-based test also tests for EMC-related sensitivity problems occurring due to the behavior of host microcomputers, actuators, and user interface devices. Because the MER-based test does not require specialized output devices and the correct reception of messages can be communicated quickly to the test station, it can be a lower cost test method than BER-based approaches.

8.4 CONCLUSION

The design of wireless sensor network node antennas is challenging due to the need to achieve low product cost and reasonable efficiency simultaneously in a physically small package suitable for volume production. It is important not to consider the antenna as a component in isolation, but instead as a component in its operating environment; the designer should be cognizant of the effect nearby materials have on antenna efficiency.

Because wireless sensor network nodes often do not produce received data as an output, nor have an input port suitable to present data to be transmitted, the measurement of basic RF performance parameters is difficult if the common definitions for data communication are used because bit errors cannot be detected. A preferred alternative is the use of message error rate, instead of bit error rate, in the parameter definition, because the network node may be kept in a normal mode of operation when under test. Standards already exist for such definitions in the field of radio paging.

References

1. Kazimierz Siwiak, *Radiowave Propagation and Antennas for Personal Communications,* 2nd ed. Norwood, MA: Artech House. 1998. p. 19.
2. As a (usually) small-signal system, for receiving antennas the output impedance of the antenna should be a conjugate match for the input impedance of the receiver for maximum power transfer; however, a mismatch is often desirable, to attain an improved system noise figure and improved system sensitivity. (See Guillermo Gonzalez, *Microwave Transistor Amplifiers: Analysis and Design.* 2nd ed. Upper Saddle River, NJ: Prentice Hall. 1997. pp. 294–322.) For transmitters, which are (usually) large-signal systems, an impedance match by the transmitting antenna to the output impedance of the power amplifier is usually not desired. Instead, an antenna input impedance that produces the desired amount of power transferred from amplifier to antenna (i.e., P_{accept}) is sought, which can result in a significant mismatch. An exception that can occur in wireless sensor network design is the use of low-transmit power (e.g., -6 dBm or less), for which even the transmitter power amplifier can be considered to be in small-signal operation.
3. Ibid., p.193.
4. John D. Kraus, *Antennas.* New York: McGraw-Hill. 1950, p. 136.
5. Ibid., pp. 136–137.
6. Siwiak, pp. 13–15.
7. $Z_0 = c\mu_0$, where c is the speed of light in vacuum (299 792 458 m/s) and μ_0 is the permeability of free space ($4\pi \times 10^{-7}$ H/m).
8. Siwiak, p. 332.
9. Alireza Zolfaghari and Behzad Razavi, "A low-power 2.4-GHz transmitter/receiver CMOS IC," *IEEE J. Solid-State Circuits,* v. 38, n. 2, February 2003, pp. 176–183.
10. Gerald F. Ross et al., *Batteryless Sensor for Intrusion Detection and Assessment of Threats.* Defense Nuclear Agency Technical Report DNA-TR-95-21. Springfield, VA: National Technical Information Service. 1 November 1995.
11. M. L. Crawford, "Generation of standard EM fields using TEM transmission cells," *IEEE Trans. EMC,* v. EMC-16, n. 4, November 1974, pp. 189–195.

12. H. Garbe and D. Hansen, "The GTEM cell concept; applications of this new EMC test environment to radiated emission and susceptibility measurements," *Seventh Int. Conf. EMC,* 1990, pp. 152–156.

13. Only payload bits are included in these calculations. Frame synchronization, destination address, and other overhead bits are not included, for reasons that will become clear in subsequent paragraphs.

14. European Telecommunication Standards Institute, Electromagnetic Compatibility and Radio Spectrum Matters (ERM); Enhanced Radio MEssage System (ERMES); Part 5: Receiver Conformance Specification. 2nd ed. Document ETSI ETS 300-133-5. Sophia-Antipolis, France: European Telecommunication Standards Institute. November 1997. Annex A.1. Hereinafter, ETSI ETS 300-133-5.

15. European Telecommunication Standards Institute, Electromagnetic Compatibility and Radio Spectrum Matters (ERM); Enhanced Radio MEssage System (ERMES); Receiver Requirements. 2nd ed. Document TBR 7. Sophia-Antipolis, France: European Telecommunication Standards Institute. December 1997. Annex A.1. Hereinafter, ETSI TBR 7.

16. International Electrotechnical Commission, IEC 60489-6: Radio Equipment Used in Mobile Services – Methods of Measurement – Part 6: Data Equipment. 3rd Edition. Geneva: International Electrotechnical Commission (IEC). July 1999. §§ 4.1.1.4, 4.1.1.5, and Annex E. Hereinafter, IEC 60489-6.

17. Ibid., Annex E.

18. Siwiak, p. 259.

19. The "20-page" method is called the "up/down" method in IEC 60489-6.

20. IEC 60489-6.

21. IEC 60489-6, Annex E.

22. For example, IEC 60489-6 assumes an unbiased test (i.e., $p(type\ I\ error) = p(type\ II\ error) = 0.5$), a 0.95 probability of passing the test if the device is 1 dB or less better than the specification threshold, and a 0.95 probability of failing the test if the device is 1 dB or less worse than the specification threshold.

Chapter 9
Electromagnetic Compatibility

9.1 INTRODUCTION

Two problems requiring special attention in the design of wireless sensor network nodes are the problems of electromagnetic compatibility (EMC) and electrostatic discharge (ESD). These problems often require a system-level view, incorporating several engineering and nonengineering specialties, if they are to be solved properly, and are particularly troublesome in wireless sensor network nodes, due to their small size, low cost, ubiquity, and need to function with a variety of other systems in a variety of environments. This chapter discusses the EMC of wireless sensor network nodes, especially the interference problems network nodes can induce in themselves due to the inherent proximity of digital and radio frequency (RF) circuits in their miniaturized designs. Chapter 10 discusses ESD.

9.2 EMC: THE PROBLEM

Wireless sensor networks by their nature are designed to be used with other circuits and assemblies — be they sensors, actuators, or computers — or installed with other systems, such as personal digital assistants (PDAs), cellular telephones, medical devices, consumer electronics, and industrial equipment. An important part of their design, therefore, is their ability to work well in a variety of electromagnetic environments, consisting of both intentional and unintentional radiation, as well as in the presence of noise on their supply, ground, and signal lines. In addition, most electronic products must meet regulatory requirements limiting their radiated emissions below a specific limit — in the United States, this requirement is in Part 15 of the Federal Communications Commission (FCC) regulations — so that other devices near them may achieve at least a modicum of interference protection.

An example of this type of EMC issue is the consumer television receiver. Bringing an NTSC television receiver near the antenna of a shortwave radio receiver can result in interference every 15.750 kHz up to 20 MHz or more; these "birdies" are harmonics of the receiver's 15.750-kHz horizontal oscillator that drives the electron beam of the picture tube across the screen.

As will be evident, the waveform of the horizontal oscillator can be a particularly troublesome one; it is a sawtooth waveform with a necessarily sharp transition as the electron beam retraces its position after filling in a line on the screen.

The effect of the television receiver on the shortwave radio receiver has been termed "desensitization" — a reduction in apparent sensitivity of the shortwave receiver, caused by the television interference. A general definition of desensitization could be "a reduction of apparent receiver sensitivity, caused by interference from nearby sources." (Interference may also affect transmitters, by corrupting, adding noise, or otherwise degrading a transmitted signal.) While the television receiver is a practical example, in modern systems desensitization usually occurs due to the presence of digital waveforms associated with microcontrollers, and the high-energy, fast-transitioning waveforms associated with switching power converters.

9.3 EXAMPLES OF SELF-INTERFERENCE

The interference sources are often within the receiving or transmitting device itself. This is likely to be more common in the future, as miniaturization of electronics and the design of systems-on-a-chip (SoCs) become prevalent. These designs will necessarily require that sources of digital and other switching waveforms be physically close to the antennas and other sensitive points of radio receivers and transmitters, exacerbating the EMC problem within the device itself. Because wireless sensor network nodes are designed to be small, highly integrated, and will of necessity incorporate both wireless transceivers and microcontrollers (and likely switching power converters), understanding self-interference is an important ability for the network node designer.

While the fundamental physics involved is often straightforward, self-interference EMC problems can present themselves in many different ways. The following is a list, taken from actual experience, of a variety of self-interference EMC problems. As one can see, the affected systems, the interference sources, the symptoms, and the solutions can be quite varied:

- Ostensibly identical 2 micron geometry microcomputers used in a radio pager were found to have different desensitization effects depending on which fabrication facility made them. Chips from some facilities (but not others) desensitized VHF pagers (i.e., pagers receiving in the 132-174 MHz band); this was cured by separating the microcomputer ground return to the battery from the voltage multiplier integrated circuit (IC) ground return on the pager's controller circuit board. Discovering this fix took two passes of the circuit board layout and two months' development effort, but it stopped the slow and expensive factory practice of sorting microprocessor ICs (with identical part numbers) by the fabrication plant code stamped on each part. As a cost reduction, these microcomputers were then shrunk from 2 to 1.5-micron geometry due to an improvement in lithography available from the chip vendor. This eliminated desensitization in the VHF band entirely (even for the original board design), but now caused desensitization of pagers in the UHF band (420–512 MHz). These microcomputers employed an integrated phase-locked loop (PLL) synthesizer to generate their system clock; the fix for the UHF desensitization was an inductor in series with the supply pin for the chip's PLL oscillator. This fix, while successful for UHF pagers, introduced desensitization on VHF pagers. Because production would be shut down, different controller circuit boards were then put into production for otherwise identical VHF and UHF pagers, greatly complicating factory flow. An additional 6 months were needed to develop and convert to production a controller board that could be used on all bands.

- A radio pager design was within six weeks of its scheduled product introduction date when a desensitization problem involving the liquid crystal display driver was discovered. It was found that, if two pages were sent to the same pager, sensitivity to the second page was about 6 dB less than to the first. This was quickly correlated to display activity: If the display were active (as it was immediately after receiving a page), desensitization occurred. After much effort, it was concluded that the root cause of the problem was the pad drivers on the display driver IC itself — a new chip just entering production. The IC design team (which was not co-located with the pager design team) had not considered desensitization in the IC design, nor was it specified in the IC specification given to them. A new revision of the chip was made, resulting in a program delay of three months.

- A capacitive switching voltage converter caused a significant desensitization problem in the receiver of a consumer electronic product. In addition to the desensitization, signals from the converter entered the voltage-controlled oscillator (VCO) of the receiver's synthesizer, broadening the spectrum of the receiver's local oscillator and resulting in a loss of adjacent channel selectivity. After much experimentation, the problems were solved by improved grounding in areas around the converter, and by the placement of the switching converter external capacitors in a particular orientation so that circuit nodes with fast, but inverse, voltage changes were placed close together.

- A service provider received field complaints of poor reception in a particular geographical area. The desired signal was strong; interfering signals present on nearby channels, while also strong, did not have the mathematical relationship necessary to produce receiver intermodulation, nor were they strong enough to cause blocking of the receiver. After much investigation, it was found that the 83-kHz switching frequency of an inductive switching converter was coupling to the receiver's first VCO. This produced a spur on the VCO 83 kHz away from the desired signal that, when mixed with an undesired received signal 75 kHz away, produced an interfering signal in the intermediate-frequency (IF) passband of the receiver that led to the desensitization. The root cause of this desensitization was found to be the coupling between the switching converter inductor and an inductor in the VCO circuit; the two were on separate circuit boards, but when the product was assembled, the two circuit boards were placed next to each other, separating the two inductors by only 2 mm. A new circuit board layout that avoided this proximity effect was required.

- Shortly after the market introduction of a pager, it was noticed that the pager's alert (its "beep") was distorted and, in some cases, too short. Because there was a large backlog of orders, this was a crisis at the worst possible time. After much investigation, the source of the problem was found to be the microcomputer clock crystal oscillator circuit. Traces on the multilayer printed circuit board leading to the liquid crystal display (LCD) were laid out underneath the oscillator circuit. The LCD lines were active when a page was received, and the switching energy of these lines was coupled to the clock oscillator, causing clock jitter. Because the microcomputer used this oscillator as the timing reference to synthesize the alert frequency, the tone of the alert was affected. A new layout of the circuit board, with the LCD lines moved away from the sensitive clock oscillator, corrected the problem. This failure resulted in a 3-week delay after shipments were announced. This was an example of a microcomputer desensitizing itself.

- A two-way portable radio in the 900-MHz band suffered from a 3-dB desensitization problem on a particular channel. The problem was finally traced to the 968th harmonic of a 972-kHz, 5-V square wave on a digital controller, desensitizing the receiver at 940.896 MHz. One of the advantages one has when dealing with such high harmonic orders is that the problem usually can be cleared by a small capacitor (in this case, 27 pF) placed on the offending source. Small value capacitors may have a minimal loading effect on the signal driver, but may significantly reduce high harmonics.

The preceding "rogue's gallery" of EMC problems is, of course, far from exhaustive. Many other types are possible; nearly all occurrences in the course of routine engineering practice seem to be unique. It is only by understanding the physics involved in the EMC problem, specifically, how interference (noise) is created and coupled into sensitive victim circuits, that trends become apparent and techniques to prevent EMC problems, and cure them when they do arise, can be identified.

9.4 THE PHYSICS ASSOCIATED WITH EMC PROBLEMS

Every period function of significance in design engineering has a Fourier series expansion. That is to say, every periodic function $f(t)$ can be formed as an infinite sum

$$f(t) = \sum_{n=-\infty}^{\infty} c_n \cdot e^{jn\omega_0 t} \tag{1}$$

where

$$c_n = \tfrac{1}{T_0} \int_{T_0} f(t) \cdot e^{-jn\omega_0 t} \, dt \tag{2}$$

are the complex Fourier coefficients
numeric T_0 is the period of $f(t)$
s and ω_0 is the fundamental frequency ($= 2\pi/T_0$) of $f(t)$, rad/s

$f(t)$ may, therefore, be constructed as an infinite sum of sinusoidal signals at multiples ("harmonics") of the fundamental frequency ω_0. These harmonics are at the root of the EMC problem.

An equivalent representation of the Fourier series is

$$f(t) = a_0 + \sum_{n=1}^{\infty} \left(a_n \cos n\omega_0 t + b_n \sin n\omega_0 t \right) \tag{3}$$

where

$a_0 = c_0,$
$a_n = c_n + c_{-n}$
$b_n = j \, (c_n - c_{-n})$

This form is often more convenient because the sine and cosine functions are available directly. Note that a_n is defined for integers $n \geq 0$, b_n for integers $n > 0$, and c_n for all integers n from $-\infty$ to ∞.

The spectrum of digital and other periodic waveforms is at the heart of the EMC problem. They may be viewed from two different scales — the "wideband view" and the "narrowband view."

9.4.1 The Wideband Spectral View

The waveform of most concern is the digital "square" wave. This waveform is actually a series of pulses with exponentially rising and falling edges associated with the charging and discharging of the distributed capacitance of (usually) a circuit board trace, through the distributed resistance of that

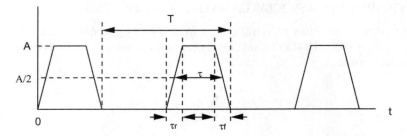

Exhibit 1.   A Digital Waveform

---

trace. Because the source driving the trace usually can be modeled as a current source, a fair model of the digital waveform is a trapezoidal wave, of variable duty cycle and rise and fall time, as illustrated in Exhibit 1.

In the practical case where the waveform rise time t_r and fall time t_f are not zero and assumed equal, it can be demonstrated that, for the real, positive frequency spectrum, the amplitude of the Fourier coefficients are[1]

$$\left|c_n\right| = 2A\frac{\tau}{T}\left|\text{sinc}\left(\frac{n\pi\tau}{T}\right)\right|\left|\text{sinc}\left(\frac{n\pi\tau_r}{T}\right)\right| , n = 0, \tag{4}$$

where

$$\tau_r = \tau_f, \text{sinc}(x) \_ \sin(x)/x$$

$$\left|c_0\right| = A\frac{\tau}{T}$$

The "wideband" view of this spectrum is a function of both the duty cycle τ/T and the rise and fall times τ_r and τ_f. Some specific examples of this spectrum can make clear the effects associated with varying these parameters.

The spectrum of a 1-MHz, 1-V peak amplitude square wave (i.e., $\tau_r = \tau_f = 0$) with 50 percent duty cycle is illustrated in Exhibit 2.

Note that only the odd harmonics are present, confirming the common wisdom about square waves.

In Exhibit 3, the same signal is used, but the duty cycle is changed to 51 percent.

Note that, with just a 1 percent change in duty cycle from a pure square wave, the even harmonics have arisen — in places, higher than the odd harmonics. Because the waveform is no longer symmetrical, valleys are now in the even and odd harmonic spectra, which, unfortunately, do not occur at the same frequencies. In practical terms, because the duty cycle of digital signals is not controlled with any precision, this implies that one

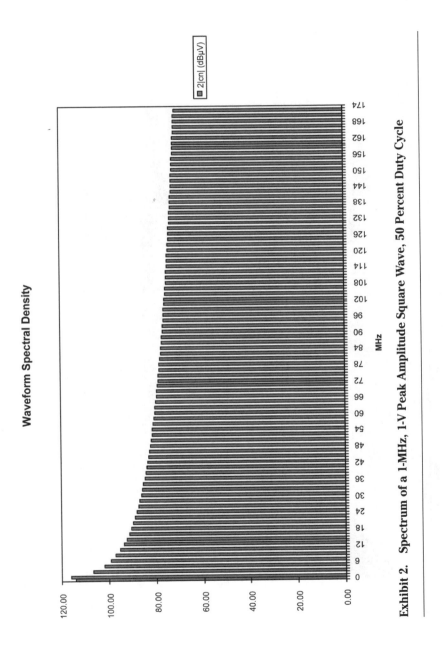

Exhibit 2. Spectrum of a 1-MHz, 1-V Peak Amplitude Square Wave, 50 Percent Duty Cycle

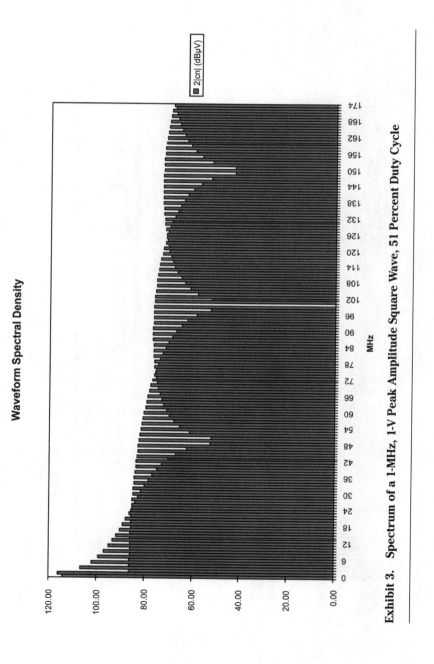

Exhibit 3. Spectrum of a 1-MHz, 1-V Peak Amplitude Square Wave, 51 Percent Duty Cycle

must expect desensitization at every harmonic — not just the odd harmonics.

The effects of rise and fall times is interesting as well. Exhibit 4 shows the spectrum of 50 percent duty cycle waveform, with $\tau_r = \tau_f = 10$ ns.

Exhibit 5 shows a spectrum of the same waveform as that in Exhibit 4, but with $\tau_r = \tau_f = 50$ ns.

Exhibit 6 shows a spectrum of the same waveform as that in Exhibit 4, but with $\tau_r = \tau_f = 100$ ns.

Finally, Exhibit 7 shows a spectrum of the same waveform as that in Exhibit 4, but with $\tau_r = \tau_f = 200$ ns.

Two things are apparent as τ_r and τ_f lengthen. The spectrum becomes less dense, meaning that fewer harmonic frequencies with significant energy are present (i.e., more nulls), and the overall power in the higher harmonics drops. Both of these are good from a desensitization standpoint. The rise and fall times of digital waveforms should be as long as possible, consistent with proper operation of their associated circuits.

In summary, although it is desirable to achieve a 50 percent duty cycle in digital circuit waveforms, the even harmonic energy builds quickly as one moves away from this ideal and it is not likely to be achieved to a level sufficient to guarantee a significant reduction in even harmonic energy under most, if not all, circuit conditions. In any event, a 50 percent duty cycle could only be achieved in clock signals because the duty cycles of general logic waveforms by design are not 50 percent. Nevertheless, an attempt to reach the 50 percent goal is worthwhile because, due to their ubiquity in digital circuits, clocks are a major source of EMC issues. Happily, this goal is also one that can be shared by designers of the digital circuits; when operating from the minimum possible clock frequency, most logic families perform best if the clock has a 50 percent duty cycle.

The importance of long rise and fall times to EMC is also apparent. Also note that, although they may be longer for pad drivers, τ_r and τ_f for internal microcomputer waveforms — even those operating at a clock rate of only a few MHz, as is common in wireless sensor networks — may be 2 ns or less. This is especially likely in single-chip, or SoC, implementations, which may require IC processes with extremely small minimum device geometries to get good RF performance; these processes usually enable very dense (i.e., low cost) logic, but logic that has very fast transitions due to the small size of the devices.

Finally, recall that the trapezoidal waveform is only a model of the physical system; the corners of the real waveform will be rounded slightly. Thus, the trapezoidal model very slightly overestimates the spectral

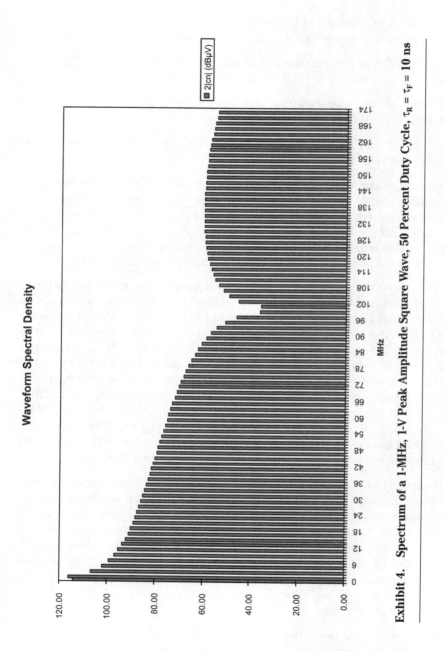

Exhibit 4. Spectrum of a 1-MHz, 1-V Peak Amplitude Square Wave, 50 Percent Duty Cycle, $\tau_R = \tau_F = 10$ ns

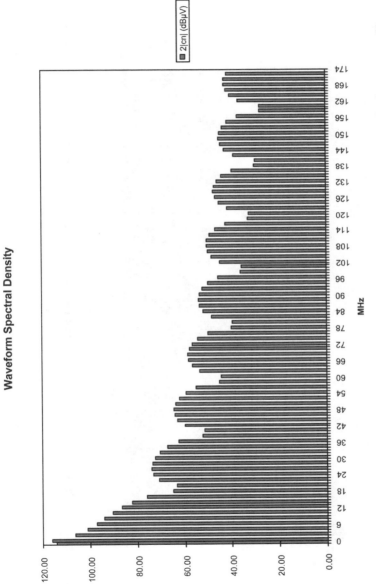

Exhibit 5. Spectrum of a 1-Mhz, 1-V Peak Amplitude Square Wave, 50 Percent Duty Cycle, $\tau_R = \tau_F = 50$ ns

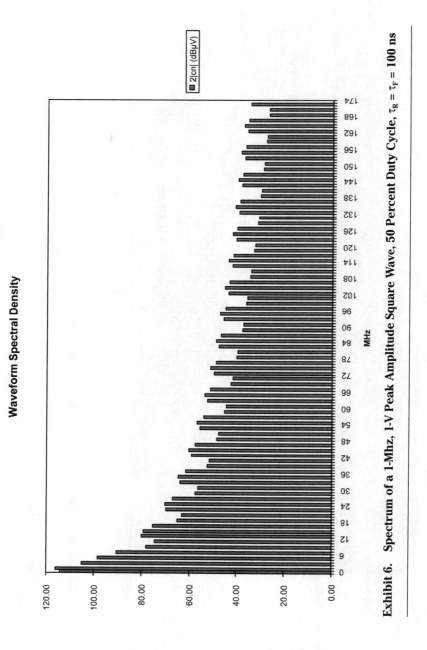

Exhibit 6. Spectrum of a 1-Mhz, 1-V Peak Amplitude Square Wave, 50 Percent Duty Cycle, $\tau_R = \tau_F = 100$ ns

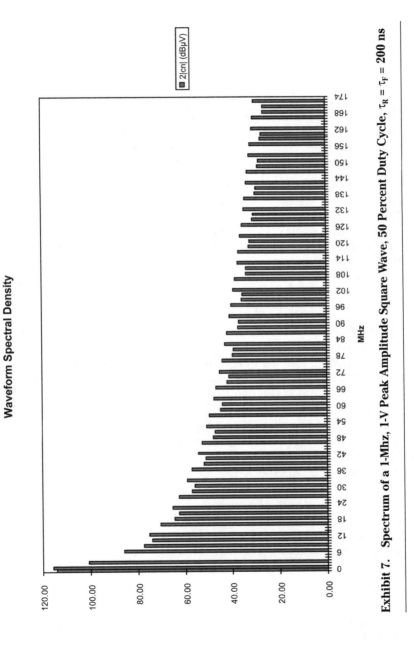

Exhibit 7. Spectrum of a 1-Mhz, 1-V Peak Amplitude Square Wave, 50 Percent Duty Cycle, $\tau_R = \tau_F = 200$ ns

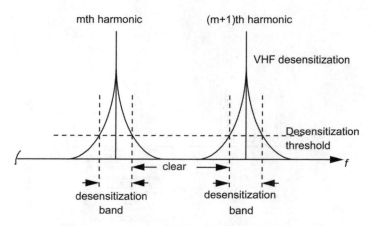

Exhibit 8. Desensitization at Low Harmonic Numbers (e.g., VHF)

energy at higher frequencies. This error is usually undetectable in practical cases, in which a far larger source of uncertainty is often the accuracy with which most circuit parameters (e.g., the resistance and capacitance (RC) model of the circuit board trace) are known.

9.4.2 The Narrowband Spectral View

The preceding analysis made it appear that the interference from digital waveforms would be isolated to discrete frequencies, albeit a large number of them. As it happens, that is not so. Due to low-frequency mixing, device 1/f noise, effects of jitter, and other factors, the interference is actually spread in frequency about the discrete frequencies mentioned. This, of course, makes the interference more difficult to avoid, because it now occurs in bands instead of at discrete frequencies. The width of the bands is determined by the amplitude of the harmonics and the "desensitization threshold," the signal level required to desensitize the nearby receiver. Even though the spectral energy is taken from a single frequency and spread in a band of frequencies, if the resulting signal level remains above the desensitization threshold, the overall effect can be deleterious. At VHF, where harmonic orders are low (and, therefore, harmonic amplitudes are high), less than half of the spectrum between two harmonics may be "clear." At higher frequencies, desensitization may occur only at individual frequencies.

Exhibit 8 and Exhibit 9 illustrate this effect. In Exhibit 8, the situation at a relatively low receive frequency, such as the VHF land mobile band, is illustrated. There, relatively wide "desensitization bands" may exist that can be larger than the clear frequency bands separating them. In Exhibit 9, the situation at a relatively high frequency, such as the 2450-MHz Indus-

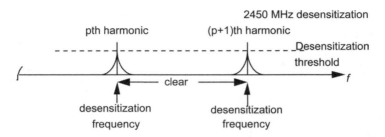

Exhibit 9.   Desensitization at High Harmonic Numbers (e.g., the 2450-MHz ISM Band)

trial, Scientific, and Medical (ISM) band, is illustrated. There, due to the lower energy of the harmonics at these frequencies, only the peak of the harmonic spectrum rises above the desensitization threshold. In this case, the desensitization bands have shrunk to become what are essentially discrete frequencies.

The spreading of spectral energy from a single frequency to a band of frequencies reduces the peak amplitude of the spectral components. If this spreading can be done over a frequency band of sufficient width, the energy will fall below the desensitization threshold of an application (or below a regulatory requirement).[2] This changing of a liability into an asset can be done by dithering the clock frequency of the digital circuits,[3,4] the duty cycle, or the rise and fall times of the clock pulses,[5] and has been employed to solve many stubborn EMC problems, such as interference from laptop computer backlights.

Note that, due to differences in coupling between the source of the digital interference and the receiver, the desensitization threshold may differ between bands, and from product to product.

9.4.3 Victim Circuits in Receivers

Although a large number of variations exist, receivers in use today may be classified into one of three types: zero-IF, low-IF, and superheterodyne.

9.4.3.1 The Zero-IF Receiver. The signal paths of zero-IF receivers are at only two frequencies — the desired RF channel, and baseband, where demodulation takes place (Exhibit 10). They are generally susceptible to interference only on these frequencies. Note, however, that baseband noise may be very difficult to filter from supply lines, etc., due to the large noise amplitudes typically found, and the large component values required for effective filtering.

233

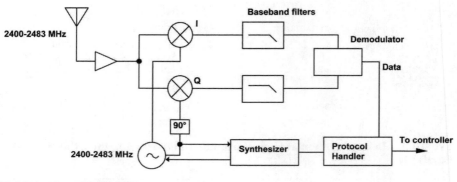

Exhibit 10.   The Zero-IF Receiver

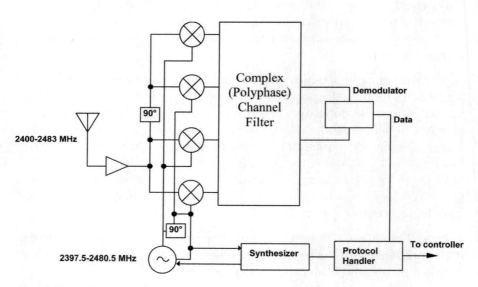

Exhibit 11.   The Low-IF Receiver

9.4.3.2 The Low-IF Receiver. From an EMC standpoint, the low-IF receiver[6] (Exhibit 11) shares many similarities with the zero-IF receiver. Both have signal paths at only two frequencies — the zero-IF receiver at RF and base-band, the low-IF at RF and a low IF, generally equal to half of the spacing between channels. The low-IF may be a few MHz, leading to a particular EMC weakness of the low-IF design for wireless sensor networks; the IF may be at or near the bus frequency of the microcontroller. Keeping energy from the microcontroller bus out of the receiver IF can be a very difficult

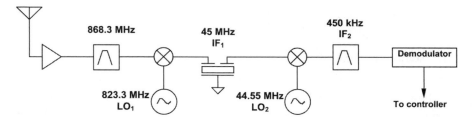

Exhibit 12.   A Superheterodyne Receiver

problem to solve, especially if the IF is relatively broad, limiting the options one has to move the bus frequency out of the receiver passband.

9.4.3.3 The Superheterodyne Receiver. The superheterodyne receiver, on the other hand, converts the received signal to one or more relatively high IFs prior to demodulation (Exhibit 12).

In addition to its RF channel, the superheterodyne receiver is also susceptible to interference at its IFs — in this example, 45 MHz and 450 kHz. Due to the amplification preceding the IF, the desired signal at an IF is stronger than the RF signal received at the antenna. Furthermore, although the IF circuits are typically integrated and, therefore, small, the antenna is probably one of the largest components in the transceiver, and can easily couple via both inductive and capacitive means. For these reasons, one is tempted to say that the probability of desensitization is less at the IFs than at RF; however, the IFs are also much lower in frequency than the RF signal at the antenna, and so digital harmonic energy typically is stronger at the IFs. In practice, the two effects counteract each other to a varying degree, and desensitization of both IF and RF signals is possible. Interference at an IF frequency is also more serious, in the sense that when it occurs it will desensitize all channels equally. When interference is arriving at the antenna, it is usually, but not always, frequency–selective, due to the discrete nature of the interfering harmonics.

It is possible to couple an interfering signal into the local oscillator (LO) circuits of any type of receiver. The contaminated LO signal then mixes with the received signal and transfers the interference to the IF or baseband signal. If the interfering signal is a relatively low frequency, the contamination will be directly placed in the receiver's passband and appear on all received channels. If the interfering signal is somewhat higher, out of the receiver's passband, it can still cause difficulty by mixing with undesired external signals to again produce an interfering signal in the passband. This can be an especially difficult type of problem to detect, because it requires a specific set of external signals to produce — a set that is usually not known before the problem is identified in the field.

9.4.4 Scope of the Problem

The wireless sensor network receiver described in Chapter 3 becomes desensitized when an interfering RF signal reaches a level of approximately $P_{desense}$ = –90 dBm (10^{-12} Watts), or $e_{desense}$ = 7.07 µVrms (17 dBµV) in a 50-Ω system. This incredibly small signal is both the magic of radio and the bane, for it means that the surrounding electrical environment must be very quiet to get maximum performance.

9.4.4.1 Digital-RF Isolation Needed. If a receiver requires an interference "floor" of –90 dBm to avoid desensitization, and it is working with a microcomputer with switching 2 V signals, the needed isolation may be computed:

$$\text{Isolation} = 20 \log (2 \text{ V}/7.07 \text{ µV}) = 109\text{dB} \tag{5}$$

or almost 11 orders of magnitude.

Fortunately, one gets several tens of dB due to the relative amplitudes of the digital signal's Fourier coefficients at the fundamental frequency and the harmonic causing the interference, but it can still be a challenge to meet this isolation requirement. For example, it was demonstrated previously that if the rise and fall times of a 1-V, 1-MHz digital waveform are 50 ns, harmonics near 150 MHz have a peak amplitude of approximately 45 dBµV. The level from the 2-V signal will be 6 dB higher, or approximately 51 dBµV. The root-mean-square (rms) value of this signal is 3 dB less than this, or 48 dBµV. Because the interfering signal, as detected by the receiver, must be less than 17 dBµV to avoid desense, there must be 48 – (17) = 31 dB of isolation between the digital signal and the receiver.

9.4.5 Coupling Mechanisms

To produce desense, the interfering signal has to reach the receiver. The process by which it does this is called coupling.

9.4.5.1 "Radiated" Coupling. "Radiated" coupling, as generally defined, means any coupling that occurs without a direct connection between threat and victim (e.g., microcontroller and receiver). This does not imply radiation in the strict physical sense; the coupling in nearly all self-desensitization cases occurs within less than one-tenth of a wavelength of the source (threat), and so occurs in the "near field" or induction region, instead of the "far field" or radiation region. Even at 2450 MHz, where a wavelength is 12.25 cm, it is rare to have the coupling distance exceed 2 cm. It becomes even less likely as the maximum dimension of a wireless sensor node approaches this length.

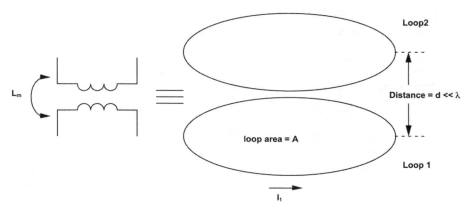

Exhibit 13.   Inductive Coupling

9.4.5.1.1 Inductive Coupling and Current Loops. It was discovered by Faraday in the early 1800s that a changing current in a coil of wire could induce a changing current in a second coil of wire. This phenomenon became known as induction, and is, among other things, the principle of operation of the transformer.

As with most physical phenomena, induction can also occur when it is not desired. This most commonly occurs in small electronic devices when a current loop is created in an integrated circuit V_{dd} and V_{ss} supply (such as a microcomputer or DC/DC converter), and couples into the device's loop antenna or, less often, an inductor in a receiver circuit.

9.4.5.1.1.1 Coupling Equation. Suppose one has two conductive loops, of negligible thickness, separated by a distance d. One of the loops, loop 1, is further assumed to carry a current I_1. The loops are so close together that all magnetic lines of flux generated by loop 1 are contained in loop 2. Further, it is assumed that d << λ, where l is the wavelength of the frequency of interest. See Exhibit 13.

It can be shown that a mutual inductance,

$$L_m = \mu A/d \qquad (6)$$

exists between the two loops, where:

L_m = mutual inductance, H
μ = permeability of the intervening medium, H/m
A = loop area, m^2
d = distance between the loops, m

If loop 2 were opened, a voltage $v_2 = L_m (di_1/dt)$ would exist between the open ends. Stated another way, the mutual inductance produces a current-controlled voltage source at the terminals of loop 2, with a value of $v_2 = L_m$ (di_1/dt). This voltage may interfere with receiver operation.

For example, suppose $A = 1$ cm^2, $d = 5$ mm (a common spacing between receiver and microcomputer circuit boards in miniaturized equipment), and the boards are in air, where $\mu = 4\pi \times 10^{-7}$ H/m. Then,

$$L_m = \mu A/d = (4\pi \times 10^{-7} \text{ H/m})(10^{-4} \text{ m}^2)/(5 \times 10^{-3} \text{ m}) = 25 \times 10^{-9} \text{ H} = 25 \text{ nH} \qquad (7)$$

This mutual inductance can couple a changing current in loop 1 into loop 2, where it may be expressed as voltage $v_2 = L_m (di_1/dt)$. Suppose that i_1 is a 50 percent duty cycle, 1 mA peak, 1 MHz trapezoidal waveform, with rise and fall times of 10 ns. If one is interested in desensitization at 151 MHz, the 151st Fourier harmonic is examined:

$$i_{1,151} = c_{151}\sin 2\pi ft. \qquad (8)$$

Then,

$$di_{1,151}/dt = (c_{151} \times 2\pi f)\cos 2\pi ft$$
$$v_2 = L_m (c_{151} \times 2\pi f)\cos 2\pi ft = (25 \times 10^{-9} \text{ H}) \times (888.3 \times 10^{-9})$$
$$\times (2\pi \times 151 \times 10^6)\cos(2\pi \times 151 \times 10^6)t$$
$$v_2 = (21 \times 10^{-6})\cos(2\pi \times 151 \times 10^6)t$$

Loop 2 now has a signal with a peak amplitude of 21 μV riding on it. Ordinarily, this would be of little concern; however, if loop 2 is the antenna, this signal (equivalent to 14.8 μV_{rms}) is 20 log (14.8/7.07) = 6.4 dB above the receiver's desensitization threshold. Note that this occurred with a loop 1 current of only 1 mA.

This coupling can be reduced in several ways. The area A may be reduced, or the distance d increased; both of these reduce the inductance L_m. The geometry may be changed, so that loop 2 is not parallel to loop 1. Although it will not reduce the coupling, it is clear that reducing the rate of change of i_1, di_1/dt, will directly reduce the voltage v_2 coupled to loop 2. All these methods may be used in product design. It is noteworthy that, with the exception of the last, all alternatives require a change in the physical layout of the components; either a relayout of circuit boards or a change in the physical design of the product (by separating the circuit boards further). This is a common theme in the prevention of EMC problems — it truly is a multidisciplinary task involving mechanical design, electrical engineering, and marketing and product design.

It should also be mentioned that arrangements of the two loops are used to minimize coupling. If two loops of radius R are close together (i.e., d << λ), it can be demonstrated that if they are offset such that their overlap is reduced to an amount D = 0.4783 R, the magnetic coupling is reduced to zero.[7] This arrangement is illustrated in Exhibit 14.

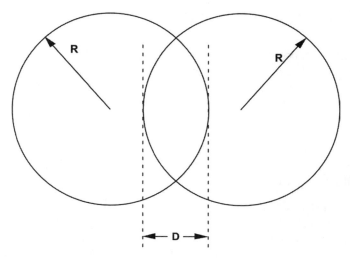

Exhibit 14.   One Method to Reduce Inductive Coupling

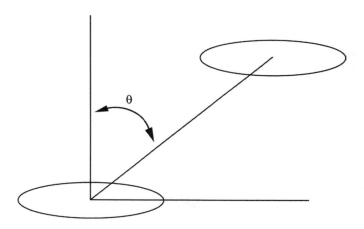

Exhibit 15.   A Second Method to Reduce Inductive Coupling

Further, if the loops are moved apart, but kept in parallel planes still in the induction region, it can be demonstrated that if the angle between the loop axis and a line drawn between the loop centers is 54.736°, the magnetic coupling is also reduced to zero.[8] This arrangement is illustrated in Exhibit 15.

Note, however, that although the magnetic coupling may be reduced to zero in these configurations, capacitive (electric) coupling between the two loops may still exist and may, in fact, be the dominant coupling mechanism.

9.4.5.1.1.2 Probably Does Not Involve Inductors. EMC problems involving inductive coupling often do not involve inductors. The main reason is that microcontroller and other digital circuits typically have few inductors; however, the ones they do have (for example, in DC/DC converters) typically carry very large currents with high harmonic content, and are important sources of desensitization energy. The usual inductive coupling problem is caused by current loops on personal computer boards, generally involving microcontroller V_{dd} and V_{ss} supplies (about which more will be said later).

9.4.5.1.2 Capacitive Coupling and Voltage Dipoles. Voltage dipoles, in the form of static electricity, have been known since antiquity. It was not until James Clerk Maxwell published his theory of electromagnetism in 1864, however, that it was predicted that a "displacement current" would flow between two conductive objects if the charge on one were allowed to vary. This displacement current is the source of capacitive coupling.

9.4.5.1.2.1 Coupling Equation. Suppose one has two conductive plates, of negligible thickness, separated by a distance d. One of the plates, plate 1, is at potential V_1. The plates are so close together that all electric field lines beginning on plate 1 end on plate 2. Further, it is assumed that d $<<$ λ, where λ is the wavelength of the frequency of interest. See Exhibit 16.

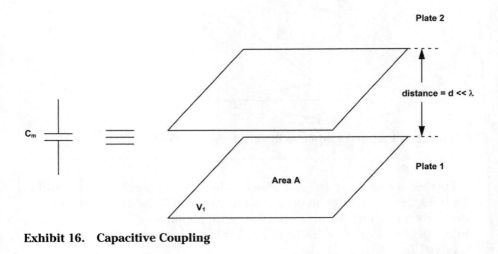

Exhibit 16.   Capacitive Coupling

It can be shown that a capacitance,

$$C_m = \varepsilon A/d \tag{9}$$

exists between the two plates, where

C_m = capacitance, F
ε = permittivity of the intervening medium, F/m
A = plate area, m^2
d = distance between the plates, m

If v_1 were allowed to vary, a displacement current $i_2 = C_m (dv_1/dt)$ would flow between the plates. Stated another way, the mutual capacitance produces a voltage-controlled current source terminating on plate 2, with a value of $i_2 = C_m (dv_1/dt)$. This voltage may interfere with transceiver operation.

For example, suppose $A = 1$ cm^2, $d = 5$ mm, and the boards were in air, where $\varepsilon \approx 8.854 \times 10^{-12}$ F/m. Then

$$C_m = \varepsilon\, A/d$$

$$= (8.854 \times 10^{-12}\ \text{F/m})(10^{-4}\ \text{m}^2)/(5 \times 10^{-3}\ \text{m})$$

$$= 177 \times 10^{-15}\ \text{F} = 177\ \text{fF} = 0.177\ \text{pF} \tag{10}$$

This capacitance can couple a changing voltage on plate 1 into plate 2, via a current $i_2 = C_m (dv_1/dt)$. Suppose that v_1 is a 50 percent duty cycle, 2-V peak, 1-MHz trapezoidal waveform, with a rise time of 10 ns. If one is interested in desensitization at 151 MHz, the 151st Fourier harmonic is examined:

$$v_{1,151} = c_{151}\sin 2\pi f t. \text{ Then} \tag{11}$$

$$dv_{1,151}/dt = (c_{151} \times 2\pi f)\cos 2\pi f t, \text{ and}$$

$$i_2 = C_m (c_{151} \times 2\pi f)\cos 2\pi f t$$

$$= (177 \times 10^{-15}\ \text{F}) \times (1.7766 \times 10^{-3}) \times (2\pi \times 151 \times 10^6)\cos(2\pi \times 151 \times 10^6)t$$

$$i_2 = (298 \times 10^{-9})\cos(2\pi \times 151 \times 10^6)t$$

Plate 2 now has a current entering it with a peak amplitude of 298 nA (or 211 nA$_{rms}$). Ordinarily, this would be of little concern; however, if plate 2 is an antenna, with an impedance of 50 Ω, the desensitization threshold in terms of current is $i_{desense} = e_{desense}/R = 7.07\ \mu V_{rms}/50\ \Omega = 141$ nA. This means the interfering signal is $20 \log (211/141) = 3.5$ dB above the receiver's desensitization threshold. Note that this occurred with a plate 1 signal voltage of only 2-V peak.

This coupling may be reduced in several ways. The area A may be reduced, or the distance d increased; both of these clearly reduce the capacitance C_m. The geometry may be changed, so that plate 2 is not parallel to plate 1. Importantly, the coupling current is directly proportional to dv_1/dt, so the coupling may be reduced by reducing the rise and fall times of the voltage waveform. The faster the rise time, the greater the coupling; this is a reason why nonsinusoidal (e.g., square) waves typically cause the greatest desense problems.

Further, a "grounded" third plate, called a "Faraday shield," may be placed between plates 1 and 2. The Faraday shield need not be actually grounded, but must be at a constant potential, so that the electric field lines (and so the displacement current) originating on plate 1 terminate on the Faraday shield, instead of on plate 2.

9.4.5.1.2.2 Probably Does Not Involve Capacitors. This coupling mechanism occurs most often due to lack of three-dimensional thinking during the physical design of the product. Commonly, a flex circuit going to a display will pass near the antenna, or the voltage-controlled oscillator on the receiver circuit board will be placed next to a fast-switching circuit on the microcontroller circuit board, when the two boards are mated together. Surprisingly, one of the most common capacitive coupling problems is the capacitive coupling between inductors: inductors tend to be large, and they carry fast-switching voltages.

9.4.5.1.3 True Radiated Coupling. To get true radiated coupling requires that the source be a relatively efficient radiator that, in turn, implies it must be at least a small fraction of a wavelength in size. This is a rare event in modern electronic equipment, and getting more so with product miniaturization. Problems can arise, however, due to true radiated coupling between separate products. Products incorporating early zero-IF receivers, for example, were often optimized so that their local oscillators (tuned to the frequency of reception) did not interfere with their own reception; however, they could interfere with nearby receivers on the same frequency. This problem, once identified, was easily corrected; however, it was often not detected until late in the product development cycle. One must also consider the possibility of interference with other services; the television receiver interfering with the shortwave radio, described earlier in this chapter, is one example of such an occurrence.

In any event, because radiation occurs when charge is accelerated, for minimum radiation opposing electric charges should be physically as close together as possible. Similarly, opposing magnetic "charges" (i.e., currents flowing in opposite directions) should be placed as close together as possible. Note that if one has minimized both inductive and capacitive coupling in a design, these two criteria are already met.

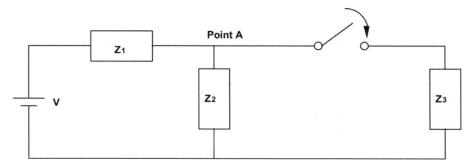

Exhibit 17.   Mutual Impedance Coupling

9.4.5.2 "Conducted" Coupling. "Conducted" coupling, as generally defined, means any coupling that occurs only with a direct connection between threat and victim (e.g., microcontroller and receiver).

9.4.5.2.1 Mutual Impedance Coupling. Suppose one has the circuit illustrated in Exhibit 17. Before the switch is closed, the voltage at point A is $V_{Aopen} = Z_2/(Z_1 + Z_2)$. After the switch is closed, the voltage is $V_{Aclosed} = (Z_2 \parallel Z_3)/(Z_1 + (Z_2 \parallel Z_3))$. The changing current through the Z_3 branch causes a change in the voltage applied to Z_2. If Z_2 were a circuit sensitive to supply voltage variations (e.g., a transceiver, sensor, or other system with significant analog circuitry), and Z_3 a circuit with large changes in supply current over time (e.g., a microcomputer, voltage multiplier coil, or transducer), it is possible to see that Z_2 circuit operation could be impaired with this arrangement.

The culprit, of course, is Z_1, the mutual impedance; if $Z_1 = 0\ \Omega$, $V_{Aopen} = V_{Aclosed} = V$ at all times. Unfortunately, all practical voltage sources have nonzero internal impedances, so if, as is often the case with wireless sensor network design, a single supply source is mandated, this problem can be minimized but not eliminated.

One is careful to state that Z_1 is an impedance, instead of a resistance. If a series inductor is placed in a line supplying current to more than one circuit and a bypass capacitor is not used, the inductor may also act as a mutual impedance, and lead to mutual impedance coupling.

9.4.5.2.2 The Importance of Current Density. The copper used in most circuit board runners has a nonzero resistivity, so the geometry of a runner has an effect on its resistance. A long, thin runner has more resistance than a short, wide one. To minimize mutual impedance coupling, it is therefore important to place current flow through the widest and shortest possible runners between source and load. Said another way, for minimum mutual impedance coupling one desires to have the current distributed through the most copper cross-sectional area (i.e., the lowest current density), for the shortest distance possible. A numerical example will illustrate the situation.

The resistivity ρ of the electrodeposited copper used in circuit boards is approximately 1.7×10^{-6} Ωcm $= 1.7 \times 10^{-8}$ Ωm. The half-ounce copper runners typically used have a thickness T of approximately 18 μm. These runners then have a sheet resistance

$$R_{sh} = \rho /T = (1.7 \times 10^{-8} \ \Omega m)/(18 \times 10^{-6} \ m) = 940 \times 10^{-6} \ \Omega \tag{12}$$

If a runner of this material has width W = 5 mil (130 μm) and length L = one inch (2.54 cm), it has a resistance of

$$R = R_{sh} \ (L/W) = 940 \times 10^{-6} \ \Omega \ (1/0.005) = 0.188 \ \Omega \tag{13}$$

If, for example, the 100 mA current of a small motor is sent through this runner, the voltage drop along the runner is E = IR = 10^{-1}A $\times$ 0.188 Ω = 18.8 mV. A receiver supplied from the motor end of this runner would see an 18.8 mV supply variation as the motor cycled on and off. If the runner is replaced with one with W = 100 mils (2.54 mm), R = 940×10^{-6} Ω (1/0.100) = 0.0094 Ω, resulting in a voltage drop of E = 10^{-1}A $\times$ 0.0094 Ω = 0.94 mV. Although this is a 20-dB improvement, the preferred method to avoid the effects of this voltage drop is to connect the receiver to the voltage source directly, avoiding the voltage drop of the motor circuit board runner entirely.

Sources of mutual impedance coupling include the internal resistance of batteries, capacitors, and voltage multipliers; battery contact resistance; and the resistance of circuit board via holes. High-current runners should have a minimum of circuit board layer changes; however, when layer changes are inevitable, multiple parallel vias should be used to minimize the resistance. Using multiple parallel vias is also good engineering practice for another reason: Vias are the most likely point of failure in a printed circuit board, due to separation between the copper of the trace and the copper in the via. This failure most often occurs due to thermal cycling or mechanical stress, and is often intermittent — the worst type of field failure to have.

9.4.6 Avoiding Coupling Problems

Avoiding desensitization begins at the beginning of product development, and continues to the end. It is a lot more than just circuit board layout. The following is a list of things to consider during product design at the system, integrated circuit, and circuit board levels.

9.4.6.1 System Level. By design, one should minimize the number and amplitude of high-frequency signals present on the circuit board and in the product. This sounds obvious, but it is remarkable how rarely this is considered. If fewer high-frequency signals are present, fewer opportunities for desensitization or other EMC problems will exist. This includes unconnected, but switching, microcontroller ports, and even internal IC nodes — all high-frequency signals should be viewed with suspicion and efforts made to eliminate them. Some methods to accomplish this are:

- *Microcontroller bus frequencies.* Two complementary strategies concerning microcontroller bus frequencies are used:
  - *Use low bus frequencies.* This is the item most often considered and, it is clear that the lower the bus frequency, the higher the harmonic needed to reach a given RF or IF frequency to cause interference. Based on the power density spectra discussed earlier, this, in turn, means a lower amplitude of the spectral component and, hopefully, less desensitization of nearby receivers.
  - *Use adjustable bus frequencies.* This can greatly reduce desensitization problems by moving bus harmonics away from desired RF channels. Again, it is important to remember that even as well as odd harmonics need to be considered when this is done. In many communication protocols, bus frequencies may need to be adjusted dynamically, in response to RF channel changes. This may be difficult to accomplish if the channel changes occur too rapidly, for example, if a protocol with a fast frequency-hopping physical layer is chosen. Systems operating in narrow frequency bands (e.g., the 868-MHz band) may have an advantage here, in that bus frequencies may be chosen such that their harmonics are outside the band completely. Nevertheless, it is often wise to make some provision for bus frequency adjustment in the initial system design, in order to respond to unexpected EMC events.
- *Minimize external bus traffic.* This is another "common sense" area in which the likelihood of EMC problems may be reduced. Power consumption also may be reduced. External bus traffic, because it is sent on relatively large circuit board runners, couples much more efficiently into the receiver than internal microcontroller bus traffic, which is confined to truly microscopic integrated circuit traces. To minimize use of external buses, attempt to keep operating system and signal processing software in internal read-only memory (ROM); signal processing data buffers (e.g., interleaving buffers) should be kept in internal randon access memory (RAM), if possible.
- *Partition integrated circuit functions intelligently.* When deciding how to partition functions between integrated circuits, a factor to consider is the speed of the interconnect signals. When possible, it is better to as-

sign functions to integrated circuits so that functions requiring high speed interconnects are placed on the same chip.

- *Consider differential, current-mode, or low-amplitude designs for particularly egregious signals.* Digital signaling does not have to be done in CMOS logic levels. As discussed previously, single-ended, voltage mode signals, although very power-efficient, are almost the worst possible choice from a desensitization standpoint. If a high-speed signal cannot be avoided, different signaling methods may be employed. Some of the options are:

  - *Differential signals.* Possibly the easiest approach is to produce a second signal inverted from the first, and place the two signals as close as possible on the circuit board. The electric field lines between the two signal lines may end on each other, instead of elsewhere on the circuit board, and so provide improved isolation. A disadvantage is the larger crowbar current (see Section 9.4.6.2) of the system, and the additional circuit runners.

  - *Current mode signals.* It is possible to use current, instead of voltage, to send information. The advantage is that the circuit board runner will keep a nearly constant voltage, so there will be little possibility for capacitive coupling to other circuits. The disadvantage is the current required in circuits to send and receive the signal — and to level shift it. In addition, one must be concerned about inductive coupling of the current loop.

  - *Low amplitude signals.* If noise margins and jitter specifications permit, it is also possible to communicate via low-amplitude signals, in which logic levels are separated by, for instance, 200 mV. This has been done quite successfully, for example, in the transmission of a 38.4 kHz clock signal from the microcontroller to the receiver in some pagers. Again, the disadvantage is the current required in circuits to send and receive the signal, to level shift it, and perhaps to buffer the signal back to CMOS logic levels. An issue with any non-CMOS logic level design is incompatibility with off-the-shelf components, especially memories. Non-CMOS logic level design is most useful in systems for which several application specific integrated circuits (ASICs) are being simultaneously designed.

- *Minimize operating voltages and currents.* If operating voltages and currents are minimized, it is apparent from the discussion on coupling that the coupled noise voltages and currents will also be minimized. Along with reduced power consumption, this is one of the strongest arguments in favor of supply voltage reduction in wireless sensor network nodes. One should not accept the minimum supply voltage requirement on an integrated circuit datasheet without discussing the matter with the vendor. Often, the datasheet value is the value to ensure operation over some performance envelope of no importance to the design at hand, and significant voltage reductions can be had with such requirements removed. For example, the supply voltage is often specified as a fixed value, yet it covers a wide temperature range. Supplying the chip with a temperature-dependent supply voltage may enable reduced-voltage operation over much of the temperature range. Similar opportunities may exist with bus frequencies — often, the devices are specified for operation at higher frequencies than the system requires, and significant reductions in the voltage specification are possible if only a lower frequency is needed. One may also consider the use of a variable supply voltage, coupled to the needed bus frequency. Often interesting opportunities become apparent after a simple characterization of a given part over supply voltage.

- *Schedule noisy events appropriately.* The behavior of the system itself can also be used to reduce the possibility of EMC problems. For example, one may schedule "noisy" events so that they occur when the receiver is off or warming up. The data packets of most wireless sensor network protocols are limited in size; it is often possible to store received data in a buffer, then process it after the receiver has gone to sleep. Processing of sensor data may be interrupted during RF transceiver operation and then resumed afterward. This strategy may not only reduce EMC concerns, it can also reduce the instantaneous peak power required from the power supply. This can extend node operation under conditions of high internal resistance in the supply (e.g., a carbon–zinc cell near end of life).

- *Consider the use of spread spectrum techniques.* For particularly difficult EMC problems, such as the cold-cathode fluorescent tubes used to backlight laptop computer displays, one should consider modulating the excitation frequency so that the interfering spectral energy is widely distributed and, therefore, lower at any given frequency of interest.

9.4.6.2 Integrated Circuit Level. The EMC characteristics of a digital integrated circuit design are very important because users of the chip often have little ability to overcome EMC deficiencies designed into it, and correction of the design is a lengthy process, with time to completion measured in months.[9] It is best if EMC is considered from the beginning of the design process; better yet if specific use cases, including frequencies of operation and the transceivers with which the digital chip will be used, are specified. Next are some of the important points to keep in mind during digital integrated circuit design.

- *Pad driver sizing.* A major cause of EMC problems is pad drivers, the output stages driving the external output pins of the chip. This is because they are often the circuits on the chip able to supply the most

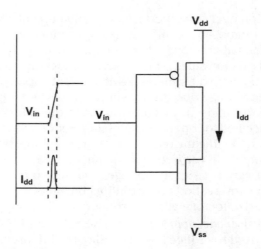

Exhibit 18.   Crowbar Current

power, and also because they are usually connected to external "antennas," in the form of circuit board runners, able to more efficiently radiate energy than are the minute traces on the chip itself. The problems with pad drivers arise from two sources.

- *CMOS shoot-through (crowbar) current.* When a CMOS gate changes state, during the transition there is a short time when both NMOS and PMOS devices are partially conducting. When this occurs, a short, high current pulse, called shoot-through, or crowbar, current, can be drawn from V_{dd}. See Exhibit 18.

  Because this current pulse has short rise and fall times, and occurs with every logic transition, it can be a major source of EMC problems. Minimizing the size (width) of the switching transistors can minimize this current. Because pad drivers typically have larger devices than standard logic blocks, minimizing their width (while still meeting drive requirements) is of primary importance, and is a major reason why EMC should be considered at the start of chip design. Sourcing and sinking crowbar current is a major function of the V_{dd} bypass capacitor, which acts as a local source for this high-frequency energy (and is recharged between pulses). If the value of this capacitor is insufficient to source the needed peak current, or if it is improperly placed, this high-frequency current may appear on the V_{dd} and V_{ss} supplies, which are typically large and effective radiators.

- *Rise and fall times; expected capacitive load.* It is, of course, important to minimize the size of switching devices, particularly pad drivers, in order to maximize the rise and fall times of the output waveform. Pad drivers, however, must still be capable of switching their maximum load capacitance at the maximum required rate. Because the capacitive load will be different on different circuit board layouts, newer microcomputers are being designed with adjustable (programmable) pad drivers. The size of the output devices may be controlled via software, so that the minimum size driver required for the design may be chosen.

The current to charge the load capacitance is drawn from V_{dd}, and is also sourced by the V_{dd} bypass capacitor. Similarly, the current discharging the load capacitance is placed on V_{ss}. As device minimum geometries continue to shrink, the effects of crowbar current will become less important, compared to that current needed to charge the load capacitance.

Rise and fall times of circuit board waveforms may be measured with an oscilloscope, but it is important to remember the impedance, especially the capacitance, of the oscilloscope probe itself when making the measurement.

- V_{dd} *and* V_{ss} *pinout.* The pinout, or physical placement, of the V_{dd} and V_{ss} pins on the integrated circuit are very important to the overall EMC performance of the device.

    - *Number of pins.* A well-designed chip may employ a number of V_{dd} and V_{ss} pins to reduce mutual impedance coupling caused by the wirebond and leadframe inductance and resistance. Usually, the analog sections, such as analog-to-digital converters and any bus synthesizers present, will have separate supply pins from the processor core and other digital sections. It is important that each of these supplies be properly bypassed to not only ensure correct operation of the chip itself, but also minimize the risk of EMC associated with high-frequency currents that may be present on the supplies (both V_{dd} and V_{ss}).

    - *Location of the* V_{dd} *and* V_{ss} *pins.* For good bypassing, it is very important that the V_{dd} and V_{ss} supplies of a given section be bonded out on adjacent pins. Because they may have a lot of high-frequency noise on them, this is especially important for digital V_{dd} supplies and their associated V_{ss} pins so that runners connecting V_{dd} and V_{ss} to the integrated circuit's bypass capacitor do not transport the noise around the circuit board. See Exhibit 19.

9.4.6.3 The Circuit Board Level. The circuit board level is often the level at which the product designer has the most flexibility regarding EMC matters. In addition to the layout of the circuit board, always a critical design for EMC, the designer often has the flexibility to specify and select discrete components used in the product. Finally, when problems arise,

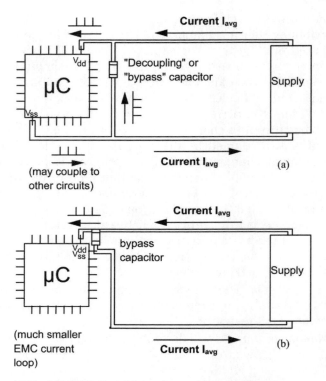

Exhibit 19.   (a) Bad and (b) Good Locations of V_{dd} and V_{ss} Pins

---

the designer has the opportunity to modify the circuit board level schematic, by the addition, respecification, or deletion of components.

9.4.6.3.1 Low-Impedance Power Sources. Low-impedance V_{dd} supplies are important to minimize mutual impedance coupling. In addition, a low impedance supply makes the development of large voltage swings on the supply lines, which may then capacitively couple to other circuits, unlikely, even when drawing large currents. Note that this desire for low impedance extends not just to the output of a voltage multiplier supplying the integrated circuits themselves, but also to the output of the power conditioning circuit and that of the primary power source itself. Each of these links in the power delivery chain may, under at least some conditions, require varying amounts of current from its supply; these current variations can produce mutual impedance coupling if the source impedance of the supply were too high.

9.4.6.3.2 Capacitor ESR and Battery R_s. As previously mentioned, both capacitors and batteries are imperfect voltage sources in that they both have internal resistance — that is to say, their terminal voltage drops when more current is drawn out of them. Due to historical reasons, the internal resistance of capacitors is called ESR (Equivalent Series Resistance), although in batteries and single cells, it is called R_s (series resistance). The internal resistance of these components makes them a possible source of mutual impedance coupling. In the case of bypass capacitors, the capacitor's ESR limits the minimum attainable impedance between the capacitors' terminals, and so limits bypassing effectiveness.

Capacitor ESR is usually specified in capacitor data sheets, and is generally on the order of a few ohms. There is a trade-off between ESR and leakage current in a given technology; capacitors with lower ESR tend to have higher leakage current. ESR varies with capacitor construction and temperature, generally being higher at lower temperatures. ESR also varies with frequency.

To get good bypassing at low frequencies, a large capacitance value is needed. In addition to their internal resistance, however, ESR capacitors also have a small amount of parasitic series inductance, an amount typically larger for the larger component packages needed for larger values of capacitance. Capacitors, therefore, can be modeled as a series RLC circuit. Similar to all series RLC circuits, above the frequency of resonance of the circuit, it appears inductive, with impedance that increases with increasing frequency. To get good bypassing at high frequencies, one therefore needs a lower value of capacitance, with an associated lower parasitic inductance and higher frequency of self-resonance.

However, one typically wants wideband decoupling, or decoupling at both high and low frequencies. This is often attempted by placing two capacitors, one a large value and the other a small value, in parallel; however, at high frequencies it is possible for the large-value capacitor to appear inductive while the small-value capacitor remains capacitive (because it remains below its frequency of self-resonance). This results in a parallel, or antiresonant, RLC circuit, with high impedance at frequency of resonance, and destroys the decoupling at that frequency. To avoid this resonance problem and achieve wideband coupling, resistors are often employed to reduce the quality factor (Q) of the parasitic resonances.[10,11] The resulting parallel RLC circuit has a lower impedance at resonance, creating a low-impedance bypass over a larger frequency range.

It has recently been demonstrated[12] that the inductance associated with circuit board vias has a strong effect on the high-frequency performance of decoupling capacitors, especially for relatively thick (30 to 60 mil) spacings between the power and ground layers of the circuit boards.

In battery-powered wireless sensor network nodes, users may place any battery they desire into the nodes when the time for battery replacement arrives. Although individual manufacturers may control their products to varying degrees, there is no global specification for the R_s of batteries of any given physical dimensions, let alone any particular chemistry, and users are certainly not equipped to return batteries to their vendor should the battery R_s be excessive. For example, although most brands of AA and AAA alkaline cells are generally quite good and have R_s values of less than 2 Ohms over most of their operating lives, some local and regional brands around the world can vary considerably. Values of 40 Ohms have been measured in some cells. Designing for battery R_s is, therefore, more difficult than capacitor ESR, where specifications are available and parts are rarely substituted. Although high battery R_s may cause many design problems (for example, microcontroller resets on transceiver activation, due to the higher current drawn from the battery at that time), in particular, it raises EMC issues.

One example of an EMC problem raised by high battery R_s is a wireless sensor network node that fails the "receive-while-actuating" test, in which the node's sensitivity is measured while the actuator (or sensor) is active. If the current drawn by the actuator (perhaps 50 mA) is drawn over a battery R_s of even 2 Ohms, 100 mV of droop in battery terminal voltage results. This voltage drop can result in a loss of sensitivity, synthesizer unlock, or other transceiver problems. In stubborn cases, when the actuator current is especially high or the RF transceiver especially sensitive, it may be necessary to power the actuator from a supply separate from that of the transceiver. A second example is the current drawn by a trasmitter, which may cause similar problems for a sensor.

One notes in passing that the R_s of Zn-O and Li cells (including those secondary cells frequently used as storage elements in power conditioning circuits) is significantly higher than that of standard alkaline AA and AAA cells, and that the impedance of some sources used for energy mining (e.g., piezoelectric elements) can be in the hundreds of Ohms.

9.4.6.3.3 Choking. Choking refers to the use of series inductance, usually in supplies, to stop ac current flow. Inductors operate in such a way as to resist current change; they, therefore, tend to pass dc and block ac. Although this is a desirable trait for a component used in the solution of EMC problems, the use of inductors has several disadvantages:

- *Inductors are relatively expensive.* The cost of a typical coil is several times that of a typical capacitor.
- *Inductors are relatively large.* Although miniaturization trends continue, monolithic chokes are not widely available in sizes below 0603 (i.e., 30×60 mils); capacitors are available down to 0402 (i.e., 20×40 mils). Chokes also have a larger z-dimension (i.e., height off of the board) — especially wirewound types.

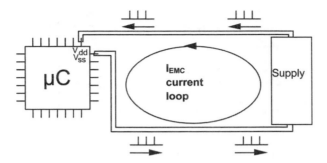

Exhibit 20. Unbypassed Circuit

- *Inductors tend to couple to other circuit components.* Because they are inductors, inductive coupling is possible; in addition, due to their size, they can capacitively couple to other components well. CMOS switching currents may shock-excite parasitic resonances in chokes at RF frequencies, exacerbating, instead of easing, EMC problems.
- *Inductor failures can be a problem.* When inductors fail, they are likely to form an open, instead of short, circuit. This is not good if they are in series with the microcomputer supply. Wirewound inductors fail more often than monolithic components.
- *The use of inductors implies that noise already exists on circuit board runners.* A much more elegant and probably cheaper design results if noise can be kept off the runners in the first place.

In applications where the DC voltage drop across an RC lowpass filter cannot be tolerated, however, an LC low-pass filter may be used effectively. One should be cognizant of the effect series elements (both L and R) may have on CMOS switching. By minimizing peak currents, they may slow CMOS operation because a portion of that current is used to charge and discharge gate and pad capacitances.

9.4.6.3.4 Bypassing. Bypassing refers to the use of shunt capacitors to pass AC signals, "bypassing" the return path to the DC source. Its importance to EMC is best understood in light of the crowbar and switching currents present in CMOS logic. These large peak currents must be sourced from somewhere and, without bypassing, the only source is the DC supply. This requires the current spikes to travel the entire DC supply path, providing limitless opportunities for coupling into other circuits. See Exhibit 20.

Using a bypass capacitor creates an RC lowpass filter by employing the internal resistance of the source. The supply then provides the average current; the bypass capacitor supplies the high CMOS peak currents and is recharged between pulses by the supply. Thinking in the frequency

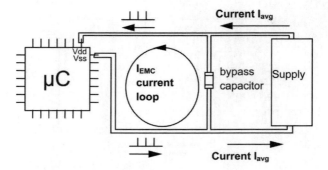

Exhibit 21. Poorly Bypassed Circuit

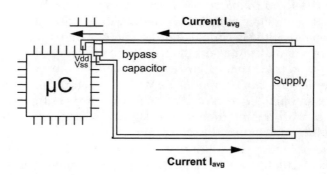

Exhibit 22. Properly Bypassed Circuit

domain, the high-frequency current passes through the low impedance of the bypass capacitor, instead of the higher impedance of the supply. In extremely stubborn cases (e.g., when bypass effectiveness is limited by the ESR of available capacitors), it may be necessary to augment the supply impedance as seen by the load by the use of series inductors or resistors placed near the load.

The physical placement of the bypass capacitor is extremely important. It should be clear from Exhibit 21 that placing the capacitor near the supply does little to reduce the resulting EMC current loop.

It is, therefore, mandatory that the bypass capacitor be placed as close as possible to the offending device (as well as the series inductor or resistor, if used), as illustrated in Exhibit 22. This minimizes the area of the EMC current loop, minimizing inductive coupling and the possibility of capacitive coupling, as well. If a series resistor is still needed, it should be placed immediately adjacent to the bypass capacitor, on the supply side.

9.4.6.3.5 RC Filters on Clocks. When low-pass RC filters are used on switching waveforms (usually clocks) to reduce the potential for EMC problems, it is important to consider the filter layout. To minimize possible capacitive coupling, the runner carrying the (unfiltered) square wave signal from its source to the filter resistor must be as short as possible. This usually means placing the resistor at the pin of the source integrated circuit. It is even more desirable, when possible, to place the resistor inside the chip — a design that both reduces part count as well as eliminates possible coupling from the IC pin and wirebond. The filter capacitor placement is less critical, but where possible, it should be placed close to the filter resistor.

9.5 PRINCIPLES OF PROPER LAYOUT

There are two principles which, when observed through the circuit board layout process, maximize the likelihood of a successful design (i.e., one with minimal EMC problems).

9.5.1 There Is No Ground

On a small, usually portable, self-contained, wireless sensor network node, there is no ground. All conductors inside are the same, and none may be described as equipotential (i.e., maintaining a constant potential). The node is often encased in nonconductive plastic, like a flashlight; no one worries where "ground" is on a flashlight schematic. Although usually the negative battery contact of a network node is defined as "ground" (when batteries are used), one may draw a correct, functional node schematic with the positive battery contact, the antenna input, or any given point in the schematic as "ground." One may just as easily draw the schematic without any ground symbols at all. From an EMC viewpoint, this means that "ground" is just another metal runner that may couple to the receiver, transmitter, sensor, or other suitable victim. The only time this rule is not valid is when the node is in a metal enclosure that is itself contacting a point of constant potential — a relatively rare occurrence in the design of wireless sensor network nodes. The "There Is No Ground" rule is meant to remind the designer that all metal is the same as far as conducting problematic EMC currents is concerned, and that there is no global point of constant potential to call "ground" in most network node designs.

9.5.2 There Is Only Return Current

The traditional European style of drawing electrical schematics is interesting. Schematics literally have two rails — V_{dd} and V_{ss} — at the top and bottom of the page, and all circuits are placed in between them. "Ground" symbols are never used, unless a connection to physical earth ground is intended.[13] (A separate symbol is used to describe connection to a metal chassis.) All currents returning to the power supplies of all integrated circuits, therefore, are identified on the schematic. This makes the identification of EMC current loops especially easy — they are all presented on the schematic. This rule is meant to remind the designer that, in light of the fact that ground does not exist in wireless sensor network nodes (rule no.1), the designer must be especially concerned with identifying and controlling not just the forward current path from the power supply to the integrated circuit (which is usually explicitly drawn on schematics), but also the return current path from the integrated circuit to its supply (which is often not explicitly drawn). Putting a via to a runner connected to the negative battery contact is not the end of the story; every microampere leaving the power supply must return to it, and many EMC problems originate in poorly controlled return currents, and lack of consideration of the complete high-frequency current loop.

9.6 THE LAYOUT PROCESS

Because no components are perfect, coupling can be minimized but not eliminated, and everything must be placed in a physically small volume, successful layout is the result of many correctly made compromises. To produce such a layout, a process should be followed to ensure that compromises are made knowingly, and that the best available options are selected.

9.6.1 Things to Look for After the Schematics Are Done but Before the Layouts Are Started

Before starting a layout, it is wise to spend a few minutes examining the electrical schematic to identify potential EMC threats and potential EMC victims. After identification, one needs to make a rough categorization of the degree to which the identified signal or component is a threat or a victim.

Among the common signals and components in a wireless sensor network node, special attention should be paid to the following.

9.6.1.1 High-Frequency Voltages and Currents. High frequency is defined as > 10 kHz or so, but the threshold varies among designs. Unfortunately, many potential sources of such signals exist; a few of the more prominent threats are:

- *Loop filter capacitors associated with integrated bus clock synthesizers.* These often have signals at a multiple of the bus frequency, and very short rise and fall times. Because synthesizer loop filters usually receive current from an up/down charge pump, it is important to also consider how these signals are bypassed to both V_{dd} and V_{ss}.
- *External buses, especially serial ports.* This includes not just the clocks, but the data line(s) as well. Because these signals can travel a relative-

ly long physical distance around the circuit board, they can be serious EMC threats; they are prime candidates for system-level EMC reduction (e.g., by not operating them when sensitive victim circuits are active). External Bus Interfaces (EBIs) and external memory buses also fall into this category. A good circuit board layout will minimize the length of these buses.

- *Voltage multipliers, switching converters, and regulators.* Often these circuits switch relatively large voltages and currents at frequencies in the tens to hundreds of kilohertz, making them important circuits to place properly in a network node layout. As with external buses, time spent in the system design phase finding ways to allow switching converters to shut down during sensor or receiver operation will pay great dividends near the end of the product design cycle.

- *The microcontroller clock itself.* It is important to consider this high-frequency circuit as both a potential threat and a potential victim. It is typically a very high-impedance circuit, subject to disruption with the application of relatively small signals; disruption of the microcontroller clock is, of course, usually fatal to device operation.

- *External signals.* Although all sources of high-frequency energy should be noted, special attention should be paid to potential sources from outside the network node itself — particularly when the node is using a host microcontroller, or is residing in another device, such as a cellular telephone. The host likely was not designed with the EMC concerns of the wireless sensor network node designer in mind, so it is good engineering practice to examine the host for the presence of high-frequency signals that, although they may not interfere with the host's performance, may seriously degrade the performance of the network node. An example is the use of switching voltage converters in the host. The host may have a system design that activates its switching converters only when the host is asleep; however, it is likely that the host and the wireless sensor network node are asynchronous, meaning that it is possible for the host's switching converter to become active while the network node is active, resulting in an EMC problem. A second example is mains noise entering through the power supply of a mains-powered node. This can be particularly difficult in industrial environments, where large amounts of noise can be placed on the mains from operating equipment. Finally, one should keep in mind that most countries regulate the conducted emissions of mains-powered devices; it is important to ensure that the network node itself does not cause interference to other services.[14]

- *Fast transitioning voltages/currents.* Almost by definition, high-frequency sources are sources of fast transitioning voltages and currents, so this may seem to be a distinction without a difference. Some signals, however, have rapid transitions at a relatively low frequency; such signals can become relevant to EMC when, for example, the transitions are rapid, not because they occur in a short amount of time, but because they change a relatively large value. In the coupling equations, it is the di/dt or dv/dt ratio that matters, not necessarily the value of di, dv, or dt alone. Circuits that commonly cause difficulty are:

- *Voltage multipliers, switching converters, and* regulators. Because they switch relatively large signals relatively quickly, these circuits can be significant EMC problems. They will need to be placed as far as possible from an internal antenna and from sensitive RF and analog circuits; the reactive element of switching converters can be one of the largest components in the network node, so its location relative to the antenna, also one of the largest components, should be considered during the early stages of product design.

- *Transducer drivers.* Transducers can require very large currents; almost any controller of a physical object, such as a heating, ventilation, and air-conditioning (HVAC) air damper, must switch hundreds of milliamperes. Particularly difficult are large audio sources, such as sirens and alarms; these often require large, square-wave input (for maximum drive efficiency), with very large harmonic content. The drivers are often in a bridge configuration, for maximum output power.

- *Display backlights.* Many technologies are suitable for display backlighting, including electroluminescent panels and cold-cathode fluorescent tubes. These can be particularly difficult EMC problems due to their relatively high drive voltages and their necessarily large physical size.

• *Areas of high current density.* Areas of high current density are important to identify because they can be sources of conducted coupling. Common points of difficulty are:

- *Voltage multipliers, switching converters, and regulators.* Particular areas to identify are the connections to the inductor in inductive switching converters, and to the capacitor(s) in capacitive switching converters, because the entire output power of the converter is transferred through these points — often in a harmonic-rich waveform. One should recognize that, in the case of the switching reactive elements, two current loops must be considered — one when the converter switch is closed, a second when it is open — and both loops need to be as small as possible, with minimal current density. Two loops that are associated with the converter input and output filter capacitors must also be considered. In the case of the input capacitor, one loop is associated with the high-frequency currents placed on it by the source (which may not be an issue if it is a constant-voltage source like a battery, but can be significant in some energy scavenging sources) and the second with the return currents of the converter itself. In the case of the output capacitor, one loop is associated with the high-frequency currents placed on it by the converter (which is nearly always significant) and the second with the return currents of the load. When the layout is done, the current paths between the converter and its input and output filter capacitors should be separate from the current paths of other circuits in the node. Note that the production of a low output current at a high output voltage must require a high input current if the input voltage is low. The current density around the input circuits of voltage multipliers is often overlooked.

- *Microcontroller* V_{dd} *and* V_{ss} *pins.* Significant current will pass through the supply pins of the microcontroller when it is active. Due to the large number of pins that must be connected to V_{dd} or V_{ss} logic levels in a typical microcontroller, it is often difficult at first glance to identify the main current sinks in a typical chip; often, several are labeled V_{dd} or V_{ss}, and some sleuthing is necessary to identify the right pins. One should ensure that the schematic has a bypass capacitor between V_{dd} and V_{ss} and recognize that current from other pins, such as those supplying onboard analog-to-digital converters, will need to be routed away from this area of high current density to avoid conductive coupling effects.

- *Transducers.* In addition to the di/dt and dv/dt concerns associated with transducers and their drivers, because they can switch large currents, a major concern is their potential for conductive coupling. This potential grows if their large currents get constricted to a small physical area on the circuit board, as can happen, for example, near a connector. If transducers are present in the network node design, the schematic should be examined to identify how their current reaches them from the supply, and how it is returned. This loop will often need to be a separate path; a common EMC error is to combine return currents from the transducer and other circuits in a common connector "ground" pin. Because the contact resistance of a connector can be significant, this is to be avoided, and a separate pin placed on the schematic for the transducer return current.

- *Transceiver* V_{dd} *and* V_{ss} *connections.* The power supply rejection ratio (PSRR) of low voltage (e.g., 1-Volt) RF and analog circuits is typically not as good as that of circuits that can operate at a higher supply voltage. In addition, a given level of noise is a higher percentage of the supply voltage, as the supply voltage is decreased. One must, therefore, pay particular attention to the transceiver supply to minimize its impedance over as wide a bandwidth as possible, by bypassing as appropriate.

9.6.1.2 Antenna Placement. Due to its physical size, and the large amount of amplification that follows it in the receive signal path, the location of an internal antenna and, to a lesser extent, the location of the connector for an external antenna, is critical in proper EMC design. The antenna is probably the component with the most contradictory requirements placed on it. It must be big enough to be effective, but small enough to fit in the product; it must be sensitive to very weak external RF signals, but not to any internally generated ones. The antenna must be kept away from lossy components and materials, like batteries, yet must be close to transceiver components to prevent undue RF losses between antenna and transceiver. It is worth spending some analysis time considering possible antenna locations, and the trade-offs involved with each. As discussed in Chapter 8, Section 8.22, an internal antenna will need a "keep out" region surrounding it in all directions, out to a distance inversely proportional to the frequency of operation. A fixed rule for the size of the keep out region cannot be given, as it depends on the physical shape and RF loss of the offending material, and the level of antenna efficiency one is willing to tolerate, but the radius of the keep out region will be on the order of $\lambda/100$ for circuit chip components, and $\lambda/10$ for larger but passive sources of loss, such as a AA cell.

9.6.1.3 Power Source Placement. The placement of the power source is important because it is likely that many different current paths will have to converge there. There should be space available so that the currents in these paths mix only at the terminals of the power source itself, instead of, for example, a narrow circuit board runner. The power source placement is also important because it is often a source of electrical noise, or at least composed of lossy materials; it should, therefore, be some distance away from the antenna.

9.6.1.4 Sensor Placement. Sensors can be sources of electrical noise, but some types can also be EMC victims if placed too close to the transmitting antenna. The designer must rely on experimentation and previous experience to determine the potential EMC problems of each sensor.

9.6.1.5 Placement of Oscillators. One source of potentially interfering signals inside a wireless sensor network node is, of course, the oscillators associated with the microcontroller and transceiver. Because they typically have high harmonic content, microcontroller clock oscillator circuits can be a problem for RF transceiver circuits and sensitive analog circuits associated with sensors. Conversely, because they are often very high impedance circuits (quartz or ceramic resonators in CMOS inverter strings), they can be corrupted by strong signals from a nearby RF transmitting antenna or power amplifier. Quartz reference oscillators associated with transceiver synthesizers have similar difficulties; although they are often higher power circuits, making them more resistant to interfering signals, they are often required to meet higher spectral purity requirements, meaning that even small degradations in performance can be unacceptable. In addition, they are typically operated at higher frequencies than microcontroller clock oscillators (in low power applications), so lower-order (and, therefore, higher-energy) harmonics of the synthesizer reference oscillator may reach RF frequencies used by the transceiver.

VCOs associated with transceiver synthesizers can cause performance degradation if not properly placed, especially when used as local oscillators driving mixers (which is to say, nearly always). In receivers, coupling of significant VCO energy into the antenna or receiver front end is especially troublesome in zero-IF designs, where it can result in DC offset and second-order intermodulation problems. In any receiver, such coupling can produce spurious responses (at frequencies related to harmonics of the VCO) and direct radiation of the VCO from the antenna. This last factor is particularly significant not only because it can be a source of EMC problems for other nearby services, but because it may exceed regulatory limits. In transmitters, coupling of transmitted energy from power amplifiers or the antenna into the VCO can result in distortion of the transmitted waveform (spectral regrowth), which may also exceed regulatory limits.

One strategy regarding oscillator placement that has been used successfully is to turn two liabilities into an asset: use lossy components, for example, a AAA battery, as RF attenuators to protect the antenna from the oscillators. One may think that the lossy component should be placed between the oscillator and the antenna, but this is often a poor trade, because it tends to move the lossy component too close to the antenna. However, simply placing the oscillator next to the lossy component appears to afford a protection of its own.

9.6.1.6 RF Filters, Low-Noise Amplifiers (LNAs), and Power Amplifiers. RF filters, receiver low noise amplifiers, and transmitter power amplifiers were saved for the end of the list, because they are some of the few circuits that should be placed close to the antenna. It is most correct to say, however, that only one port of these circuits (the port connected to the antenna) should be placed close to the antenna. A very great reduction in filtering effectiveness can result if the opposite port of an RF filter is visible by the antenna; instability can result in both LNAs and power amplifiers, if sufficient energy can couple from the LNA output into the antenna, or from the antenna into the power amplifier input. Good engineering practice is to bring these circuits radially outward from the antenna, so that their connections to the antenna are as short as possible, for minimum loss, although their other ports are as far from the antenna as possible, for minimum coupling. When both LNA and power amplifier are integrated, as is usually the case in wireless sensor network design, the job is to place the chip in the proper orientation so that the LNA input and power amplifier output are as close as possible to the antenna. The external signal path from IC to antenna, which may include an antenna switch, balun, RF filter, and perhaps even antenna tuning or other circuits, should be as short and as straight as possible.

9.6.2 EMC-Aware Layout Procedure

When modifying an existing design, with components of known characteristics and systems of known performance, it is usually not necessary to make an exhaustive EMC analysis; however, when starting a new design, possibly in a new market, incorporating a new transceiver and maybe new sensors, it is wise to have a structured approach to component placement and layout. A successful strategy has been to employ the organized signals concept,[15] which, although originally designed to optimally arrange pins in a connector for minimum cross talk, can be applied more generally:

1. The first step is to list all potential threats and victims present in the design. This list can be drawn from the preceding discussion, plus any signals or components particular to the design that are known to have caused problems in the past.

2. From this list, identify the greatest potential threat and the greatest potential victim. Usually the greatest potential threats will be the reactive elements of switching voltage converters, loop filter capacitors associated with integrated bus clock synthesizers, and the V_{dd} and V_{ss} pins of all CMOS integrated circuits. The greatest potential victim is usually the receiving antenna, due to its size, but perform the analysis on each design. For capacitive coupling, high-impedance nodes tend to be victims and low-impedance nodes threats; for inductive coupling, low-impedance nodes tend to be victims and high-impedance nodes threats. The evaluation of potential threats is the area in which an engineer's judgment most comes into play; general principles, similar to the preceding principles, can be stated, but there truly is no substitute for experience.

3. The next step is to floor plan the product so that the greatest potential threat and the greatest potential victim are as far apart as possible, given the additional constraints of desired product size and shape. This minimizes both inductive and capacitive coupling. If multiple circuit boards are to be used in the design (e.g., if a separate plug-in sensor board is to be used), do not forget that the boards will be mated together, and that one should think three-dimensionally.

4. The following step is to return to the threats and victims list, and select the greatest potential threat and the greatest potential victim from those components remaining on the list. Those components are then placed on the floor plan. The process then repeats until the list is empty. If an unequal number of threats and victims exist, the last remaining victim is used for the remaining threats (and vice-versa). In the highly integrated designs common to wireless sensor network nodes, the organized signals process is usually a short one because often, only two or three potential victims (the antenna and a sensor), and relatively few threats (voltage converters and oscillators) exist.

5. The next step in the EMC-aware layout procedure is to place components and circuit board traces so that the area of the return current loops of EMC-relevant circuits are minimized. This minimizes inductive coupling. Prominent circuits and components to consider here are the reactive elements and filter capacitors of switching voltage converters, loop filter capacitors associated with integrated bus clock synthesizers, and the bypass capacitors on the V_{dd} and V_{ss} pins of all CMOS integrated circuits. Do not place all return currents on a single "ground" plane; follow the current from each source to each load and back to the source, and ensure that that loop is as small as the component geometry and other design considerations can make it.

6. Next, minimize the length of circuit board runners carrying high-frequency signals. Prominent among these are the external buses, both serial and parallel, including any connections to external memory. This minimizes possible capacitive coupling.

7. Next (if not done so already as part of an earlier step), separate the current loops for high current loads from those of other circuits. Of principal concern here are any large currents going to actuators, sensors, or user interface items that may switch on and off. These loops should be separated from those of other circuits, joined only at the terminals of the power source. This minimizes possible conducted coupling.

8. Finally, if any connectors are used in the design, employ the organized signals concept to assign signals to the pins in a manner that will minimize cross talk.

9.7 DETECTIVE/CORRECTIVE TECHNIQUES

Despite one's best efforts, EMC problems will still occur on occasion. The product engineer can therefore benefit from knowledge of ways to analyze EMC problems and possible correction techniques. The following list of analysis and correction techniques is not exhaustive; other methods, concepts, and viewpoints are presented in Kultsar,[16] Muccioli, Catherwood, and Stevens,[17] Greb,[18] and Montrose.[19]

9.7.1 The "Hole in the Bucket" Principle

The "hole in the bucket" principle refers to a leaky bucket — all holes must be repaired before the bucket stops leaking. Similar to the bucket, a network node may have more than one EMC problem. For example, a design may have both inductive and capacitive coupling to a noise source. The engineer tasked with producing a working design should be aware that multiple fixes may have to be applied simultaneously to eliminate the problem(s), and that the fix may be more than the sum of its parts; two fixes that show no effect when tried separately may cure the problem when applied together. The best strategy is often to produce a functioning unit by any means possible, and then begin a process of subtraction to identify the minimum set of changes required.

9.7.2 Substitution

Substitution is a powerful method to isolate EMC problem causes and cures. For example, one of the most telling substitutions to make to resolve a receiver desensitization problem is to remove the antenna and measure sensitivity "conducted" (i.e., with the receiver not in a TEM cell, but directly connected to a signal generator). If the desensitization is not present when conducted sensitivity is measured, it is an indication that the interference pathway passes through the antenna. This, in turn, is a likely indication that the root cause of the problem is inductive or capacitive coupling, instead of conductive coupling.

9.7.3 Software to Control Specific MCU Functions

Many times, the source of the difficulty can be understood, or at least the testing made easier, if the microcomputer controlling the network node can be programmed to independently control specific functions. These functions may then be performed repetitively or exclusively, or not performed at all, and the effect on the EMC problem noted. Common functions useful to EMC problem investigations are

- Constant toggling of actuators; activation and deactivation of sensors
- Writing to displays
- Reading and writing to memory (especially external memory)
- Controlling pad drivers on microcontroller ports (turning them on and off; varying their drive capability)
- Toggling external buses with 1-0 data patterns
- Continuous synthesized bus, alarm, and display backlight operation

This topic should be discussed at the beginning of the product development program because "throwaway," or nonshippable, code will have to be written.

9.7.4 Physical Separation

Because both inductive and capacitive coupling are functions of distance, distance can also be used to aid the investigation of an EMC problem. If the product design includes multiple circuit boards, a ribbon cable connection between two boards may allow study of the EMC problem as the physical geometry of the two boards is varied.

9.7.5 The Fallacy of Shields

Shields have their uses, although they are unpleasant in most factory production processes, expensive, and often a quality problem; however, the engineer should remember that there is no ground in most wireless sensor network nodes. In most cases, "shields" do not function as their designers believe they do — as Faraday shields to stop capacitive coupling. Usually, shields function as another circuit board layer to redistribute return currents in a more favorable way. It is nearly always true that, with proper circuit board design, shields are unnecessary to meet the performance requirements typical of a wireless sensor network node. An exception can be discrete frequency generation circuits (VCO, synthesizer, etc.) of a transmitter, which may receive energy coupled from the antenna sufficient to cause improper operation if shielding is not used. In most wireless sensor network nodes, however, these components are largely integrated, so this exception rarely occurs.

9.7.6 Get to Know the IC Designer

Much insight can be gained into many EMC problems if the operation of the integrated circuits involved is understood and appreciated. Knowing, for example, which internal function is supplied by which V_{dd} and V_{ss} pins on a microcontroller is a powerful advantage to both understanding what is going on, and fixing it. The integrated circuit may have undocumented test modes that may be employed. If nothing else, the integrated circuit designer should be made aware that the chip is causing an EMC problem, so that future designs (possibly including a redesign of the chip now having the problem) may be improved. If at all possible, get to know the IC designer and design group.

9.7.7 Simulation

Simulation of EMC problems is getting better. With newer field solver software, it has become possible to verify that a proposed coupling mechanism indeed produces an EMC problem, and that a proposed fix will help.

What is still beyond simulation is to take an existing physical design, include all the possible waveforms that may exist on the circuit board runners, and determine if a problem exists before the product is built.

9.8 CONCLUSION

Because wireless sensor network nodes must coexist with many other electronic devices, and in fact work closely with them as part of their normal operation, EMC is an important consideration in their design. Further, because network nodes are true mixed-signal devices, often including digital, analog, and RF circuits on the same integrated circuits, the designer must be aware of coupling mechanisms that may transfer harmonic energy from digital signals into sensitive analog and RF circuits. The harmonic energy is a function of the frequency, rise and fall times, and duty cycle of the digital waveform, and can easily reach frequencies and powers necessary to desensitize nearby receivers or disrupt other sensitive circuits. Such coupling may occur via inductive, capacitive, conducted, or (more rarely) true radiated means.

Fortunately, specific design techniques are available at the system, circuit board, and integrated circuit levels to minimize this problem. At the system level, one attempts to minimize the number and level of high-frequency signals present, and to time them so that they do not occur during times when potential victim circuits are active. At the circuit board level, layout is of fundamental importance; the need to minimize the size of high-frequency current loops by placing bypass capacitors close to digital circuits, where they may operate as local sources of high-frequency energy, is first on the list. At the integrated circuit level, the design should consider this need, and place V_{dd} and V_{ss} on adjacent package pins.

An EMC-aware circuit board layout procedure is available that employs the organized signals concept to maximize the separation of the most sensitive potential victims from the most egregious threats. Finally, detective and corrective techniques do exist, so that when, despite the best efforts of all involved, an EMC problem does develop, the coupling mechanism may be quickly identified and the problem corrected.

References

1. William T. Presley, Spectral density of digital waveforms, *EMC Test & Design,* May 1994, pp. 14–17.
2. Donald R. Bush, John T. Fessler, and Keith B. Hardin, Spread spectrum clock generation for the reduction of radiated emissions, *IEEE Int. Symp. on Electromagnetic Compatibility,* 1994, pp. 227–231.
3. Keith B. Hardin et al., *Spread Spectrum Clock Generation and Associated Method.* U.S. Patent No. 5,488,627. Washington, D.C.: U.S. Patent and Trademark Office. 30 January 1996.

4. John B. Berry et al., Design considerations of phase-locked loop systems for spread spectrum clock generation capability, *IEEE Int. Symp. on Electromagnetic Compatibility*, 1997, pp. 302–307.
5. Mark L. Servilio, Michael J. DeLuca, and Edgar H. Callaway Jr., *Method and Electronic Device Using Random Pulse Characteristics in Digital Signals*. U.S. Patent No. 6,169,889. Washington, D.C.: U.S. Patent and Trademark Office. 2 January 2001.
6. J. Crols and M. S. J. Steyaert, A single-chip 900-MHz CMOS receiver front-end with a high-performance low-IF topology, *IEEE J. Solid-State Circuits*, v. 30, n. 12, December 1995, pp. 1483–1492.
7. Pantelis Angelidis, Kostas Vassiliadis, and George D. Sergiadis, Lowest mutual coupling between closely spaced loop antennas, *IEEE Trans. Antennas and Propagation*, v. 39, n. 7, July 1991, pp. 949–953.
8. Kazimierz Siwiak, *Radiowave Propagation and Antennas for Personal Communications*, 2nd ed. Norwood, MA: Artech House. 1998. p. 17.
9. M. Thompson, Microcontroller offers reduced EMI/RFI, *EMC Test Design*, December/January 1995, pp. 22–25.
10. C. Dirks, Wideband supply voltage decoupling for integrated circuits, *EMC Test Design*, November/December 1992, pp. 21–25.
11. Jiro Oouchi, Chikara Igarashi, and Kenichi Itoh, Study of noise reduction by 3 multiple decoupling capacitors with series resistors, *Int. Symp. Electromagnetic Compatibility*, 1999, pp. 232–235.
12. Jun Fan et al., Quantifying Decoupling Capacitor Location, *IEEE Int. Conf. Electromagnetic Compatibility*, 2000, v. 2, pp. 761–766.
13. When, in fact, it is called "earth."
14. Clayton R. Paul and Keith B. Hardin, Diagnosis and reduction of conducted noise emissions, *IEEE Trans. Electromagnetic Compatibility*, v. 30, n. 4, November 1988, pp. 553–560.
15. William T. Presley, Organized signals concept and cross talk, *EMC Test Design*, January/February 1992, pp. 23–24.
16. Emery Kultsar, EMC modeling tutorial, *EMC Test Design*, November 1993, pp. 22–23.
17. James P. Muccioli, Mike Catherwood, and Dale Stevens, Near-field emissions from microcontrollers, *EMC Test Design*, November 1993, pp. 14–18.
18. Vincent W. Greb, An intuitive approach to EM fields, *EMC Test Design*, January 1994, pp. 30–33.
19. M. I. Montrose, *Printed Circuit Board Design Techniques for EMC Compliance*. Piscataway, NJ: IEEE Press. 1996.

Chapter 10
Electrostatic
Discharge

10.1 INTRODUCTION

This chapter reviews the problem of electrostatic discharge (ESD) as it affects wireless sensor network nodes, and the importance of proper electrical, mechanical, and product design in its control. Similar to electromagnetic compatibility (EMC) problems, ESD problems often require a multidisciplinary approach for proper resolution. Unlike most EMC problems, however, ESD problems can be difficult to identify in the field; often, the first indication of an ESD weakness in a design are sporadic reports of "dead" nodes not attributable to other causes, or reports of very specific failure types, such as corruption of certain register contents in a microcontroller. ESD problems are difficult enough to detect in wireless sensor networks designed for consumer and home automation applications, where an ESD event may be instigated by a charged individual, but become even more troublesome in networks for industrial, agricultural, and military applications, many of which operate essentially autonomously, without direct human contact. Lack of human contact may lower the overall failure rate, but can result in a lack of information on the failures that do occur. It is important, therefore, to understand the ESD problem, so that it may be "designed out" of the finished product.

10.2 THE PROBLEM

10.2.1 Examples

Most people are familiar with static electric discharges on cold winter days or other especially dry conditions. These discharges, although only unpleasant to humans, may be fatal to modern electronic products; further, weak discharges that are undetectable to humans can still damage electronic products.[1] What follows is a selection of ESD-related events associated with consumer electronic products, illustrating the importance of the problem and the variety of guises in which it may appear.

- A new product was developed one spring, and volume shipments began the following summer. That winter, however, field failures, in the form of random access memory (RAM) errors, were noted in

269

temperate regions. The problem was believed to be ESD-related, and became serious when the RAM errors could not be duplicated in the manufacturer's laboratories, because the problem could not be corrected until the product was made to fail under laboratory conditions. Although many attempted design improvements were tried in the field, the field failures did not stop until spring. It became apparent that the ESD test procedure as performed by the manufacturer at that time was inadequate, and a task force was organized to identify improvements. The recommendations of the task force, including an environmentally controlled ESD testing room and a new testing procedure based on internationally accepted standards, were adopted after the new test accurately duplicated not only the field failure of this product (which was cured by a bypass capacitor on a RAM integrated circuit [IC]), but also the warranty failures of several other products. (ESD had not even been considered as a possible root cause of the failures in the other products.) By the next winter, products of modified design were in the field, and there was no seasonal rise in ESD-related failures. The cost of the new ESD testing room and equipment was more than repaid by the improvement in the manufacturer's reputation among its customers.

- During development of a product, several problems were uncovered during ESD testing. It was noted that the switching voltage converter IC on a microcontroller circuit board would go into a test mode when an ESD discharge reached a circuit board runner through a gap in the housing around the "5" key on the product's keypad. An added capacitor and diode to the runner cleared this problem. It was next noticed that the RAM memory locations storing trim values for analog circuits would become corrupted when a discharge occurred on the serial port contacts used to program the microcontroller, through another housing hole. Placing a diode between the two contacts and placing spark gaps (pointed circuit board runner shapes) between the runners to the contacts fixed the RAM problem; instead, a discharge to the serial port contacts would now cause the microcontroller to reset. Making the serial port runners thin to increase their inductance and modifying the ground metal beneath them, fixed this problem.

Now, after a discharge to the serial port contacts, the product's real time clock (RTC) appeared to stop — but the product would continue to work normally in every other way. It was found that the microcontroller had two programming bits that controlled the RTC increment rate. The RTC could be set to increment either once per second, once per minute, once per hour, or to not increment at all. The ESD event was somehow corrupting these bits so that the RTC would not increment, and, therefore, appear to stop. Working with the designers of the microcontroller, the product engineer assigned to this problem determined that the RTC was supplied power through a particular supply pin, the V_{rr} pin. A 150-Ω resistor in series with the V_{rr} pin (made possible due to the very low current drain of the RTC) fixed it. To speed testing, special product firmware was written during this part of the investigation that constantly checked the value of the sensitive bits. If a change were noted, a light-emitting diode (LED) on the product would turn on.

10.2.2 Failure Modes

Product failures resulting from ESD events may take many forms. They can, however, be placed into two general categories, long-term and short-term failures, depending on the time scale of the failure. Short-term failures are generally found immediately after the ESD event. Long-term failures take time, perhaps even months or years, to become apparent.

- *Long-term failures and latent defects.* Long-term failures, by definition, do not become apparent until some time after the ESD event.[2] The most common long-term failure involves CMOS ICs, which may have a shorter operating life after an ESD event, due to gate oxide degradation. The noise figure of radio frequency (RF) transistors may also increase over time after ESD exposure. It has also been demonstrated that ESD events during manufacturing processes produce latent defects in Schottky diodes used in microwave transceivers, leading to a loss of sensitivity after a few months of operation.[3]

- *Short-term failures.* Short-term failures become apparent immediately following the ESD event. They can have a confusing, application- or market-specific taxonomy; in particular, the distinction between catastrophic ("hard") versus noncatastrophic ("soft") failures can be vague. Although one may debate the classification of a particular product failure, the point to remember is that all failures are failures, from a quality and customer satisfaction point of view (the only point of view that matters). It is rare to find a user interested in entering into a lengthy discussion with a manufacturer on the precise categorization of a product failure due to ESD.

  - *Catastrophic failures.* A catastrophic failure is, simply put, any failure from which the network node cannot be made to recover by action taken over the air or at the user interface. This includes transceiver parametric failures, such as loss of receiver sensitivity, and failure of any internal sensors and actuators. Typically, catastrophic failures involve physical component damage, and require the replacement of hardware (e.g., integrated circuits) to return the network node to its proper operating condition. These types of failures are those most commonly encountered when reading the ESD literature.

- *Noncatastrophic failures.* Noncatastrophic failures, also called system upsets, include such transient phenomena as inadvertent turn-off, improper operation of the transceiver (partial or complete transmission of inappropriate messages, for example), improper operation of any internal sensor or actuator, low battery indication, or system reset (which can include loss of microcontroller state, loss of volatile memory, and system turn-off). Of these undesirable events, system turn-off is particularly undesirable. If the user (or network, if automatic network monitoring is in place) does not notice the ESD event and its effect on the node, the user may not turn the node back and so miss future messages. Because such future messages could be to inform the user of a fire alarm, breaking window, or other safety-related event, the loss of future messages may not be "noncatastrophic." Even if the wireless sensor network is not performing such a critical function, if the node suffering the ESD failure is the master of a star network, or performing some other centralized function, loss of the node can lead to a failure of the network. Noncatastrophic failures should be considered just as seriously as catastrophic failures.

- *Data loss.* Loss of data falls in the gray area between catastrophic and noncatastrophic failure. The loss of messages queued for transmission, unprocessed sensor data, or received data not stored in nonvolatile memory, although unfortunate and undesirable, probably falls in the "noncatastrophic" category. The loss of factory tuning values for the transceiver or calibration values for a sensor, which renders the network node inoperable, probably falls in the "catastrophic" category. The loss of security keys needed to communicate with other network nodes via symmetric-key cryptography may or may not be considered catastrophic, depending on whether or not a mechanism exists by which the user may enter keys into the node. (A failure of a memory chip itself, so that it is no longer capable of storing information, is clearly a catastrophic failure.)

10.3 PHYSICAL PROPERTIES OF THE ELECTROSTATIC DISCHARGE

The phenomenon of electrostatic discharge has been studied extensively in the last three decades, largely due to its increasingly important economic effect on the semiconductor industry, and several excellent texts are available on the subject from the viewpoint of integrated circuit design and protection (e.g., Amerasekera and Duvvury[4]). The following is a view of the ESD phenomenon from the perspective of a product containing integrated circuits.

10.3.1 The Triboelectric Effect

It was discovered in antiquity that, when certain dissimilar materials were rubbed together, what we now call an electrostatic field was generated between them. One of the materials becomes more positive with respect to the other. This phenomenon is now known as the triboelectric effect, or triboelectric charging. Modern work has led to the development of the "triboelectric series," an ordered list of materials from higher to lower electron affinities. The greater distance between two materials on the list, the easier it is to generate a charge redistribution resulting in the generation of an electrostatic field.

Triboelectric charging is usually the mechanism responsible for the generation of electrostatic voltages relevant to electronic products, but others occur, including the direct charging of a device by a charged second device and charging by the improper use of air ionizers (usually in factory settings).

10.3.2 Air Breakdown

The charge redistribution associated with an ESD event usually, although not always, includes an air discharge because the event itself lasts only nanoseconds, but most physical movement of electronic products occurs on much longer time scales. It is, therefore, useful to discuss air breakdown.

If an increasing voltage is developed between two fixed spheres in air, a point will be reached beyond which the air ionizes between the two spheres, and a spark travels between them. The electric field strength at which this initial breakdown occurs is dependent upon several factors, including the humidity and motion of the air and the surface roughness of the spheres, but may be conservatively estimated at approximately 1 million volts per meter. (In most cases, it is somewhat less, but it varies greatly and this value is sufficiently accurate for the purposes at hand.) If the distance between the two spheres is one meter, it requires 1 million volts to jump the gap. Viewed another way, if there is a difference of 15 kV between the spheres, they must be separated by at least 15 mm to prevent a discharge.

Note that once ionization occurs, the path between the spheres requires much less voltage differential to conduct because the gases in the path have already been ionized. If a current source exists to continue the spark (discharge), an arc results. This is familiar to horror movie fans as the Jacob's ladder — the vertical, diverging wires with the arc rising across them. The voltage is set just high enough so that the spark can initially form at the bottom (because that is where the wires are closest). Because it is connected to a high voltage transformer, it has a continuous supply of current, so a (continuous) arc forms. The arc rises because it heats the

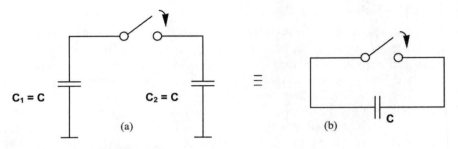

$C_1 = C$ $C_2 = C$ (a)   (b)   C

Exhibit 1.   Charge Redistribution

---

ionized channel of air through which it is passing. Eventually, the distance between the wires becomes too much to sustain the arc, and it stops; the process then repeats.[5]

It is important to distinguish the concept of the spark, which is an electrostatic effect (i.e., it does not conduct DC current), from that of the arc, which is an electrodynamic effect (and does conduct a DC current).[6]

10.3.3 Charge Redistribution

Because a spark is an electrostatic effect, and does not conduct a DC current, a return DC path to the source is not needed to support it. See Exhibit 1a.

Suppose a voltage V made by charge Q exists on capacitor C_1, and capacitor C_2 is initially uncharged. There is no energy stored in capacitor C_2, so the total energy of the two capacitors is the energy stored in capacitor C_1, or QV/2. If the switch is closed, the charge originally present on C_1 is redistributed between C_1 and C_2. Because Q = CV is conserved, and the capacitance is now doubled to 2C, the voltage on the capacitors is now halved, to V/2. The total energy stored in the two capacitors is now Q(V/2)/2 = QV/4. Where did the energy go? The "missing" energy, QV/4, was dissipated in the spark jumping the switch terminals as it closed.

The interesting point in this thought experiment is one that is rarely mentioned; it is not necessary for the "bottom" plates of the capacitors to be connected together, or to anything else, to see this effect, because it is an electrostatic effect — a physical manifestation of charge redistribution. No return path is needed because no direct return current exists— only the displacement current associated with the capacitance between the two capacitors. This is clear to those performing helicopter rescues on ships at sea; triboelectric charging associated with the helicopter moving through the air can result in a significant ESD event when the helicopter lowers its rescue basket to the deck of a ship, especially in dry, fair weather. Sailors are told to let the basket touch the deck first, before they touch it.

Perhaps a more intuitive (and in some ways, more physically correct) view is to remove the ground completely as illustrated in Exhibit 1b, and consider the ESD event to be the discharge of a single capacitance between two objects, one of which is neutral and one of which has a mobile charge Q placed on it. (Because charge is conserved, an object with a charge –Q on it exists somewhere in the universe, but this object will be ignored.) Without loss of generality, one may assume the neutral device as a reference; it is clear then that the only currents involved in the ESD event are the current through the discharge itself, and the displacement current through the electric field of the discharging capacitance.

10.4 THE EFFECTS OF ESD ON ICs

Generally, ESD can damage an IC in two ways: thermal effects due to the energy of the discharge applied to the small geometries of active devices or interconnects, and overvoltage effects, in which a MOS gate or other dielectric may break down. Additional information on damage mechanisms may be found in Amerasekera et al.,[7] Amerasekera and Duvvury,[8] and Duvvury and Amerasekera.[9]

The AC current density during an ESD event may be large enough to cause metal vaporization, resulting in metalization opens and circuit failure. Many ESD events can produce sufficient heating to melt the silicon of active devices, a feature exacerbated by the fact that the resistivity of silicon drops as it is heated. A section of an active device that begins to overheat decreases its resistance, encouraging more current to flow through that section, increasing the heating still further until failure occurs.

In addition to catastrophic device failures, an ESD event may degrade semiconductor junctions by the introduction of traps. These traps are a source of noise, and can raise the noise figure of RF devices. In wireless sensor network nodes, this is most troubling when an antenna (a large, easy target) is hit by an ESD discharge, which travels to the RF amplifier and may damage the RF amplifier transistor. Such effects can be immediate, or they can produce latent defects that may appear after weeks or months of operation.

The AC voltages present during an ESD event may cause dielectric breakdown of the gate oxide of MOS devices. As the minimum feature size of CMOS devices is reduced, the gate oxide thickness of the devices is also reduced, making them increasingly susceptible to this effect.

In addition to physical damage, an ESD event can result in system upset. One way in which this can occur is by the inversion of supply voltages. A positive discharge onto a "ground" plane may raise the potential of V_{ss} far above V_{dd}. A similar effect occurs with a negative discharge onto a V_{dd} circuit board runner. These events, of course, usually inhibit proper IC operation during the ESD event, and may have lasting effects afterward due to the loss of system state information. This can be caused, for example, by the loss of the microcomputer program counter value or activation of a system power-on reset. Similar effects can occur due to a discharge to the microcomputer reset line itself; unlike V_{ss} and V_{dd}, it is not necessary for the reset line to exceed the supply rails to cause an upset — a simple change of logical state (e.g., momentarily transitioning from a logical "1" to a logical "0") is all that is required. For this reason, the protection of internal reset lines is key to producing an ESD-resistant design.[10]

Because RAM operates by stored charge on integrated capacitors, it is sensitive to ESD events, and RAM memory values may be partially or completely erased by them. Note that it is possible to have memory erasures without an actual air discharge; the presence of a strong electric field is sufficient to change the charge on RAM cells.

10.5 MODELING AND TEST STANDARDS

For design, development, and testing of ESD-resistant devices and systems — just a part of a complete ESD control program[11] — the ESD event must be modeled. Two basic approaches are used: an analytical model, describing the event as a double exponential pulse, and an equivalent circuit model, describing pulse generation in terms of discrete circuit elements.

10.5.1 The Double Exponential Pulse

The waveform of an electrostatic discharge current has a 10 to 90 percent rise time of perhaps 150 ps to a peak amplitude of perhaps 10 Amperes, then falling off gradually with time.[12] The length of the pulse, as measured at the two 50-percent points, may be 150 ns. The discharge can be modeled as a double exponential pulse of the form[13]

$$y = S\,[e^{-at} - e^{-bt}] \tag{1}$$

where:

S is a scaling (amplitude) constant
a and b (with $a < b$) are two modeling parameters that define the shape of the pulse

For a typical ESD event, a is in the tens of MHz, and b is in the tens of GHz.

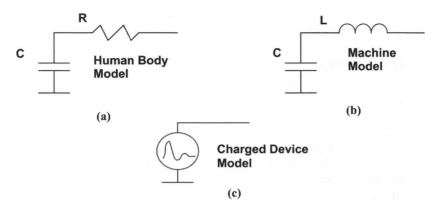

Exhibit 2.   ESD Models

10.5.2 Human Body, Machine, and Charged Device Models

Another way of modeling the ESD event is by an equivalent circuit model (Exhibit 2). To model a discharge from a human body, an RC network consisting of a capacitor discharging through a resistor is used. One may see differing component values used in the literature, reflecting different assumptions on the geometry and location (e.g., whether the discharge occurs when sitting or standing) of the discharge being modeled. For example, the IEC 61000-4-2 standard[14] recommends a 150-pF capacitor and a 330-Ω resistor; although the ESD STM5.1 standard from the Electrostatic Discharge Association (an industry organization)[15] recommends a 100-pF capacitor and a 1500-Ω resistor. To model a discharge from a small metal object, such as a conductive robotic arm, the so-called machine model has been used; however, there is some controversy over how well it represents ESD events encountered in factory equipment settings. Similar to the human body model, different component values have been proposed for this model. The IEC 61000-4-2 standard recommends a 200-pF capacitor and a 0-Ω resistor (i.e., no inductor or resistor); although the ESD STM5.2 standard[16] recommends a 200-pF capacitor discharged through a 500-nH inductor.

A third model, the charged device model (CDM), is more widely used today.[17] This model represents the discharge associated with a charged component, such as an integrated circuit, through one of its leads. It is viewed as more relevant than the machine model to simulate discharges that occur due to handling of devices in manufacturing processes prior to insertion into products. It is defined by a discharge waveform, not circuit elements, and so strictly speaking is not an equivalent circuit model, but is included here because the waveform greatly resembles the discharge of a capacitor through a very small series inductance and resistance.

10.5.3 Detailed Requirements of an ESD Standard

A large number of ESD standards and test methods are available.[18,19] It is instructional to consider the requirements of a particular standard in more detail. The IEC 61000-4-2 standard, from the International Electrotechnical Commission (IEC) is typical. This document "defines the immunity requirements and test methods for equipment which must withstand electrostatic discharges, from operators directly, and to adjacent objects. Several severity levels are defined which relate to different environmental and installation conditions."

The tests are to be made in a temperature (15–35°C) and humidity (30–60 percent) controlled environment. The IEC 61000-4-2 standard specifies the test set-up to include a metallic (copper or aluminum) ground reference plane (GRP) on the floor of the test area. A wooden table 0.8 m high is placed on the GRP. A horizontal coupling plane (HCP) is placed on the table, with the device under test separated from the HCP by an insulating support that is 0.5 mm thick.

The standard lists several types of discharge tests, and several voltages for each type of discharge. All voltages listed should be checked. It is quite possible for a device under test to pass at a high voltage, and fail at a low voltage. One way, for example, could be if an IC had an improperly designed ESD protection circuit that activated between the high and low voltage, but the IC itself could fail at the lower voltage, where the protection circuit did not protect it. It is also possible for a high voltage ESD discharge to jump to a new path that bypasses an IC, where a low voltage ESD discharge may take a different path through an IC, damaging it.

Field effect test. This test should be performed before the discharge tests. The standard states, "The EUT [equipment under test] shall be placed on the insulator on the HCP. The discharge probe, charged to 15 kV, shall be passed over the entire exterior (top, bottom, front, back, sides) at a distance of one-half inch."

Direct contact discharge. The tip of the discharge electrode should touch the EUT before the discharge switch is operated. The test voltages should begin at 4 kV, and continue in order through 6 kV and 8 kV, with at least ten discharges per polarity per voltage.

Direct air discharge. Air discharge is the most common, and the least repeatable, of the ESD test methods. The test voltages should begin at 4 kV, and continue in order through 8 kV, 10 kV, 12 kV, and 15 kV with at least ten discharges per polarity per voltage.

Indirect discharge. In the indirect discharge test, 8-kV ESD events are made to occur to the HCP and to a vertical coupling plane placed next to the EUT. The purpose is to identify failures caused by the induced current from the nearby discharge.

10.5.4 Performance Standards

ESD resistance may be specified at two levels: the product level and the component level. Product level performance standards are typically much higher than those of the components from which the product is made. Product level standards may reach 15 kV, although most integrated circuits are specified at 2 kV. Achieving 15-kV performance in a product built with 2-kV components requires good product design technique.[20]

10.6 PRODUCT DESIGN TO MINIMIZE ESD PROBLEMS

Several strategies can be employed to improve the robustness of a wireless sensor network node to an ESD event. Many of them concern the mechanical and industrial design of the node, and only tangentially involve its electrical design. Because many network nodes employ internal antennas and, therefore, must use plastic or other nonconductive housings, the use of grounded metal enclosures to protect the node are usually not possible. Nodes inside plastic housings can be made resistant to ESD events, however, if properly designed.[21,22]

10.6.1 Prevent Discharges from Entering or Exiting the Housing

The fundamental rule, and paradoxically the rule most easily overlooked by electrical design engineers, is to prevent electrostatic discharges from entering or exiting the housing of the network node. This can be accomplished by several techniques.[23]

10.6.1.1 Avoid Holes in the Housing. Eliminating housing holes is the first rule in the book. In fact, a waterproof network node would be ideal. It may seem strange to describe ESD as a mechanical engineering problem, but in this regard, it truly is so. Network node development programs that start from the first day worrying about openings in the housing, and how to keep conductors away from them, are far ahead of those who do not think of ESD until the first customer shipment. Holes in plastic housings of wireless sensor network nodes can be greatly reduced or eliminated by the use of a few design techniques:

- *Use tang-and-clevis joints.* As noted earlier, an air gap of 15 mm is needed to prevent a 15-kV discharge. This is obviously impractical for miniature wireless sensor network nodes that may be only 25 mm wide. The solution is to make the discharge travel an extended, nonlinear path. Where housing openings cannot be eliminated, create hairpin curves so that the path of the discharge through the opening doubles back on itself. This problem occurs often where two sides of a housing meet. The first impulse of most designers is to employ a butt joint (i.e., a joint in which the housing sections are simply abutted together). The butt joint inevitably leaves a gap, however, through which a discharge may travel to reach the sensitive circuits inside. The solution is to use a tang-and-clevis joint. See Exhibit 3. In the tang-and-clevis joint, the path of the discharge into the housing is serpentine and greatly extended, protecting the circuit board components inside.

- *Use elastomeric buttons and switches.* Because they represent holes in the housing, the number of buttons (and other user interface components) should be minimized. When buttons are required, however, elastomeric (e.g., silicone rubber) buttons and switches have many benefits for ESD protection. They are nonconducting, and they seal the housing switch opening with a layer of elastomer that can stay flat against the housing for a considerable distance, offering excellent ESD protection.
- *Avoid metallic external connections.* Wireless sensor network nodes typically connect to external sensors and actuators, often by cables. When the cables are in place, they can be significant ESD problems because they are large, conductive, and typically travel directly to the most sensitive circuits of the node (e.g., the microcomputer). When the cables are not in place, their external connectors represent metallic contacts leading directly from the outside to the sensitive circuits. The node must be protected in both conditions (i.e., with and without the cable present); however, it is very difficult to provide good ESD protection for these external metallic contacts. As an alternative, consider infrared (IR) or other noncontact communications. Using IR eliminates this problem, as long as there is an ESD-proof seal (i.e., a seal without any gaps or other openings in the housing) around the lens covering the IR LED and the phototransistor.

10.6.1.2 Locate Circuit Boards and the Metal on Them Away from Housing Holes. Due to product requirements other than ESD protection (e.g., customer requests, need for compatibility with other systems, etc.), almost every housing design is a compromise and has at least one hole. When holes are inevitable, locate circuit boards as far as possible from them. If moving the circuit board is not possible, at least move all metal runners (ground included) away from the holes. There is no need to invite a discharge, and discharges will not land on bare (i.e., etched) circuit board material. Do not forget to move internal metal on multilayer circuit boards back away from the edge of the board, as well — discharges can travel from the edge into the board, along the layer laminations.

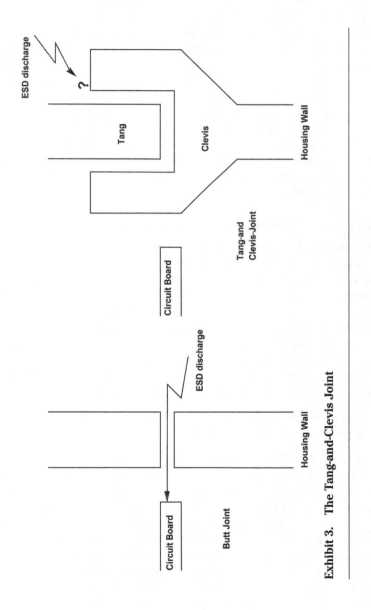

Exhibit 3. The Tang-and-Clevis Joint

10.6.1.3 Eliminate Metal Points and Burrs. Electric fields are particularly high near the points of sharp conductive objects — that is why lightning rods look like they do. The same effect happens anywhere in a wireless sensor network node where metal burrs, the cut ends of wire battery contacts, or edges of chrome connectors can get exposed to an electrostatic discharge. The resistance of a network node to a discharge can be improved to a remarkable degree by simple mechanical changes to metallic components. For example, turn the ends of a wire loop antenna inward, away from the edge of a circuit board. This makes it less likely that any metal burrs on the end of the antenna, produced when it was cut and formed in the manufacturing process, will be the destination of a discharge from outside the node. Continuing a wire battery contact spring for an additional one-half turn, so that the end of the spring is near the inside of the node, instead of the outside, is another example. Use round wire and large radii wherever possible, instead of stamped metal, due to the sharp points that can exist along stamped metal edges.

10.6.2 Once Inside, Design Paths for the Discharge to Travel

If, despite the best efforts of the designer, the discharge arrives on the circuit board, one is in a significantly more difficult position. The charge will want to redistribute on all metal it can reach; however, it still obeys Ohm's law, and the most current and the least voltage will develop across the lowest impedance. Thus, one must steer the ESD current pulse down a low impedance path that avoids sensitive components. Because the discharge is a high-frequency event, one should consider the impedance at high frequency of paths leading from the point of entry of the discharge. The electrostatic discharge, which has major spectral components above 1 GHz, will tend to avoid series inductance and will capacitively couple very easily.

With this behavior in mind, several circuit board layout design features can increase protection of integrated circuits:

- Use wide and short V_{dd} and V_{ss} circuit board runners to integrated circuits. Keep these runners close to each other, for maximum capacitance. This design method produces low-impedance structures that encourage more of the discharge energy to remain on the circuit board, and less to travel inside the integrated circuit. In addition, wide, short runners produce smaller voltage drops when large ESD currents pass through them. This reduces the likelihood of voltage upset. Finally, it minimizes the area and, therefore, the radiation efficiency of any high-frequency current loop resulting from an ESD event.
- Place V_{dd} and V_{ss} decoupling capacitors close to their integrated circuit, with wide, short connections. This policy allows the discharge to reach the V_{dd} and V_{ss} pins of the integrated circuit nearly simulta-

neously, resulting in little or no change in their voltage differential. This reduces the likelihood of voltage upset.

- Place RC networks on sensitive digital inputs. Series R-shunt C networks can protect sensitive digital inputs of integrated circuits, by making the entry into the integrated circuit a higher impedance path than an alternative on the circuit board. The series resistor should be placed as close to the integrated circuit as possible, to reduce the possibility that the discharge may travel around the resistor by another path and still enter the integrated circuit. The shunt capacitor should be placed as close as possible to the point of discharge entry onto the input runner, so that the discharge is controlled as early as possible in its path on the circuit board.

- Make circuit board runners to external connector contacts as narrow as possible. This includes the ground runner to the connector. Narrow runners have high series inductance and, therefore, higher impedance to a high-frequency ESD event.

- Consider the use of diodes between digital signal lines and both V_{dd} and V_{ss} to clamp the voltage swing possible on the signal lines during an ESD event. Keep in mind, however, that, many times, the diode switching time is too slow to be effective, and that the real mechanism at work is likely to be the diode capacitance, which acts as a bypass to carry the discharge current away from the integrated circuit. If an experiment is performed and diodes are found to be successful, try replacing the diodes with capacitors, which are usually smaller and cheaper.

- Use care in the placement of "influential" microcontroller runners, such as reset and interrupt request lines. Keep them away from the edges of circuit boards and other ESD entry points, such as housing openings. Small capacitors from the influential runners to V_{dd} and/or V_{ss} can be used to bypass electrostatic discharges. When multilayer circuit boards are used, runners may also be buried on their inner layers.

- Keep high-impedance circuits away from ESD entry points. High-impedance circuits are the circuits most susceptible to ESD charge injection. An example is the microcontroller clock circuit. The impedance of the crystal oscillator circuit may be 10^6 Ohms or more; a small amount of charge placed there can be sufficient to upset the circuit.

- Keep voltage management and voltage or current references away from ESD entry points. Many times, ESD failures result not from the corruption of signal lines, but from corruption of the references to which they are compared. This frequently happens in the case of external analog sensors using an internal analog-to-digital converter. Corruption of the voltage reference used by the analog-to-digital converter is just as damaging as corruption of the sensor data itself.

- Consider spark gaps for troublesome entry paths. Spark gaps, or field effect structures, can be used to encourage electrostatic discharges to follow desired paths. These structures, which are sharp shapes of circuit board metalization, can produce low-impedance paths for high-voltage signals, while remaining high-impedance paths for low-voltage signals. These can be helpful in particularly difficult ESD situations, such as those involving external connectors and other direct connections from the outside to integrated circuits. They should be placed as close as possible to the discharge source. Other nearby metal runners and edges, associated with undesired ESD paths, should be as round and smooth as possible.

10.6.3 Reaching the Integrated Circuit

Protecting an integrated circuit from an electrostatic discharge is difficult because the structures in an integrated circuit have very limited power dissipation capabilities and overvoltage survival abilities. Once a discharge reaches the package of an integrated circuit, however, it is likely that the discharge has been weakened by both wide distribution and attenuation. The integrated circuit designer has several design techniques available to further attenuate the discharge and to design paths for the discharge to travel away from sensitive circuits within the chip itself. There has been a great amount written in the literature on ESD-resistant design of integrated circuits; exemplary resources are Amerasekera et al.,[24] Amerasekera,[25] Amerasekera and Duvvury,[26] Duvvury and Amerasekera,[27] and Duvvury.[28]

The first step in integrated circuit protection is to employ structures at the pads of the chip to protect internal circuits from incoming ESD discharges. Protection structures are widely used for both CMOS[29] and BiCMOS[30] integrated circuits, although their design necessarily changes greatly with the minimum feature size and other details of the fabrication process.[31] Generally, however, pad-level ESD protection is usually accomplished by the same impedance-control methods used at the circuit board level. These impedance-control methods provide low shunt impedances to the V_{dd} or V_{ss} supply of the circuits to which the pads are connected, while providing high series impedances on the signal lines leading to the protected circuits. For digital input and output pads, the series impedance is often just a series resistor. The mechanism by which the low impedance to the V_{dd} or V_{ss} supply is generated can vary greatly, from passive systems such as integrated capacitors and diodes, to relatively sophisticated active circuits that detect the high incoming ESD voltage (or current), then couple the signal pad to one of the V_{dd} or V_{ss} supplies for the duration of the event. The design of ESD protection circuits is nontrivial because the circuits must survive the ESD event themselves, and most circuit modeling software does not adequately model the behavior of semiconductor

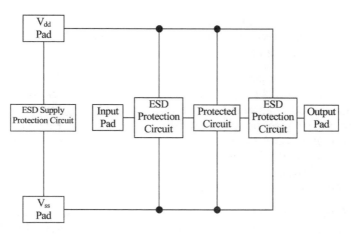

Exhibit 4. IC Protection

---

devices at the extreme levels of voltage and current commonly encountered during a discharge. As the minimum feature size of integrated devices shrinks with improving lithographic techniques, circuit devices become more sensitive to the effects of ESD, and the design of their associated ESD protection systems becomes more difficult. One advantage of small-geometry processes, however, is that their switching devices can be fast; this can be useful in some active protection circuit designs if the devices are not directly exposed to the ESD event itself.

The second step in ESD protection is to protect the supplies of the integrated circuit. ESD supply protection circuits are two-terminal structures that generally have the dual role of reverse-voltage protection, so that V_{ss} does not exceed V_{dd} by a significant amount, and forward-voltage clamping, so that V_{dd} does not rise so high relative to V_{ss} that damage may occur. The final protection scheme is illustrated in Exhibit 4.

The design of ESD protection structures is made still more difficult when it is desired to protect RF, instead of digital, signal pads. Because the purpose of such pads is to pass high-frequency energy, and the circuit impedance is relatively low, it is usually not possible to place fixed resistors in series with them without affecting their performance. Further, the parasitic capacitance associated with most ESD protection circuits often provides an unacceptably low impedance path to the V_{dd} or V_{ss} supplies for the desired signal, even when the circuit is not active. The parasitic capacitance is often high due to the size of the ESD protection circuit required to ensure that it can pass the discharge current safely, without damage occurring, and that its resistance is low enough to avoid the generation of an IR drop sufficient to damage the protected circuit. In addition, the ESD protection structure itself can degrade the noise figure of low-noise amplifiers, due to capacitive coupling to noise present in the substrate or to noise generated within the protection structure itself.[32]

Nevertheless, it is possible to achieve a level of ESD protection to RF pads. One approach is to incorporate the parasitic capacitance of the ESD protection structure as part of the impedance matching structure of the RF circuit, and to employ needed matching components for ESD protection.[33] An example of this is the use of the series matching inductor in a CMOS low noise amplifier as the series input impedance element providing ESD protection to the circuit.[34,35] Recently, a 900-MHz receiver that survives 8-kV Human Body Model (HBM) ESD events has been reported.[36]

An additional level of complexity arises when systems-on-a-chip (SoCs) are considered. For EMC reasons, SoCs often employ several multiple V_{dd} and V_{ss} supply pins, each supplying specific circuits. For example, the supplies for analog and digital circuits are often separated, and supplied from separate pins on the integrated circuit. It then becomes necessary to place additional integrated ESD supply protection circuits between the supplies. The resulting system is illustrated in Exhibit 5. An important consideration when such multiple-supply systems are used is to ensure that the supply-to-supply ESD protection structures will not conduct under all possible conditions of operation (short of an actual ESD event, of course), and for all possible variations in the supply voltages. For example, suppose V_{dd2} is nominally 2.0 V, generated by a switching voltage converter from V_{dd1}, supplied by removable primary battery, with a nominal voltage of 1.5 V. Further, assume that the ESD supply protection structure between the two supplies is a simple reverse-biased diode. Under normal operation, V_{dd1} is always less than V_{dd2}, so no current is drawn by the protection structure; however, when a battery is first inserted in an inactive device, V_{dd1} immediately appears, although V_{dd2} is still at or near zero volts because the voltage converter has not yet become active. This can result in a very large current being drawn by the now forward-biased ESD protection structure. Under these conditions, an anomalous high current spike will occur when the battery is inserted. In the best-case scenario, this current does not cause damage to the protection structure, and the spike ends when the voltage of V_{dd2} rises to the level needed to reverse-bias the ESD protection structure again. In the worst-case scenario, the current spike is large enough to cause the terminal voltage of the battery to drop significantly (due to the nonzero ESR of the battery). The terminal voltage may drop far enough, in fact, that the voltage converter will be unable to generate its nominal 2.0 V, latching the node in this undesirable state. Often, the exit from this inoperative but battery-draining state occurs when the ESD protection device fails in an open circuit; alternatively, the battery may quickly discharge. The problem of undesired conduction can also occur at a

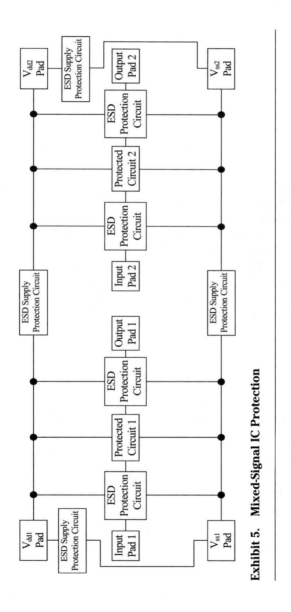

Exhibit 5. Mixed-Signal IC Protection

temperature extreme, if V_{dd1} and V_{dd2} have significantly different temperature coefficients. In this example, the problem could be prevented by placing not one, but three series diodes in the ESD supply protection circuit. With three diodes, the 1.5 V of the replaced battery is insufficient to cause conduction of the ESD supply protection circuit. It also means that the ESD supply protection circuit between V_{dd1} and V_{dd2} does not begin to protect the V_{dd1} supply during an ESD event until V_{dd1} rises above the V_{dd2} voltage plus three forward diode voltage drops, or about four volts in this example. Until then, the V_{dd1} supply pin must be protected by the ESD supply protection circuit to V_{ss1}. Whether or not this is sufficient depends on the integrated circuit technology and type of discharge at hand; but for these and other reasons, ESD supply protection circuits more sophisticated than a simple reverse-biased diode are often used in modern integrated circuit designs.

An additional point that is often overlooked in ESD protection is the assignment of signals to package pins (or pads, in the case of ball grid arrays [BGAs] and other pinless integrated circuit packages). In practice, ESD events on integrated circuits tend to strike the corner pins most often because they are the most exposed. On BGAs, the outer row of pads, and especially the corner pads on the outer row, are the most exposed. One may take advantage of this effect and assign the most sensitive signal lines, such as those leading to RF and analog inputs and other minimally protectable CMOS gates, to pins or pads in the center of the package. In the case of RF input pins, this approach has the added benefit of (usually) minimizing the length of input wire bonds, providing the minimum possible parasitic package inductance and, therefore, the minimum variation in the input impedance match associated with manufacturing variations of bond wire length.

10.6.4 Once an ESD Event Occurs, Limit Discernable Effects

Despite all efforts to the contrary, an ESD event may still occur. If it produces physical damage, little can be done in most cases except replacement of the affected component or the entire node. One of the few exceptions to this is the use of memory error detection and correction techniques, which can help identify and avoid permanently damaged memory elements; however, few other portions of a wireless sensor network node contain redundancy comparable to that found in a memory. So, generally, a node cannot automatically recover from a hardware failure.

The same is not true, though, for system upsets. The effects of a nondestructive ESD event are greatly affected by system firmware and software.[37] The analogy is to the engineering definitions of stress and strain: stress is the disturbing force; strain is the response of the system to the applied stress. In the case of ESD-resistant firmware and software design, stress is often the corruption of a register by an ESD event, and strain is how the system firmware and software respond to the incorrect register value. ESD-resistant firmware can reduce the consequences of ESD events, should they occur.

The design of ESD-resistant firmware and software is often done at increased cost, both in development time and in size of the implemented code. The code size is a particularly sensitive point, due to the large effect memory costs have on total network node cost. This cost must be balanced with the "mission-critical" nature of the wireless sensor network application (i.e., the effect losing a network node or having it perform an improper operation due to an ESD event will have on the overall application) and the cost of additional hardware efforts to reduce the rate of ESD events. Often it is less expensive to reduce or eliminate the effect of an ESD event than it is to eliminate the ESD event itself.

Register corruption can affect all parts of microcontroller operation. Corruption of data registers is an obvious problem. In addition, because the Program Counter is an internal register, it can be corrupted by an ESD event, leading to incorrect program flow. Finally, interrupt masks and other status information are also stored in internal registers; corruption of these can lead to improper behavior.

One of the basic tenets of ESD-resistant programming is to not trust the existing state of the microcontroller. Bits set or cleared at some time in the past may have become corrupted due to an ESD event. To protect against this possibility, the desired values should be restored immediately prior to their use, regardless of their present state. This minimizes the time during which the system is vulnerable to upset. Other techniques are specific to the type of information stored:

- *Data integrity checks.* When possible, the data coming in from external sources should be checked for its integrity. Digital data should employ an error-correcting code (ECC), such as a Reed–Solomon or other block code, added at the source and checked upon receipt to detect and correct data corruption during the transmission process (regardless of the transmission medium, i.e., whether the data was received via a connected cable or via the RF link). Such a method can even correct corruption due to an ESD event occurring at another network node. If the implementation of an error-correcting code is not possible, at a minimum, a parity bit or cyclic redundancy check (CRC) block should be checked. Data produced by an analog-to-digital converter should have simple checks performed for "reasonableness." For example, if the process is known to be slowly varying, like the room temperature of a house, the last value of the data sequence (in degrees Celsius) of 29, 29, 30, −1 (in hex, 1D, 1D, 1E, FF) should be viewed with suspicion — especially if the samples are taken once per second. Incoming data that cannot be corrected by the ECC employed, or fails the CRC or other incoming integrity check, should not be used, of course; depending upon the application, it may be simply ignored, a default value used in its place, or a retransmission request may be made. The integrity of stored data should be rechecked immediately prior to its use, to avoid the use of data corrupted during storage. Critical data should be stored in multiple places in RAM; at least two copies should be compared and found identical, prior to use of the data.

- *Program flow checks.* These checks confirm that the program flow is correct, and that the program has not improperly entered some area of memory, due perhaps to corruption of the program counter. The first line of defense here is the watchdog timer, which resets the microcontroller should it not receive regular setting from the running program. The second line of defense is checks within the running code itself. Progression through the program can be monitored by the use of flags set when a segment of code is entered, then checked against the set value when the segment is exited. If, upon exiting, the flag is not set, it indicates that that segment of code was entered improperly.

- *Status information checks.* Modern microcontrollers have several registers dedicated for specific purposes. One example cited at the beginning of this chapter was the register that stored the increment rate of the microcontroller's real time clock. As in this example, when these registers get corrupted, unexpected software events (USEs) can occur. These registers should have their values regularly refreshed as part of the program's regular housekeeping subroutine, especially those values, such as the RTC increment rate, that are adjustable by design of the microcontroller, but fixed for any particular application.

Boxleitner[38] has an excellent discussion on principles useful in ESD-resistant programming.

10.7 CONCLUSION

This chapter reviewed the problem of ESD as it affects wireless sensor network nodes, and the importance of proper electrical, mechanical, and product design in its control. ESD is a pervasive problem in the design of network nodes, due, in part, to their use of high levels of integration with state-of-the-art integrated circuit processes, and also to their need to be inexpensive, yet well-connected to the outside world via sensors and actuators. The need to be inexpensive often leads to the need for internal antennas, which leads to the need for a plastic (i.e., nonconductive) housing. The outside connections can be sources of ESD events that can be difficult to predict, and, therefore, to design against.

Interestingly, although an electrostatic discharge is clearly an electrical event, many design techniques to protect network nodes from its effects involve the mechanical design of the node, including the need to minimize the number and size of holes in the product housing, how external connections are made to the node, and the design of buttons and other user interface devices. These efforts reduce the possibility of an ESD event reaching the node's internal components. Should that happen, however, in addition to hardware design points to minimize the likelihood of physical damage, firmware and software techniques can be employed to minimize the likelihood and severity of system upset resulting from an ESD event.

References

1. Steven H. Voldman, Lightning rods for nanoelectronics, *Scientific American,* v. 287, n. 4, October 2002, pp. 90–97.
2. Owen J. McAteer, *Electrostatic Discharge Control.* Upper Falls, MD: MAC Services Incorporated. Available from the Electrostatic Discharge Association, Rome, NY. 1990. Chapter 11.
3. Yogi Anand and Dana Crowe, Latent ESD failures in Schottky barrier diodes, *Proc. Electrical Overstress/Electrostatic Discharge Symp.,* 1999, pp. 160–167.
4. Ajith Amerasekera and Charvaka Duvvury, *ESD in Silicon Integrated Circuits.* 2nd ed. Chichester, West Sussex, England: John Wiley & Sons. 2002.
5. The arc has another interesting property — negative resistance. Once it is formed, an increase in current across the arc makes the arc thicker, resulting in a lowering of the voltage drop across the arc. This phenomenon was used to make continuous wave radio transmitters of up to 1 MW of output power by 1918, before the advent of short waves (and the vacuum tube) rendered them obsolete.
6. This distinction also holds for the arc transmitter, which is not to be confused with the spark transmitter, popular at about the same time but which transmitted damped, instead of continuous, waves.
7. Ajith Amerasekera et al., ESD failure modes: characteristics, mechanisms, and process influences, *IEEE Trans. Electron Devices,* v. 39, n. 2, February 1992, pp. 430–436.
8. Amerasekera and Duvvury, *ESD in Silicon Integrated Circuits.*
9. Charvaka Duvvury and Ajith Amerasekera, ESD: a pervasive reliability concern for IC technologies, *Proc. IEEE,* v. 81, n. 5, May 1993, pp. 690–702.
10. William D. Kimmel and Daryl D. Gerke, Three keys to ESD systems design, *EMC Test Design,* September 1993, pp. 24–27.
11. Electrostatic Discharge Association, ANSI/ESD-S20.20-1999: ESD Association Standard for the Development of an Electrostatic Discharge Control Program for Protection of Electrical and Electronic Parts, Assemblies and Equipment (Excluding Electrically Initiated Explosive Devices). Rome, NY: Electrostatic Discharge Association. 1999.
12. It is interesting to note that the stated rise time of an ESD event has been decreasing in the literature over the years, as measurement equipment improves.
13. A. M. J. Mitchell, Analyzing the double exponential pulse, *EMC Test Design,* September 1994, pp. 27–30.
14. International Electrotechnical Commission, IEC 61000-4-2: Electromagnetic Compatibility (EMC) Part 4-2: Testing and Measurement Techniques — Electrostatic Discharge Immunity Test. Edition 1.1. Geneva: IEC. May 1999.
15. Electrostatic Discharge Association, ESD-STM5.1-2001: ESD Association Standard Test Method for Electrostatic Discharge Sensitivity Testing — Human Body Model (HBM) Component Level. Rome, NY: Electrostatic Discharge Association. 2001.

16. Electrostatic Discharge Association, ANSI/ESD-STM5.2-1999: ESD Association Standard Test Method for Electrostatic Discharge Sensitivity Testing — Machine Model — Component Level. Rome, NY: Electrostatic Discharge Association. 1999.

17. Electrostatic Discharge Association, ANSI/ESD-STM5.3.1-1999: ESD Association Standard Test Method for Electrostatic Discharge Sensitivity Testing — Charged Device Model (CDM) Component Level. Rome, NY: Electrostatic Discharge Association. 1999.

18. Stanley Weitz, New trends in ESD test methods, *EMC Test Design,* February 1993, pp. 22–26.

19. P. Richman, Handheld MIL-STD-883C ESD PCB testing, *EOS/ESD Technol.,* February/March 1992.

20. Warren Boxleitner, The ESD threat to PCB-mounted ICs, *EOS/ESD Technol.,* October/November 1991.

21. Warren Boxleitner, *Electrostatic Discharge and Electronic Equipment: A Practical Guide for Designing to Prevent ESD Problems.* New York: IEEE Press. 1989.

22. M. Mardiguian, ESD hardening of plastic housed equipment, *EMC Test & Design,* July/August 1994, pp. 21–24.

23. Boxleitner, *Electrostatic Discharge and Electronic Equipment.*

24. Ajith Amerasekera et al., Modeling MOS snapback and parasitic bipolar action for circuit-level ESD and high current simulations, *Proc. 34th Annu. IEEE Intl. Reliability Physics Symp.,* 1996, pp. 318–326.

25. Ajith Amerasekera, Addressing ESD for microprocessors and ASICS in 21st-century technologies, *Symp. on VLSI Circuits Dig. of Technical Papers,* 2000, pp. 84–87.

26. Amerasekera and Duvvury, *ESD in Silicon Integrated Circuits.*

27. Charvaka Duvvury and Ajith Amerasekera, ESD: a pervasive reliability concern for IC technologies, *Proc. IEEE,* v. 81, n. 5, May 1993, pp. 690–702.

28. Charvaka Duvvury, ESD protection device issues for IC designs, *Proc. IEEE Custom Integrated Circuits Conf.,* 2001, pp. 41–48.

29. Charvaka Duvvury and Ajith Amerasekera, Advanced CMOS protection device trigger mechanisms during CDM, *IEEE Trans. Components, Packaging, and Manufacturing Technology — Part C,* v. 19, n. 3, July 1996, pp. 169–177.

30. William D. Mack and Robert G. Meyer, Protecting BiCMOS circuits, *EMC Test Design,* July/August 1992, pp. 26–31.

31. Amerasekera and Duvvury, *ESD in Silicon Integrated Circuits.* Chapter 9.

32. Ke Gong et al., A study of parasitic effects of ESD protection on RF ICs, *IEEE Trans. Microwave Theory and Techniques,* v. 50, n. 1, January 2002, pp. 393–402.

33. Frederic Stubbe et al., A CMOS RF-receiver front-end for 1 GHz applications, *Symp. VLSI Circuits Dig. of Technical Papers,* 1998, pp. 80–83.

34. P. Leroux and M. Steyaert, High-performance 5.2 GHz LNA with on-chip inductor to provide ESD protection, *Electronics Lett.,* v. 37, n. 7, 29 March 2001, pp. 467–469.

35. Paul Leroux, Johan Janssens, and Michiel Steyaert, A 0.8-dB NF ESD-protected 9-mW CMOS LNA operating at 1.23 GHz [for GPS receiver], *IEEE J. Solid-State Circuits,* v. 37, n. 6, June 2002, pp. 760–765.

36. Ming-Dou Ker et al., ESD protection design for 900-MHz RF receiver with 8-kV·HBM ESD robustness, *Proc. IEEE Radio Frequency Integrated Circuits Symp.,* 2002, pp. 427–430.

37. Boxleitner, *Electrostatic Discharge and Electronic Equipment.*

38. Ibid.

Chapter 11
Wireless Sensor Network Standards

11.1 INTRODUCTION

The value of wireless sensor networks rests on their low cost and distribution in large numbers. To achieve the economies of scale needed to reach a large market at low cost, certain features of wireless sensor networks need to be standardized, so that products from many manufacturers may interoperate. This synergy will add utility to wireless sensor networks, and therefore encourage their use.

To this end, efforts are under way to standardize the various layers of wireless sensor network communication protocols, including the data payloads sent by their sensors. The success of wireless sensor networks as a technology rests on the success of these standardization efforts to unify the market, leading to large numbers of low cost, interoperable devices, and avoiding the proliferation of proprietary, incompatible protocols that, although perhaps optimal in their individual market niches, will limit the size of the overall wireless sensor market. Two of these standardization efforts are the IEEE 802.15.4 Low Rate Wireless Personal Area Network (WPAN) standard[1] (with the ZigBee Alliance, its marketing and compliance certification organization), and the IEEE 1451.5 Wireless Smart Transducer Interface standard.

11.2 THE IEEE 802.15.4 LOW-RATE WPAN STANDARD

As noted in Chapter 1, the scope of the IEEE 802.15.4 task group, as defined in its original Project Authorization Request, is to "define the PHY and MAC specifications for low data rate wireless connectivity with fixed, portable and moving devices with no battery or very limited battery consumption requirements typically operating in the Personal Operating Space (POS) of 10 meters." Further, the purpose of the project is "[t]o provide a standard for ultra low complexity, ultra low cost, ultra low power consumption and low data rate wireless connectivity among inexpensive devices. The raw data rate will be high enough (maximum of 200 kbps) to satisfy a set of simple needs such as interactive toys, but scaleable down to the needs of sensor and automation needs (10 kbps or below) for wireless communications."[2] The maximum and minimum raw data rates were later raised to 250 and 20 kb/s, respectively.[3]

This diverse set of goals requires the IEEE 802.15.4 standard to be extremely flexible. Unlike file transfer protocols such as IEEE 802.11 that are designed for a single application, the IEEE 802.15.4 standard supports a nearly infinite variety of possible applications in the POS. These applications vary from those requiring high data throughput and relatively low message latency, such as wireless keyboards, mice, and joysticks, to those requiring very low throughput and able to tolerate significant message latency, such as intelligent agriculture and environmental sensing applications. The IEEE 802.15.4 standard supports both star and peer-to-peer connections, and is therefore able to support a wide variety of network topologies and routing algorithms. When security is used, the AES-128 security suite[4] is required.

The standard employs beacons, although their use is optional. The beacon period is variable in binary multiples of 15.36 ms, up to a maximum of 15.36 ms $\times 2^{14}$ = 4 minutes, 11.65824 seconds, so that the optimum trade-off can be made between message latency and network node power consumption for each application. Beacons may be omitted for applications that have duty cycle limitations, as can happen on networks in the 868 MHz band (which has regulatory limits on network node duty cycle), or applications that require network nodes with constant reception. Channel access is contention based, via a carrier sense multiple access mechanism with collision avoidance (CSMA-CA); the beacon is followed by a contention access period (CAP) for devices attempting to gain access to the channel. The length of the CAP is adjustable as a fraction of the period between beacons. A "battery life extension" mode is also available that limits the CAP to a fixed time of approximately 2 ms. To address the needs of applications requiring low message latency, the standard supports the use of optional guaranteed time slots (GTSs), which reserve channel time for individual devices without the need to follow the CSMA-CA access mechanism.

The standard has a 16-bit address field, meaning that up to $(2^8 - 2) \times (2^8 - 2)$ = 64,516 devices may be assigned logical addresses (two values in each byte are reserved); however, the standard also includes the ability to send messages with 64-bit extended addresses, allowing an almost unlimited number of devices to be placed in a single network. Message transmission can be fully acknowledged; each transmitted frame (with the exception of beacons and the acknowledgments themselves) may receive an explicit acknowledgment. This produces a reliable protocol; the overhead associated with explicit acknowledgments is acceptable given the low data throughput typical of wireless sensor networks. The use of acknowledgments is optional with each transmitted frame, however, to support the use of passive acknowledgment techniques. These techniques are used in some ad hoc routing schemes, for example, the gradient routing (GRAd) algorithm discussed in Chapter 5.

The IEEE 802.15.4 standard incorporates many features designed to minimize power consumption of the network nodes. In addition to the use of long beacon periods and the battery life extension mode, the active period of a beaconing node can be reduced (again by powers of two), allowing the node to sleep between beacons.

Coexistence with other services using the unlicensed bands with IEEE 802.15.4 devices was also a major factor in the protocol design, and is evident in many of its features. For example, dynamic channel selection is required; should interference from other services appear on a channel being used by an IEEE 802.15.4 network, the network node in control of the network (the personal area network [PAN] coordinator) scans other available channels to find a more suitable channel. In this scan, it obtains a measure of the peak energy present in each alternative channel and then uses this information to select a suitable channel. This type of scan can also be used prior to the establishment of a new network. Prior to each frame transmission (other than beacon or acknowledgment frames), each IEEE 802.15.4 network node must perform two clear channel assessments (CCAs) as part of the CSMA-CA mechanism to ensure the channel is unoccupied prior to transmission.

A link quality indication (LQI) byte is attached to each received frame by the physical layer before it is sent to the medium access control layer. The receiving node expects to used this information for a number of purposes, at the discretion of the network designer:

- It can be used as an indication of channel impairment, perhaps leading to the need to perform the dynamic channel selection process and move to another channel.
- It can be used for power control of its own transmitter, under the assumption of a symmetrical channel.
- It may be used as part of a relative location determination algorithm, to estimate the location of each network node relative to its peers.
- It may be used as part of a network routing algorithm to establish packet routes based only on link quality between network nodes.

The LQI may be generated from a signal level determination, a signal-to-noise determination, or a combination of the two, at the discretion of the network node implementer. This enables both received signal strength indication (RSSI) and correlation-based signal quality estimators to be used. Although a byte (8 bits) is reserved for the LQI, to ease the burden on implementers that do not desire to make use of it (an appropriate decision for some applications), the standard specifies that at least eight unique values shall be used in the LQI, including 0×00 and $0 \times FF$. The values 0×00 and $0 \times FF$ are to be associated with the lowest and highest quality IEEE 802.15.4 signals detectable by the receiver, respectively.

To maximize the utility of the standard, the IEEE 802.15.4 task group had to balance the desire to enable small, low-cost, and low-power network nodes with the desire to produce a standard that met a wide variety of market applications. The resulting standard includes three types of network node functionality:

- *PAN coordinator.* The PAN coordinator is the node (strictly speaking, the coordinator node) that initiates the network and is the primary controller of the network. The PAN coordinator may transmit beacons and can communicate directly with any device in range. Depending on the network design, it may have memory sufficient to store information on all devices in the network, and must have memory sufficient to store routing information as required by the algorithm employed by the network.
- *Coordinator.* The coordinator may transmit beacons and can communicate directly with any device in range. A coordinator may become a PAN coordinator, should it start a new network.
- *Device.* A network device does not beacon and can directly communicate only with a coordinator or PAN coordinator.

These three functions are to be embodied into two physically different device types:

- *Full function device (FFD).* An FFD can operate in any of the three network roles (PAN coordinator, coordinator, or device). It must have memory sufficient to store routing information as required by the algorithm employed by the network.
- *Reduced function device (RFD).* An RFD is a very low cost device, with minimal memory requirements. It can only function as a network device.

The archetypical RFD is the wireless light switch. It must be as inexpensive to produce as possible, is likely to be battery-powered, and has very limited functional requirements, needing to communicate only with a light. The light itself, however, may be the archetypical FFD because it can be (slightly) more expensive, has access to mains power, and can have additional network functions as a more permanent feature of the building.

As noted earlier, the IEEE 802.15.4 standard supports multiple network topologies. In the standard, two general types are discussed — star networks and peer-to-peer networks. In the star network, the master device is the PAN coordinator (an FFD), and the other network nodes may either FFDs or RFDs. In the peer-to-peer network, FFDs are used, one of which is the PAN coordinator. RFDs may be used in a peer-to-peer network, but they can only communicate with a single FFD belonging to the network, and so do not have true "peer-to-peer" communication.

Similar to all IEEE 802 wireless standards, the IEEE 802.15.4 standard standardizes only the physical and medium access control (MAC) layers. The IEEE 802.15.4 standard, in fact, incorporates two physical layers:

- *The lower band:* 868.0–868.6 MHz (for Europe), plus the 902–928MHz (for much of the Americas and the Pacific Rim)
- *The upper band:* 2.400–2.485 GHz (substantially worldwide)

The channel numbers and their center frequencies are defined as follows:

$$Fc = 868.3 \text{ MHz, for } k = 0$$

$$Fc = 906 + 2 \ (k - 1) \text{ MHz, for } k = 1, 2, \ldots, 9, 10$$

$$Fc = 2405 + 5 \ (k - 11) \text{ MHz, for } k = 11, 12, \ldots, 25, 26$$

where k is the channel number.

Both lower and upper bands employ a form of direct sequence spread spectrum (DSSS). In the lower band, binary phase shift keying (BPSK) with raised-cosine pulse shaping is employed. In the 868-MHz band, a data rate of 20 kb/s and a chip rate of 300 kc/s are used, while in the 902–928-MHz band, a data rate of 40 kb/s and a chip rate of 600 kc/s are used. In the upper band, a modified version of the scheme described in Chapter 3 is used;[5] offset quadrature phase shift keying (O-QPSK) with half-sine pulse shaping is employed at a chip rate of 2 Mc/s, along with a 16-ary orthogonal symbol scheme sent at 62.5 ksymbols/s, resulting in a data rate of 250 kb/s. The PN sequences of each of the orthogonal symbols are related to each other through cyclic shifts and/or conjugation (inversion of chips transmitted on the Q-channel). Because all of the possible received symbols may be easily derived from a single PN sequence, this scheme simplifies the design of the receiver in comparison with other 16-ary DSSS methods that would require the storage of sixteen PN sequences.

11.3 THE ZIGBEE ALLIANCE

As noted previously, the IEEE 802.15.4 standard does not standardize the higher communication protocol layers, including the network and application layers. To assure interoperability between devices operating the IEEE 802.15.4 standard, the behavior of these layers must be specified. The creation of such a specification has been taken up by the ZigBee Alliance,[6] San Ramon, California, an industry consortium of chip manufacturers, OEM manufacturers, service providers, and users in the wireless sensor network market, many of which worked to develop the IEEE 802.15.4 standard itself.

In addition to the creation of higher layer specifications, the ZigBee Alliance is also the marketing and compliance arm of IEEE 802.15.4, in a manner analogous to the relationship between the Wi-Fi Alliance, Mountain View, California and the IEEE 802.11 WLAN standard (marketed as "Wi-Fi®").

Development of the ZigBee network specification, which will include both star and peer-to-peer topologies, along with the first application profiles, is due for completion in the first half of 2004. The development of additional application profiles is expected to be a continuing process, due largely to the widening use of wireless sensor network technology.

11.4 THE IEEE 1451.5 WIRELESS SMART TRANSDUCER INTERFACE STANDARD

The development of wireless sensor network technology and standards is not being driven solely by the communications interface. Manufacturers, implementers, as well as users of sensors have expressed a desire for wireless connectivity.[7] This need has been recognized by the IEEE, and has led to the formation of the IEEE 1451.5 Wireless Sensor Working Group.[8]

The IEEE 1451 family of standards (e.g., Std. 1451.1-1999[9] and Std. 1451.2-1997[10]) is sponsored by the Technical Committee on Sensor Technology of the IEEE Instrumentation and Measurement Society. The development of these standards followed a meeting in 1993 that was cosponsored by the IEEE and the National Institute of Standards and Technology, a division of the U.S. Department of Commerce, to address the issue of transducer compatibility. Manufacturers were finding it difficult to produce transducers compliant with the increasingly large number of network communication protocols. The solution was the development of a smart transducer interface, which is a single communication protocol usable by all sensors; this effort became IEEE 1451.

One of the major benefits provided by the IEEE 1451 standardization process was the creation of a standard Transducer Electronic Data Sheet (TEDS) in IEEE 1451.2.[11] This standard facilitated the use of smart sensors by standardizing the interface between the sensors and the "Network Capable Application Processor (NCAP)" (i.e., the protocol-handling processor between the sensor and the network). As part of this standardization, the TEDS was created to provide a way in which the transducers can describe themselves to measurement systems, control systems, and, in general any device on the network. Representation of almost every conceivable parameter associated with the transducer, including its manufacturer, calibration, and performance parameters, can be represented in its TEDS. This greatly enhances interoperability between sensors and actuators, and makes the communication network sensor-agnostic; it can communicate equally well with thermometers, psychrometers, and robotic actuators. Because the transducer-NCAP interface is standardized, the network need only have a single NCAP design for the communication protocol it uses, which can then be duplicated for each sensor.

IEEE 1451.3 is standardizing the protocols and architectures associated with a distributed multidrop transducer bus (i.e., having several transducers attached to the network via a single NCAP). IEEE 1451.4 is adding analog transducers to the standard interface and is proposing a mixed-mode approach, in which both analog and digital signaling is performed, but not simultaneously.

The latest 1451 standard, the wireless sensor standard 1451.5, is now under development. Balloting for this standard is expected in 2004.

References

1. Institute of Electrical and Electronics Engineers, Inc., IEEE Std. 802.15.4-2003, IEEE Standard for Information Technology — Telecommunications and Information Exchange between Systems — Local and Metropolitan Area Networks — Specific Requirements — Part 15.4: Wireless Medium Access Control (MAC) and Physical Layer (PHY) Specifications for Low Rate Wireless Personal Area Networks (WPANs). New York: IEEE Press. 2003.
2. Sean Middleton, *IEEE P802.15 Wireless Personal Area Network Low Rate Project Authorization Request.* Document No. IEEE P802.15-00/248r4. 2000. Sections 9 and 10. http://grouper.ieee.org/groups/802/15/pub/2000/Nov00/00248r4P802-15_LRSG-PAR.doc.
3. http://ieee802.org/secmail/msg02790.html.
4. U.S. Department of Commerce, National Institute of Standards and Technology, Information Technology Laboratory, *Specification for the Advanced Encryption Standard (AES).* Federal Information Processing Standard Publication (FIPS PUB) 197. Springfield, VA: National Technical Information Service. 26 November 2001. http://csrc.nist.gov/publications/fips/fips197/fips-197.pdf.
5. Jose A. Gutierrez, Edgar H. Callaway Jr., and Raymond Barrett, *IEEE 802.15.4 Handbook.* New York: IEEE Press. 2003.
6. http://www.zigbee.org.
7. Michael R. Moore, Stephen F. Smith, and Kang Lee, "The next step — wireless IEEE 1451 smart sensor networks," *Sensors Mag.,* v. 18, n.9, September 2001, pp. 35–43.
8. http://grouper.ieee.org/groups/1451/5/; see also http://ieee1451.nist.gov.
9. Institute of Electrical and Electronics Engineers, Inc., IEEE Std. 1451.1-1999, IEEE Standard for a Smart Transducer Interface for Sensors and Actuators — Network Capable Application Processor Information Model. New York: IEEE Press. 1999.
10. Institute of Electrical and Electronics Engineers, Inc., IEEE Std. 1451.2-1997, IEEE Standard for a Smart Transducer Interface for Sensors and Actuators — Transducer to Microprocessor Communication Protocols and Transducer Electronic Data Sheet (TEDS) Formats. New York: IEEE Press. 1997.
11. Ibid.

Chapter 12
Summary and Opportunities for Future Development

12.1 SUMMARY

Wireless sensor networks are an emerging technology, poised for rapid market growth. The combination of multiple user applications, the development of communication protocols for self-organizing ad hoc networks, high levels of product integration, and standardization is expected to lead to high manufacturing volumes and their associated economies of scale. It truly is an exciting time for the wireless industry, which has suffered lately due to the maturity of its existing markets. The development of wireless sensor networks will be its next major growth area.

Six major market classifications for wireless sensor networks were presented:

1. Industrial control and monitoring
2. Home automation and consumer electronics
3. Security and military sensing
4. Asset tracking and supply chain management
5. Intelligent agriculture and environmental sensing
6. Health monitoring

This list is surely limited only by the imagination, however, and will grow as potential users become aware of the technology and its capabilities. One need only ask someone in almost any field of endeavor, "What could you do with a collection of small, autonomous, low power communication devices?" to expand the list. Business opportunities exist at every level, from device design and manufacture through the mining of the data produced by the networks.

The development of self-organizing ad hoc communication protocols is an important step in the development of the wireless sensor network industry. The design of the physical, logical link control, and network layers of a wireless sensor network can trade lower data throughput and higher message latency, when compared with a conventional Wireless Personal Area Network (WPAN), for lower cost and lower power drain.

A physical layer design for wireless sensor networks was presented that includes the use of code position modulation, an efficient direct sequence spread spectrum modulation method first proposed for wireless sensor networks in "A Communication Protocol for Wireless Sensor Networks."[1] Code position modulation combines a low symbol rate and a high over-the-air data rate (for long battery life) with the possibility of low-cost digital implementation. A high data rate was demonstrated as desirable to minimize total transceiver active time and, therefore, maximize battery life. In addition, since code position modulation employs orthogonal signaling, it retains good detector sensitivity while sending multiple bits per symbol. Performance was simulated in Additive White Gaussian Noise (AWGN) and a Bluetooth™ interference environment. Since code position modulation is a form of direct sequence spread spectrum, it enjoys a degree of interference rejection from narrowband interfering signals, such as Bluetooth, due to the processing gain of the modulation. This physical layer design was proposed and accepted as the basis for the physical layer of the Institute of Electrical and Electronics Engineers (IEEE) 802.15.4 LR-WPAN standard, approved in May, 2003.

A data link layer design was presented that employs the Mediation Device (MD) protocol, a novel method by which low-cost devices may temporarily synchronize to exchange information, while maintaining low duty cycle and, therefore, power efficient operation. The MD protocol enables low duty cycle devices to communicate easily and does not require a high accuracy synchronization reference. In fact, by randomizing the times that different devices send query messages, a certain amount of synchronization reference instability is actually beneficial. The MD protocol also introduces the concept of "dynamic synchronization," which extends the random channel access feature of ALOHA to peer-to-peer networks, and adds the logical channel reservation of time division multiple access (TDMA) channel access methods. By combining the use of lower-cost hardware with good battery life, the MD protocol improves the practicality of wireless sensor networks. This low duty cycle protocol is compatible with the IEEE 802.15.4 LR-WPAN standard.

Also presented were simulation results of a cluster tree network employing the distributed MD protocol, including message throughput, message delay, effective node duty cycle, and channel collision performance. This network was designed to address the problem inherent in networks of clusters, routing to non-neighboring clusters without keeping a global image of the network in every node (which would entail significant memory costs). A discrete-event network simulator was employed to evaluate the throughput, message latency, and number of packet collisions of the network, and the network node duty cycle. The network was found to have good performance even with node duty cycles as low as 0.19 percent, and able to handle an offered data throughput of 0.5 messages/node/hour. It was demonstrated that the node duty cycle did not vary significantly with position in the cluster tree hierarchy; this is due to the fixed (but low) duty cycle of the node beacons, which are simply replaced by data messages when sent, keeping overall duty cycle nearly constant with variations in message traffic. Packet collisions were found to be rare, validating the medium access function of the distributed MD protocol.

Message latency was found to be a strong function of the MD period, rising from an average of 241 seconds with a 500-second MD period, to an average of 1321 seconds with a 2000-second MD period. This is as expected because communication between nodes can only take place after MD activity of a nearby node. The advantage of the distributed MD protocol in dense networks is also apparent; even though the average message made at least two hops, the average message latency is approximately half of an MD period, indicating that multiple potential MDs were available at each hop in the path of the average message.

Implementation of these protocols must be practical for the product to be successful, and the design of practical wireless sensor network nodes is the subject of the second half of this book.

The partitioning decision, or the determination of which functions in the block diagram of the network node get placed in which integrated circuit, represents the first implementation decision faced by the designer. Factors to consider include the application computational requirements, the ability to use host resources, the desired market flexibility, product physical size goals, the required time to market, and the degree of network heterogeneity. Additional factors include the integration of digital logic, radio frequency (RF) and analog performance, the need for high voltage and/or high current operation, the method of packaging and chip-to-chip interconnection, and the location of high-speed signals, due to their power consumption and electromagnetic compatibility (EMC) issues. With such a large list of factors to consider — and this list is certainly not exhaustive — there are many possible designs, each optimum for a given weighting of the relevant factors. A related issue is the transducer interface. The selection of the transducer interface is a balancing of the need for market flexibility with the need for high levels of integration and their associated cost reductions.

The practicality of a wireless sensor network node rests, to a large degree, on its ability to meet the low power consumption goals typically set by the network application. Low system power consumption depends on a good match between the power source and the load itself, so that significant amounts of power are not lost in conversion processes. The selection of the power source itself is often the key to a successful implementation; the use of energy scavenging is possible only with the low power consumption that is typical of wireless sensor networks.

Antenna selection is a design decision that is often overlooked, but can be critical to the overall performance of the wireless sensor network node, since the antenna can have a large effect on product cost, market flexibility, and range of the node. The antenna design can also affect the external appearance of the node, and certainly has a large influence on the internal circuit board layout.

Due to their inherent miniaturization and interfaces with a variety of other systems, EMC is an important topic for the design of wireless sensor network nodes. The nodes must be protected, to the degree possible, from external interfering sources, while producing little interference themselves. Most important, they must not produce self-interference, implying that the designer should be aware of the potential EMC threats and victims of the proposed design, and the coupling mechanisms that may exist between them.

Since wireless sensor network nodes are typically small, often exposed to the environment, employ small geometry CMOS integrated circuits, and may have plastic housings to enable the use of internal antennas, electrostatic discharge (ESD) protection is a major design point. For products in nonconductive housings, ESD protection can be largely mechanical engineering task, with the goal of eliminating any holes or gaps in the housing that may present a pathway for the discharge to reach the sensitive circuits within.

Finally, much of the success of the wireless sensor network market will depend on standards, so that nodes from multiple vendors can be interoperable. This will rescue the industry from its present state of very successful but incompatible proprietary communication protocols existing in a number of niche markets, with their combined volume too low to reach economies of scale necessary for significant growth.

12.2 OPPORTUNITIES FOR FUTURE DEVELOPMENT

One is tempted to state that, since the technology is only now emerging from academic research into commercial application, all areas of wireless sensor networks represent opportunities for future development. However, some areas merit particular attention.

Opportunities for future work on the physical layer include the addition of higher layers of the protocol stack to the simulations presented in Chapter 3, so that package error rate and message error rate may be simulated and the whole 2.4-GHz coexistence question may be considered. If the duty cycle of a wireless sensor network node is 0.2 percent, can it be a significant interference source to a Bluetooth piconet? Other services, such as IEEE 802.11b Wireless Local Area Networks (WLANs) and microwave ovens, exist in the 2.4-GHz band; the performance of wireless sensor networks in their presence is an interesting study, hinging first on the definition of performance used and then on the type of application to which the wireless sensor network is applied. Similarly, the performance of other services in the presence of wireless sensor networks is an interesting subject.

The related problem of multiple access protocols in the Industrial, Scientific, and Medical (ISM) bands also is waiting to be solved. Methods to quickly determine the nature of signals present in the channel, not just their existence, would greatly help the coexistence problem now facing 2.4 GHz, and certain to arrive at 5 GHz.

As semiconductor processes improve, one may turn to the 24- and 60-GHz ISM bands where, due to the reduced wavelength, interesting features such as integrated antennas[2] may be considered in wireless sensors. This would further reduce their size, and perhaps create new applications for them, but the optimum design of such antennas remains an open issue.

Turning to the mediation device protocol, at present, the design assumes a simple random process to determine the MD period. The MD period could be made dynamic, and a function of the number of node neighbors, the amount of recent message traffic, or the amount of energy remaining in the node's power source. The effect of this dynamic MD period on the node duty cycle, network throughput, and message latency are open questions. The MD behavior could be tied to network quality of service (QoS) provisions, and respond to a QoS field in the message header or separate control messages.

One can envision that, if the offered message rate were low enough, a lower-power algorithm could eliminate all regular MD activity in the network altogether, and replace it with the "emergency mode" MD described in Chapter 4, Section 4.3.3. As the offered message rate rose, however, the constant scanning of the neighborhood before every message would result in power consumption higher than the MD protocol. A protocol could then be devised that changed modes as the traffic rate varied, to maintain optimum power consumption. This would include, perhaps, operating in "emergency mode" but maintaining a record of neighbors' beacon timing so that, for multiple messages sent in succession, the receive period could be discarded for the follow-up messages.

The physical topology of the network needed for maximum MD throughput is not known. What is the optimum number of neighbors, or children, or cluster size? This is of special importance when the nondistributed form of the MD protocol is used, and special-purpose MD nodes are placed in the network.

MD nodes (both distributed and dedicated) could be given a message store-and-forward capability, instead of the simple node synchronization function described in Chapter 4. This would be analogous to the more common use of the telephone answering machine, wherein the caller does not suggest a time for the return call, but actually leaves a message. How this would affect network performance is an interesting question. It may increase message latency in some cases by requiring two hops where nodes could have transferred the message directly, but it could also improve the connectivity of the network, by linking two nodes that otherwise could not hear each other.

Although it was designed for operation in wireless sensor networks in which nodes are assumed to be static, the use of the dedicated MD protocol in mobile networks presents some interesting possibilities. The plasma metaphor could be used to describe a network of MD nodes with store-and-forward capability in a network of highly mobile nodes. The heavy, positive ions of the plasma (the MDs) are relatively stationary, while the light, negative electrons (network nodes) move at a high velocity, storing messages with the nearest MD as they pass by for later forwarding by the intended recipient when it is next in range. This leads to the use of MDs as data fusion/aggregation devices, in which redundant data is eliminated, metadata is created, and only a minimum number of messages are transmitted.

Further, the fault tolerance of any hierarchical tree is always a question; in this case, it is not known if the dependence of the network on the MD for communication link establishment affects the fault tolerance of the network. It is possible that the failure of one or more MD could partition the network.

The development of routing algorithms for wireless ad hoc networks is perhaps the most active area of research related to wireless sensor networks; many approaches, from those based on the behavior of ants to those based on the hierarchical structures of military organizations, have been proposed. Does a single algorithm exist that is flexible enough to meet the needs of both wireless mice and military sensing applications, while simple enough to be implemented in the most inexpensive applications?

Chapter 5 described a network with the "monocrystalline" network association model (i.e., growth of the network starts with the DD). A network designed after the "polycrystalline" network association model, in which network growth starts with ordinary nodes, would make an interesting comparison.

One may argue that the most useful feature of wireless sensor networks is the low power consumption of their nodes, which makes them practical for a variety of new uses. In this area, perhaps the most intriguing areas of research are the development of true low power mixed-signal CMOS processes (i.e., processes designed for low power operation rather than maximum speed), and the development of practical energy scavenging systems.

Although low-power CMOS processes have been in production for many years, their use has, in general, been limited to specific low-leakage applications, such as digital watches. Their performance, although adequate for that task, is, in general, insufficient for mixed-signal designs at 900 MHz and above. However, much of the improvement in high-performance CMOS has come from the reduction of parasitic capacitance and the reduction of supply voltages, two performance improvements that also reduce power consumption, and a true low-power, mixed-signal CMOS process, perhaps on SOI, would be a boon to the wireless sensor network market.

The development of a wireless sensor network node with a practical energy scavenging system would certainly receive publicity similar to the wide acclaim received by Reutter with the introduction of his perpetual clock in 1928.[3] In addition to the favorable publicity, the development of a practical energy-scavenging system would almost certainly enlarge the total available market for wireless sensor networks by making new applications practical.

Finally, probably the greatest opportunity for future development lies in the actual application of wireless sensor networks. The opportunity for the development of creative uses for these small, inexpensive, low-power, self-organizing communication systems appears limitless.

References

1. Edgar H. Callaway Jr., A Communication Protocol for Wireless Sensor Networks, Ph.D. dissertation, Florida Atlantic University, Boca Raton, FL, August 2002.
2. K. Kim and K. K. O, Characteristics of integrated dipole antennas on bulk, SOI, and SOS substrates for wireless communication, *Proc. IITC,* 1998, pp. 21–23.
3. Jean Lebet, *Living on Air: History of the Atmos Clock.* Le Sentier, Switzerland: Jaeger-LeCoultre. 1997. pp. 29–31.

Appendix A
Signal Processing Worksystem (SPW)

The Signal Processing Worksystem (SPW) from Cadence Design Systems, Inc., San Jose, California, (http://www.cadence.com) is a system-level design tool originally designed for the development of digital signal processing (DSP) algorithms and hardware. The increased use of DSP in both wired and wireless communication systems has expanded the utility of SPW to include the simulation of complete communication systems, including modulators, demodulators, and communication channels. The software employs a graphical "connect-the-blocks" format to speed design, enable sharing among members of design groups, and provide an intuitive understanding of the simulated system.

The SPW simulator has three main parts:

1. Systems are entered into SPW via the Block Diagram Editor, which, as its name suggests, constructs the simulated system by connecting low-level functional blocks, such as adders and delay elements. Blocks may be hierarchical, to capture higher-level functionality.
2. The Simulation Manager defines and controls the simulation(s) to be run; the length of the simulation and parameters to be swept, for example, are controlled here.
3. Finally, the Signal Calculator edits and analyzes both input and output signals of the simulated system.

Appendix B
WinneuRFon
Dr. Yan Huang, May 2001

B.1 INTRODUCTION

WinneuRFon is a simulation tool for multi-hop neuRFon™ device networks using the distributed Mediation Device (MD) protocol. This appendix documents the WinneuRFon simulation program, which was initiated and developed with the objective of observing the behavior and collecting performance data of multi-hop neuRFon device networks that use the distributed MD protocol. In addition to giving the current system requirement, features, status, achievement and results, this document also identifies many important potential capabilities of this simulation method, which may be exploited by extensions of the program.

B.2 MOTIVATION

With the research on neuRFon device networks and the development of the Institute of Electrical and Electronics Engineers (IEEE) 802.15.4 standard, it becomes very desirable to find a method to observe the behavior and performance of the distributed MD protocol in large-scale multi-hop neuRFon device networks. Existing simulation methods using MATLAB, OPNET, or other network simulation tools can only provide us limited information. For example, currently these tools are only used to simulate the distributed MD protocol in a one-hop neuRFon device network and a dedicated MD. It is very difficult to use these tools' facilities to simulate more complicated situations such as the distributed MD and very large multi-hop neuRFon networks. Moreover, the information provided by these tools is limited by their design, which makes it impossible to obtain some types of needed information, such as message trace, device duty cycle, and all possible events in the network. Although a large amount of new ideas and protocols are still being created with neuRFon networks, like tree network formation, optimization of the distributed MD protocol, and location-based routing, it is even more difficult to manipulate the network with these new ideas and judge their performance and feasibility.

Observing the difficulty to simulate large-scale, multi-hop neuRFon device networks using general tools, it was decided to create a dedicated simulation program for it, temporarily called WinneuRFon (Version 1.0).

Because WinneuRFon is developed specifically for simulating behaviors of neuRFon networks, currently using a tree topology and the distributed MD protocol, we are able to collect any type of information we need and extend the network to any scale. In the future, it will also be very easy to extend this program to simulate any new ideas for the network.

B.3 SYSTEM REQUIREMENTS

The objective of WinneuRFon is to simulate and evaluate the performance of the distributed MD protocol using distributed MD in multi-hop neuRFon tree networks. Required input and output information follow:

Input:

Network Information:
- A network's physical area: (X, Y)

Device Information:
- Number of nodes in the area
- Period of each node's Rx/Tx cycle in seconds
- Device Rx/Tx duty cycle in percentage
- Range of each device
- Average number of messages per device per hour, with constant inter-arrival time

Dedicated Device (DD) Information:
- The coordinate of DD

MD Information:
- Receiving period in seconds
- Restart period range (START, END), in seconds

Simulation Information:
- Length of the simulation, in number of 10 minutes
- The pace of the graph view of the simulation procedure

All messages will be sent at random times from random nodes to a single node, DD. Current routing algorithm uses neighbor list/cluster tree routing: nodes check their neighbor list first; if destination is not on the lis, the message is sent to the node's parent. The phase of each node's Rx/Tx cycle and each MD's activity are randomized.

Output:

Message:
(a) Trace
- Time and node of origination of each message
- Time between each message's origination and delivery
- Number of nodes traversed by each message
- Path taken by each message

- Time each message waited at each node

(b) Statistics
- Average transmission time of all delivered messages
- Average transmission time of all delivered messages by their originating level
- Percentage of messages that are delivered
- Percentage of messages that are delivered by their originating level
- Average one-hop message transmission time
- Average one-hop message transmission time by hop level

Node and MD:
(a) Trace
- Number of messages facilitated by MD for each MD cycle
- Number of messages generated and relayed by each node

(b) Statistics

- Average duty cycle of each node
- Average duty cycle of each node by their level
- Average number of relayed messages each node
- Average number of relayed messages each node by their level

B.4 SUPPORTED FEATURES

The current version of WinneuRFon (Version 1.0) supports the following features:

- It provides a friendly graphical user interface (GUI) for users to provide all input information.
- It provides a graph view of the simulation procedure that allows users to observe the network behavior more intuitively.
- It provides the output of all required trace and statistics information of messages and devices of each simulation run to text files.
- It generates statistics information for a user-defined serial of simulation runs to text files for comparable analysis of the network performance.
- All statistical data are generated in a way that can be easily imported to Excel™ files to create a performance graph for analysis.

B.5 CURRENT STATUS AND ACHIEVEMENT

The current version of WinneuRFon (Version 1.0) is a stable system running on Microsoft Windows™ and supporting the features listed previously. It simulates the message delivery behaviors in a tree network formed by devices with given parameters under the distributed MD protocol. It allows the user to choose from a graph view (animated) simulation (with simulation delay for graph view of the procedure) or a nongraph view simulation. (The simulator runs faster without graph delay.)

The graph view of simulations shows the following:

- The tree topology of uniformly distributed neuRFon devices
- The events of MD receiving (node becomes black)
- The events of each device's communication request (black links) while there is an MD in its range
- The events of each device's communication query (green links) while there is an MD in its range
- The events of real message delivery (red links) among devices after the synchronization assisted by MD
- The situation of increasing number of messages queued at a device (increase the darkness of a node)

Both graph view and nongraph view generate trace and statistic results in the following text files:

- **result:** message traces, device and MD traces
- **tot_avg_msg_arr_rsp:** average transmission time of all delivered messages
- **level_avg_msg_arr_rsp:** average transmission time of all delivered messages by the originating level
- **tot_percent_div_msg:** percentage of messages that are delivered
- **level_percent_div_msg:** percentage of messages that are delivered by their originating level
- **tot_hop_rsp:** average one-hop message transmission time
- **level_hop_rsp:** average one-hop message transmission time by hop level
- **tot_avg_duty_cycle:** average duty cycle of each node
- **level_avg_duty_cycle:** average duty cycle of each node by their level
- **tot_num_rel_msg:** average number of relayed messages each node
- **level_num_rel_msg:** average number of relayed messages each node by their level.

The following enumerates some important assumptions of the current simulation system. Performance results collected by this version are highly dependent on these assumptions:

- The tree network formation procedure is not performed in this version because the current focus is the performance of the distributed MD protocol. In this version, the tree is formed using a breadth-first-traversal of the graph constructed based on given devices' number, location, and range (there is an edge between two devices that are in range of each other in the graph) with a trivial constraint of letting all nodes choose the nearest node as its parent among all nodes that could be parent in the traversal.

- Each node has an infinite buffer for queuing messages that need forwarding. No message dropping is used. This is a reasonable assumption for low throughput networks, although care must be used to consider the case in which message density is high, such as in nodes near the DD.
- If multiple communication requests are made to a single device in a MD period, we assume only one communication between MD and the destination device is needed to notify the receiver about all senders' time offsets. After this, all communication is simulated in the order of the senders' RTS sequence. This is a reasonable approximation because the additional node time offset information is short compared to the size of a transmitted packet (due to the packet header overhead).
- If one node in an MD period needs to both send message and be sent message, only the earlier of these events can happen.
- No more than one MD device is permitted to be in the range of any one node at any given moment. As discussed in the text, methods (such as having an MD transmit a warning beacon at the end of its MD receiving period) exist to ensure this in practice; however, WinneuRFon, at present, ensures this by design. This is probably the largest behavioral deviation at the network layer from a physical system, and the inclusion of the MD warning beacon will probably be the first addition to WinneuRFon. The effect of this behavior will be observed first in very dense networks.
- Any communication request, communication query, and message delivery can be completed using one device Tx/Rx period, without considering these entities' format and length. Because these transmissions are very short, this is equivalent to saying that WinneuRFon assumes no packets are lost in the wireless channel other than by packet collisions. This is a reasonable assumption for a first-generation network-layer simulator, although the wireless channel is certainly not reliable and the effects of an unreliable physical layer must be included in later versions of WinneuRFon.
- Timing among all nodes is assumed to be ideal; that is, although the phase offset of the nodes is randomized, all nodes operate from the same time base so that their relative phase offset remains constant. This is clearly not the case for a physical network; however, the distributed MD protocol requires synchronization only for short periods (e.g., the MD receive and transmit periods) during which timing offsets should be small. Nevertheless, this will also be added to the next release of WinneuRFon.

The major achievement of current WinneuRFon is being able to show a clear picture of how the distributed MD protocol works in multilevel neuR-Fon tree networks. The simulation data collected by WinneuRFon have been used to draw performance graphs. By changing the input configuration and generating statistic data and pictures, it can help determine the performance of the distributed MD protocol in different situations, and what can be a feasible configuration in realistic applications. Moreover, it can also help determine which variables should or can be optimized to provide the best solutions.

B.6 SIMULATION METHOD AND MORE POTENTIAL FUNCTIONALITIES

WinneuRFon is implemented using C++ for the simulation and Microsoft® Foundation Class MFC Visual C++ for its interface. The simulation uses some essential data structures and an event-driven simulation method.

As presented in Exhibit 1, WinneuRFon is divided into the following major components:

- Input module
- Output module
- Tree formation module
- Protocol module
- Event generation/handling/management module
- Data collection module
- GUI and graph view module

The following defines the major events in WinneuRFon:

- **DEV_TX:** Device Tx starts
- **DEV_TXEND:** Device Tx ends
- **DEV_RV:** Device Rv starts
- **DEV_RVEND:** Device Rv ends
- **AMD_STR:** MD Rv starts
- **AMD_END:** MD Rv ends
- **AMD_TX:** MD Tx starts
- **AMD_TXEND:** MD Tx ends
- **DEV_MSG:** Device message generation

The simulation starts by generating all basic events for all devices including DEV_TX, AMD_STR, and DEV_MSG. All events are inserted into an event queue based on the event time. The time line moves ahead based on the progress of all events. When a new event is handled, it will trigger another event based on the neuRFon device's behavior and the distributed MD protocol. All event-triggering behavior is enumerated in the following manner:

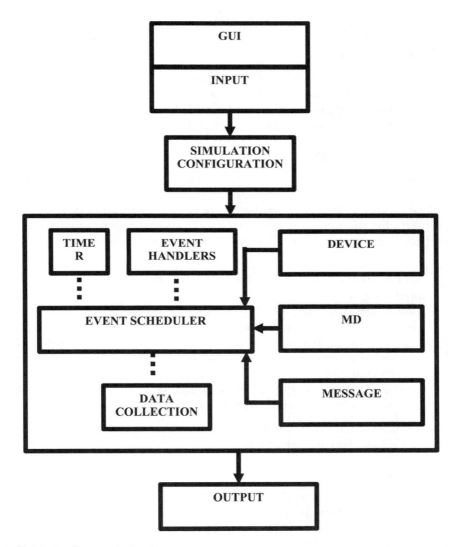

Exhibit 1. WinneuRFon Components

- DEV_TX triggers DEV_TXEND.
- DEV_TXEND triggers DEV_RV.
- DEV_RV triggers DEV_RVEND.
- DEV_RVEND triggers DEV_TX.
- AMD_STR triggers AMD_END.
- AMD_END triggers AMD_TX or next AMD_STR.
- AMD_TX triggers AMD_TXEND.
- AMD_TXEND triggers next AMD_TX or next AMD_STR.

The event-handling procedures are the most important part of the simulation. They are also based on the neuRFon device behaviors and the distributed MD protocol.

The event-driven method can provide many potential functionalities of WinneuRFon. If we want to simulate more new ideas or protocols using WinneuRFon, more new events can be added and event-handling procedures can be programmed based on new solutions.

B.7 PROPOSAL FOR FUTURE WORK

WinneuRFon is a very efficient method to simulate multi-hop neuRFon networks with the distributed MD protocol. Because of the method it uses, it is more flexible than other simulation tools, and can also very efficiently extend its functionalities and features to support simulating more ideas and protocols for neuRFon networks. Some possible simulation proposals are:

- Use this method to simulate the tree network formation protocol. By simulating the message delivery procedure in this protocol and evaluating its performance, the tree network can be formed.
- Based on the results collected by current version of WinneuRFon, some optimization schemes can be designed for the distributed MD protocol. They can then be integrated into WinneuRFon for performance evaluation.
- Using a simplified version of MD, location-based routing can all be integrated or simulated separately using this method.

It would be much more difficult and inefficient to use general simulation tools to evaluate the preceding items.

B.8 SUMMARY

This appendix summarized the advantages, system requirements, features, development status, achievements, results and methods used in the WinneuRFon simulation tool, developed for large-scale neuRFon networks using the distributed MD protocol. The appendix also describes other opportunities to extend WinneuRFon, to simulate more issues, and identify solutions for neuRFon networks.

Appendix C
An Example Wireless Sensor Network Transceiver Integrated Circuit (IC)

This appendix describes a short-range, low-power 2.4-GHz Industrial, Scientific, and Medical (ISM) band transceiver designed for the emerging IEEE 802.15.4 wireless standard supporting star and mesh networking (Exhibit 1). When combined with an appropriate microcontroller (MCU), it provides a cost-effective solution for short-range datalinks and networks. Applications include remote control and wire replacement in industrial systems such as wireless sensor networks, factory automation and control, heating and cooling, inventory management, and radio frequency identification (RF ID) tagging. Potential consumer applications include wireless toys, home automation and control, human interface devices, and remote entertainment control.

The receiver includes a low-noise amplifier, 1.0-mW PA, VCO, full spread-spectrum encoding and decoding compatible with Institute of Electrical and Electronic Engineers (IEEE) 802.15.4, and buffered transmit and receive data packets for simplified use with low-cost microcontrollers (Exhibit 2). The device supports 250-kbps O-QPSK data in 5.0-MHz channels, per the IEEE 802.15.4 2.4-GHz physical layer specification. A Serial Peripheral Interface (SPI) to the MCU is used for RX and TX data transfer and control (Exhibit 3). This allows MAC, Network, and application software to be supported by an appropriately sized MCU for the use model. In many applications, a larger read-only memory (ROM) version of the existing MCU can be used resulting in embedded solutions.

Features include:

Power supply range: 2.0 to 3.6 V (Exhibit 4)
16 channels

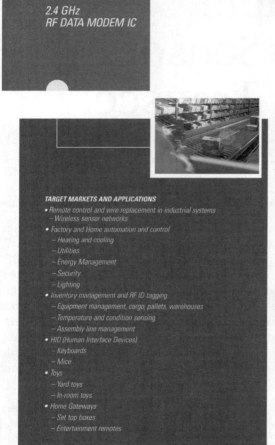

OVERVIEW

Motorola is a leading member of the IEEE 802.15.4 Standards Body. The 802.15.4 specification is a cost effective low data rate (< 250 kbps), 2.4 GHz and 868/928 MHz wireless technology designed for short-range and personal area networking. Target markets for the 802.15.4 Standard include industrial control and networking, home automation and control, inventory management, human interface devices, as well as wireless sensor networks.

The 802.15.4 Standard is the basis of an application and network layer protocol known as ZigBee™. The ZigBee™ Alliance is an association of companies working together to create software interoperability certification and testing for 802.15.4 systems. The alliance's website address is www.ZigBee.com

Motorola will be offering comprehensive system solutions comprised of the RF data modem designed to support the 2.4 GHz band of this standard along with reference designs and software. The software is targeted to support a broad suite of Motorola 8-bit microcontrollers, allowing the end user maximum flexibility in application design, memory requirements, and peripheral interfaces.

With more than 8 billion processors currently in use, Motorola embedded electronic solutions are everywhere and permeate virtually all advanced technology applications. Motorola is number one in Embedded Microprocessors, Microcontrollers, and 8-bit MCUs.

Ratification of this IEEE standard is expected in January of 2003. Engineering samples of the RF data modem are expected to be offered in early 2003 to development partners to coincide with ratification of this standard.

Exhibit 1.   An Example Transceiver

Low power drain
Power-down modes for power conservation
RX sensitivity of −90 dBm at 1.0 percent Packet Error Rate (PER)
Full spread-spectrum encode and decode compatible with 802.15.4

Transition times from power down modes to active modes are short, to allow for maximum power conservation (Exhibit 5).

An Example Wireless Sensor Network Transceiver Integrated Circuit (IC)

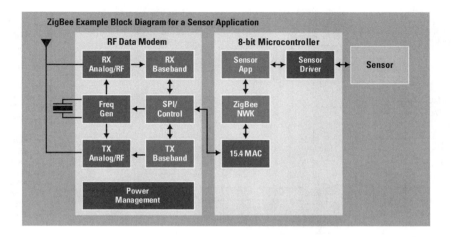

RF DATA MODEM KEY FEATURES

- Designed for the IEEE 802.15.4 and ZigBee™ standards

- Operates in the 2.4 GHz ISM band available worldwide

- Cost effective CMOS design. Low external components. Includes on-chip low noise amplifier, 1.0 mW PA, VCO, full spread-spectrum encoding and decoding compatible with 802.15.4

- RX sensitivity of -90 dBm at 1% PER, well above specification

- Buffered transmit and receive data packets for simplified use with low-end microcontrollers

- Engineered to support 250 kBit/s O-QPSK data in 5.0 MHz channels, per the IEEE 802.15.4 specification

- No line-of-sight limitations as with infrared (IR)

- Multiple power down modes enabling standard Alkaline battery lifetimes from months to years

- Power Supply Range: 2.0 to 3.6 V with on-chip voltage regulator

- SPI data and control interface, operates up to 8MHz

MAC LAYER SOFTWARE KEY FEATURES

- Designed to support peer to peer and star topologies

- Optional guaranteed time slots for low latency transfer

- Will support optional Zigbee™ Network layer software

- Power saving modes: Doze and Hibernate, application configurable

Exhibit 1.   (Continued)

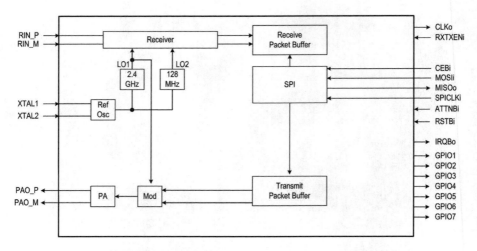

Exhibit 2.   Simplified Block Diagram

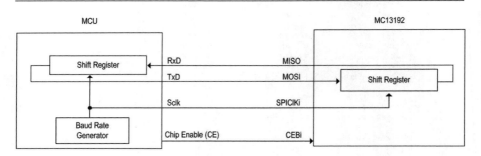

Exhibit 3.   Serial Peripheral Interface

Exhibit 4.   Recommended Operating Conditions

Characteristic	Symbol	Min	Typ	Max	Unit
Power supply voltage	V_{DD}	2.0	2.7	3.6	Vdc
Core supply voltage, logic interface	V_{CORE}	—	1.8	—	Vdc
Digital interface supply	V_{DDINT}	2.0	2.7	3.6	Vdc
Input frequency	f_{in}	2.405	—	2.480	GHz
Ambient temperature range	T_A	−40	25	85	°C
Logic input voltage low	V_{il}	0	—	30 percent V_{DDINT}	V
Logic input voltage high	V_{ih}	70 percent V_{DDINT}	—	V_{DDINT}	V
SPI clock rate	SPICLKi	—	—	8.0	MHz
RF input power	P_{max}	—	−20	0	dBm
Ref Osc frequency ($\pm$ 40 ppm over temperature range)	f_{ref}	16 MHz only			

An Example Wireless Sensor Network Transceiver Integrated Circuit (IC)

Exhibit 5. Mode Definitions and Transition Times

Mode	Definition	Transition time
Off	All IC functions Off, Leakage only. Responds to RSTBi	23 ms to Idle
Hibernate	Reference Oscillator Off. IC. Responds to ATTENBi	15 ms to Idle
Doze	Ref Osc On but no CLKO output. Responds to ATTENBi and TC2	3.3 ms to Idle
Idle	Ref Osc On with CLKO output available. SPI active.	
Receive	Ref Osc On, Receiver On, SPI should not be accessed.	144 µs from Idle
Transmit	Ref Osc On, Transmitter On, SPI should not be accessed.	160 µs from Idle

RAM = random access memory.

323

About the Author

Edgar H. Callaway Jr. received a B.S. in mathematics and an M.S.E.E. from the University of Florida, Gainesville in 1979 and 1983, respectively; an M.B.A. from Nova (now Nova-Southeastern) University, Davie, Florida, in 1987; and a Ph.D. in Computer Engineering from Florida Atlantic University, Boca Raton, in 2002.

Dr. Callaway joined the Land Mobile Division of Motorola in 1984 as an RF engineer working on 800-MHz and (later) 900-MHz trunked radio products. In 1990, he transferred to Motorola's Paging Products Group, Boynton Beach, where he designed paging receivers for the Japanese market.

From 1992 to 2000, Dr. Callaway was engaged in paging receiver and transceiver system design and was the lead receiver designer of Motorola's paging platform. In 2000, he joined Motorola Labs, Plantation, Florida, where his interests include the design of low-power wireless networks. He is a Registered Professional Engineer (Florida). He has published several papers and has had more than 20 U.S. patents issued.

Index